The Patrick Moore Practical Astronomy Series

Series Editor

Gerald R. Hubbell
Mark Slade Remote Observatory, Locust Grove, VA, USA

More information about this series at http://www.springer.com/series/3192

The Patrick Moore Practical Astronomy Series

Philip M. Bagnall

Atlas of Meteor Showers

A Practical Workbook for Meteor Observers

 Springer

Philip M. Bagnall
Wallsend, UK

ISSN 1431-9756 ISSN 2197-6562 (electronic)
The Patrick Moore Practical Astronomy Series
ISBN 978-3-030-76642-9 ISBN 978-3-030-76643-6 (eBook)
https://doi.org/10.1007/978-3-030-76643-6

© The Editor(s) (if applicable) and The Author(s), under exclusive license to Springer Nature Switzerland AG 2021
This work is subject to copyright. All rights are solely and exclusively licensed by the Publisher, whether the whole or part of the material is concerned, specifically the rights of translation, reprinting, reuse of illustrations, recitation, broadcasting, reproduction on microfilms or in any other physical way, and transmission or information storage and retrieval, electronic adaptation, computer software, or by similar or dissimilar methodology now known or hereafter developed.
The use of general descriptive names, registered names, trademarks, service marks, etc. in this publication does not imply, even in the absence of a specific statement, that such names are exempt from the relevant protective laws and regulations and therefore free for general use.
The publisher, the authors, and the editors are safe to assume that the advice and information in this book are believed to be true and accurate at the date of publication. Neither the publisher nor the authors or the editors give a warranty, expressed or implied, with respect to the material contained herein or for any errors or omissions that may have been made. The publisher remains neutral with regard to jurisdictional claims in published maps and institutional affiliations.

Cover image: © solarseven/Getty Images/iStock

This Springer imprint is published by the registered company Springer Nature Switzerland AG
The registered company address is: Gewerbestrasse 11, 6330 Cham, Switzerland

To Joshua, Sophia and Eden

Key Events in Meteor Astronomy

687 Mar 15 BCE	First known observations of a meteor shower—the Lyrids—by Chinese court astrologers
340 BCE	Aristotle in *Meteorologica* attempts first scientific explanation of meteors, describing them as '…hot, dry exhalations rising from the Earth when it is heated by the Sun'
CE 36 July 15	First record appears of the Perseid meteor shower in Chinese annals, which note that 'At dawn more than 100 small meteors flew in all directions'
CE 855	Leonids' first recorded appearance by Imam Ibn al-Jawzī
CE 902	Possibly the first record of the Orionids according to Adolphe Quetelet
1066	Earliest depiction of Halley's Comet in the Bayeux Tapestry following the Battle of Hastings
1676	Geminiano Montanari estimates the height and speed of the Great Meteor observed over Italy on March 31, which he calculates at 34–40 Italian miles (55–65 km) and at least 160 miles per minute or 4.3 km/s
1741 Dec 6	Georg W. Krafft, St. Petersburg, Russia, discovers the Andromedids
1762	Petrus van Musschenbroeck notes in *Introductio ad Philosophiam Naturalem* that 'Falling stars are more numerous in August than at any other time of the year', referring to the Perseids
About 1763	Prof. Thomas Clap, Yale, suggests the brightest meteors were caused by friction with the air at speeds of up to 500 miles per minute (30,000 mph or 48,276 km/h = 13.4 km/s)
1783	Nevil Maskelyne encourages people to record their observations of meteors as soon as possible after the event and to send them to him for analysis
1794	Ernst Florens Friedrich Chladni, having studied various accounts of meteorite falls, concludes that rocks did indeed fall from the sky
1795 Dec 20	Probably first reference to the Ursids in Chinese annals
1798	Brandes and Benzenberg, two students at the University of Göttingen, use a 16-km baseline to triangulate the heights of meteors
1799	Alexander von Humboldt witnesses the Leonid meteor storm from Cumaná, Venezuela

1803 April 26	The townsfolk of L'Aigle in Normandy, France, witness the fall of more than 3000 stony meteorites, proving Chladni's hypothesis of rocks falling from the sky to be correct
1803 Apr–May	First modern observations of the April Lyrids from Richmond, Virginia (see 687 BCE)
1825 Jan 2	First recorded observation of the Quadrantids made by Antonio Brucalassi from Tuscany, Italy
1830 Dec 12/13	Ludvig F. Kämtz discovers the Geminids
1833 Nov 13	The Great Leonid Storm indicates that nearly all the meteors appeared to radiate from a particular part of the sky: the *radiant*
1834	Prof. Denison Olmsted, Yale, explains the Leonid radiant as being an optical illusion and states that the meteors actually travel in parallel paths. He also suggests the shower is an annual event, having recalled a shower of meteors appearing in November 1832
1836	Adolphe Quetelet, Director of the Brussels Observatory, is the first person to recognize the annual nature of the Perseids
1839	Adolph Erman is the first person to calculate the orbits of meteors by careful observation of five Perseids
1858 Aug	The Southern δ-Aquariids are discovered by Eduard Heis and Georg Balthazar von Neumayer and recorded in their book *On Meteors in the Southern Hemisphere*, published in 1867
1861	Prof. Daniel Kirkwood speculates on the origin of meteors, suggesting that they are the remains of ancient disintegrated comets
1862 Jul 16–19	Lewis Swift and, on July 19, Horace P. Tuttle discover a new comet later shown to be the parent body of the Perseids
1863	H.A. Newton proposes the existence of the η-Aquariid meteor shower based on ancient observations
1866	Giovanni Schiaparelli confirms Kirkwood's 1861 hypothesis by demonstrating that the orbit of Comet Swift-Tuttle was very similar to the meteors that appeared in mid-August, which he called the 'Perseids', thereby inadvertently introducing a method of naming annual meteor showers
1867	Eduard Heis discovers the Taurids based on observations made between 1839 and 1849
1868	Italian astronomers observe the η-Aquariids but do not recognize it as a shower.
1868	R. Falb suggests a link between Halley's Comet, the η-Aquariids and the Orionids
1869 Apr 29	Lieutenant-Colonel George L. Tupman in the Mediterranean makes the first planned observations of the η-Aquariids
1870	The British Association for the Advancement of Science establishes a committee to study meteors
1878	Prof. Herschel discovers that the η-Aquariids radiant drifts eastward
1879	W.F. Denning probably discovers the Northern δ-Aquariids, which he calls the β-Piscids
1885 Nov 27	Ladislaus Weinek photographs a meteor for the first time—an Andromedid
1890	The British Astronomical Association establishes a Meteor Section with David Booth as its first Director
1893 Aug 9	William L. Elkin, Yale, captures a meteor on two cameras using a 6-km baseline
1899	William F. Denning publishes a catalogue of more than 4000 radiants in the *Memoirs of the Royal Astronomical Society*. Many turned out to be spurious
1911	Charles P. Olivier notes similarities between the η-Aquariids and the Orionids

Key Events in Meteor Astronomy

1911	Charles P. Olivier establishes the American Meteor Society
1911–1912	After years of declining numbers, the Perseids reach an all-time low of just 1 or 2 meteors per hour
1915	M. Davidson predicts the existence of the Draconid meteor shower based on calculations of the orbit of Comet Giacobini-Zinner
1916 Jun 28	June Boötids discovered by William F. Denning from England
1916	Denning mentions observing the Ursids for the first time in *The Observatory*
1929	Hantaro Nagaoka suggests that random bursts of electron density observed at night by radio astronomers are due to the appearance of meteors momentarily ionizing the atmosphere
1929 Nov	Ronald A. McIntosh publishes first detailed report on the η-Aquariids in the *Monthly Notices of the Royal Astronomical Society*
1932 Nov 14–16	John Schafer, William Goodall and Albert Skellett correlate radio bursts with visual observations of the Leonids
1933	The Great Draconid Storm occurs, with an estimated ZHR of 10,000
1935	Ronald A. McIntosh publishes a list of 320 radiants in the Southern Hemisphere but, like Denning's catalogue of 1899, it also contains many phantom radiants
1936	Harvard College Observatory sets up all-sky camera stations in Massachusetts and New Mexico. The project results in the orbits of 144 meteors being calculated by the time the project is terminated in 1951
1946 Oct 7–11	James Hey, Gordon Stewart and John Parsons successfully bounce 5-m wavelength signals off Draconid meteors, detecting a peak of almost 300 meteors per hour. They also succeed in estimating the meteoroids' geocentric velocity for the first time: 22.9 km/s
1950s	Zdeněk Ceplecha introduces a systematic method of calculating meteor shower rates. The zenithal hourly rate (ZHR) calculation becomes popular as a standard means of indicating the strength of shower activity
1952–1954	The Harvard Meteor Project, the successor to the 1936 initiative, produces 2529 orbits from meteors photographed over the skies of New Mexico
1963	Richard Southworth and Gerald S. Hawkins, using data from the Harvard Meteor Project, develop a mathematical method of determining whether two orbits are related, a method they called the D-criterion
1966 Nov 17	A Leonid storm with an estimated ZHR of 144,000 reported over the western United States
1988	International Meteor Organization formed
1999	David J. Asher, in collaboration with Robert H. McNaught, develops a model of the evolution of the Leonid meteoroid stream. They demonstrate that filaments within the stream, caused by planetary perturbations, mean that the 1998 outburst was caused by debris laid down in 1333 by Comet Tempel-Tuttle, and the 1966 storm was caused by the comet's 1899 return. Asher also makes predictions about future Leonid activity
2019	Quanzhi Ye predicts that a new shower would be found in the constellation Sculptor, caused by the disintegration not of a comet but of an active asteroid, (101955) Bennu

About this Atlas

Meteor observation is one of the more dynamic aspects of astronomy. There is nothing more exciting than watching a piece of cosmic debris blaze its way through the Earth's atmosphere. Far from being rare events, as popularly believed, countless meteors appear every second of every day. At certain times of the year, the chances of witnessing a meteor event greatly increase as the Earth plunges headlong into one of the ten major meteoroid streams that cross our planet's orbit. Knowing when and where to look for these spectacular events is a key element of this *Atlas*.

The *Atlas* contains all the information an observer will need to locate, track, record and analyse meteor activity from all of the major showers and a selection of the more interesting minor showers. Beyond explaining the fundamentals of meteor astronomy, it contains charts, observation record sheets, details of shower visibility and activity from almost any location on Earth, as well as practical checklists and wind-chill charts. The text is data-rich but highly readable.

Several helpful charts can be downloaded for printing and use from the Springer website (https://doi.org/10.1007/978-3-030-76643-6_4):

D1: Meteor Observation Record Sheet
D2: Magnitude Comparison Charts
D3: Gnomonic Plotting Charts
D4: Major Meteor Shower Radiant Charts
D5: Minor Meteor Shower Radiant Charts
D6: ZHR Calculation (Excel Spreadsheet)

Preface

I first became interested in meteor observation as a teenager in the early 1970s. They were exciting times. Astronauts were walking on the Moon and the Universe was ours. There was a group of us, eight or nine in all, and over the years we built up a sizeable collection of observations, mainly of the major meteor showers but also a few minor showers.

We used to get into countless arguments as to where a radiant would be on any particular night or what the zenithal hourly rate was likely to be. In the 1970s, reliable information was not so easy to come by. It often involved a trip to the local library and then a request to the British Library for a journal or book that was not available locally. The service could be slow at times and often the publication did not give us the detail we so desperately needed. It was probably towards the late 1970s, or perhaps the early 1980s, that I decided that a meteor shower atlas would be useful. Partly because of a lack of good-quality data, and partly because life gets in the way, I shelved the idea until fairly recently, when the arrival of COVID-19 left me with more time on my hands than was healthy.

The one nagging doubt I had about this project was whether it was worth spending hours writing a book when there is so much information available on the Internet. A preliminary search soon settled that question. Information on the Internet basically falls into two categories, at least as far as science is concerned. Much of the information about meteor showers is, frankly, so bland as to be almost pointless. The really useful information is often tucked away in academic journals, many of which are inaccessible to anyone who is not a member of a university or other institution. And those journals that are readily available are written in the sort of language that the layperson would struggle to understand—that is the nature of scientific journals. So there were two very good reasons why a meteor shower atlas would be worth the effort. And there was a third driver. No one has ever produced a similar publication. Sure, there have been a few books on observing meteors, but from a practical let's-go-out-tonight-and-watch-the-Geminids stance, none have provided all the charts, report sheets and calculations that an observer needs.

This *Atlas* is intended primarily for visual observers. Technology, particularly over the past decade or so, has become cheaper and easier to use and provides an avalanche of invaluable data. But technology has its limitations, and the visual observation of meteor showers complements, supplements and puts into context the information obtained by photography, video, radio and radar. And apart from that, watching a meteor tear through the atmosphere, live, is on par with observing the aurora or experiencing a total solar eclipse. It's exciting!

A book such as this has a limited shelf life of about 10 years. Activity from meteor showers evolves. New information gives us better insight into meteoroid streams. Nothing stands still. And the text and diagrams in this *Atlas* will one day have to be revised.

Wallsend, UK   Philip M. Bagnall

Acknowledgements

In preparing this *Atlas*, I have sought the help, opinions and observations of many people, all of whom have been generous and often insightful.

I would like to thank in particular Harry Burke, David Norton and Malcolm Young for their honest critique and for helping me focus on the salient points in a sea of what is often conflicting data and views.

I am also grateful to the following for access to their images and for their guidance:

Sean Adams, Pete Almond, Keith Asher, Lynne Banks, Paul Bennet, Cyril Blount, Alun Boyce, Hannah Bush, David Clark, Norman Clegg, Martin Collinson, Alan Cook, Kevin Cook, Carol Corn, Bill Davidson, Paul Davies, Shawn DeGroot, Kaz Downing, Charlie Drummond, Keith Edwards, Natalia Eriksson, Mark Evans, Peter Fairley, Colin Faulkner, Iain Fischer, Victoria Ford, Frank Forster, Alice Fox, Mike Frost, Chris Gates, Claire Geddes, David Hale, Debbie Hamilton, Eric Hargrove, Steve Harris, Liam Jones, Peter King, Gary Kronk, Patrick Lambert, Carl Locke, David Marsh, Richard Miller, Fred Milton, William Morgan, Eric Murphy, Robert Murray, Tom Newton, Izzie Nilsson, Val Norman, Phil Oates, Ozzie Parker, Andrew Parsons, Simon Perry, Harold B. Ridley, Xavier Ronan, Freddie Roy, Lance Russell, George Savill, Ilya Shaporev, Brian Thompson, Thame Vincent, John Walker, John Wallis, Brian Watkins, David Welch, Morag West, Mike Williams and Paula Wood.

Much of the raw data for this *Atlas* came from the various individuals mentioned above, various publications, including long-defunct magazines, such as *Ruat Coelum, The Meteor Observer* and *Meteor!*, and databases, such as those held by the International Meteor Organization and the IAU Meteor Data Centre. Where a single database or individual has supplied the information used, they are acknowledged in the text or with an image credit. Usually, however, information comes from various sources as this often helps with gaps in the data. Images without a credit line are my own work.

Finally, I would like to thank Hannah Kaufman, associate editor for Astronomy, Astrophysics, Astronautics, and Space Studies at Springer, for having faith in this project and for guiding me through the various aspects of publication and her colleague, Dinesh Vinayagam and his team, for technical assistance.

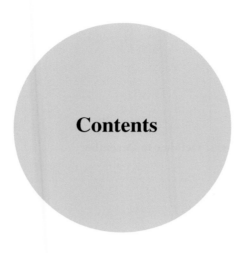

Contents

1 Introduction 1
    About Time!. 2
    Radiants 3
    Major Showers and Minor Showers 5
    Shower Names 6
    The Zenithal Hourly Rate (ZHR) 6
    Magnitude Distribution Graphs 8

2 Meteors 11
    A Meteor Event 15
    The Zodiacal Dust Cloud 18
    The View from Earth 20
    References 23

3 Meteoroid Steams and Meteor Showers 25
    The Structure of the Solar System 25
    The Structure and Behaviour of Comets 30
    Comets and Meteoroid Streams 31
    Active Asteroids and Meteoroid Streams 33
    Types of Meteor Showers 33
    Summary 35
    References 36

4 Observing Meteor Showers 37
    Before You Start 38
    Essentials 43
    The Observing Session 44
    Fireballs and Bolides 48
    Adding a Camera to Your Meteor Watch 48
    Analysing Your Observations 48
    Mathematical Approach 50
    And Why Not Try… 52

5 The Major Meteor Showers: January to August	53
Introduction	53
Activity and Additional Data Tables	53
Charts, Graphs and Diagrams	55
The Quadrantids	58
The April Lyrids	78
The η-Aquariids	100
The Southern δ-Aquariids	120
The Perseids	142
References	168
6 The Major Meteor Showers: October to December	171
The Orionids	171
The Taurids	196
The Leonids	224
The Geminids	246
The Ursids	270
References	285
7 Discovering Minor Showers	287
Type 1: Remnant Shower	287
Type 2: Close Approach	290
Type 3: One-Off Display	290
Type 4: New Shower	290
Type 5: Developing Shower	296
Type 6: Variable Activity Shower	299
Ones to Watch	304
α-Centaurids (ACE)	304
June Boötids (JBO)	306
Aurigids (AUR)	310
Piscis Austrinids (PAU)	312
September ε-Perseids (SPE)	315
November Orionids (NOO)	318
Southern χ-Orionids (ORS)	318
σ-Hydrids (HYD)	318
Phoenicids (PHO)	320
December Monocerotids (MON)	323
Puppid-Velids (PUP)	324
Observing Minor Showers	326
References	327
Appendix A: Greek Alphabet	329
Appendix B: Constellation Abbreviations and Meteor Shower Names	331
Appendix C: Right Ascension to Degrees Conversion Table	335
Appendix D: Glossary, Symbols and Abbreviations	337
Appendix E: Wind Chill Chart	345

Appendix F: Visual Meteor Observation Form	347
Appendix G: Magnitude Comparison Charts	349
Appendix H: Gnomonic Plotting Charts	367
Appendix I: Minor Meteor Showers	397
Appendix J: Major Meteor Showers	401
Author Index	403
Subject Index	407

Chapter 1

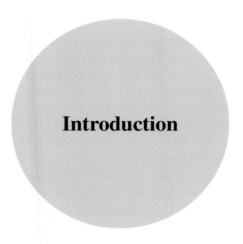

Introduction

> *By watching I know that the stars are not going to last.*
> *I have seen some of the best ones melt and run down the sky.*
> ***Eve's Diary***
> Mark Twain

Every second of every day, countless numbers of cosmic dust particles fall into the Earth's atmosphere and burn up, producing a momentary streak of light in the night sky. Often referred to as falling or shooting stars, these *meteors* are a constant reminder that, far from being empty, the Solar System is pervaded by a cloud of dust, which the Earth passes through continuously during its orbit around the Sun. About ten times a year the Earth encounters a thick stream of dust. When this happens, the number of meteors can suddenly jump as a *meteor shower* takes place. More rarely, meteors can be seen to fall like rain, with several visible at once. These events are called *meteor storms* and, although not common, some can be predicted with some degree of certainty.

The purpose of this *Atlas* is to provide a data-rich but easy to use practical reference for meteor observers. Using this *Atlas*, an observer should be able to quickly check the observing prospects for a particular shower on a particular night, see at a glance how high the radiant will be during the course of their meteor watch, grab a map of the radiant, a star chart for estimating magnitudes and a report form. Weather permitting, they will have a good, productive night. Chapters 5 and 6 are the core of this *Atlas* and provide much of the practical information observers need for a successful meteor watch.

To get the most out of this *Atlas*, you will need to be aware of a few things. Even if you are an experienced meteor observer, you might still benefit by taking a few moments to read this short chapter.

About Time!

The Earth does not spin once on its axis in exactly 24 h. Nor does it orbit the Sun in exactly 365 days. And everything in the Solar System is gyrating. That makes time-keeping a little tricky because there are no invariable fixed reference points. One of the most obvious problems that this causes is its effect on the date and time when a meteor shower will reach maximum activity. The April Lyrids, for example, will peak on either April 22 or April 23 over the next decade; two thousand years ago, the date was April 19. Clearly, the civil calendar isn't sufficiently accurate for meteor observation. There's no point looking for the April Lyrids' maximum on April 23 if, this year, the shower peaks on April 22.

To overcome this anomaly, meteor astronomers use an alternative time system based on *solar longitude,* which has the symbol $\lambda_\odot$ and is measured in degrees. This is a more accurate method of calculating time, determined by where the Sun is in the sky in relation to the Vernal Equinox. The April Lyrids reach maximum at $\lambda_\odot$ 32° regardless of the year. Of course, the Sun's apparent path across the sky is not constant over a long period of time, so its longitude is not stable and does gradually drift, but the drift happens over thousands of years so can be disregarded over the timescale of a human life. Throughout this *Atlas,* the civil date is given alongside the solar longitude. Should you need to convert solar longitude to a civil date, the International Meteor Organization provides conversion tables on its website at www.imo.net, or search the Internet for 'convert solar longitude to date'.

Another consequence of our continuously gyrating Solar System is that the so-called 'fixed stars' gradually appear to drift out of position. The amount is tiny, currently just 50.29 arcseconds per year, but the effect is noticeable as time progresses. As a result, star charts and star atlases must be redrawn every 50 years to realign the celestial coordinate grid of Right Ascension and Declination. The current version is usually referred to as *Epoch 2000.0*. The previous version was *Epoch 1950.0*. For accuracy, always ensure you use the right Epoch.

You will find reference to local time, or LT, throughout this *Atlas*. Essentially, if a meteor shower radiant (see below) rises at, say, 8 pm LT, then it will be 8 pm wherever you are in the world. Well, almost! If you are in a country that applies Daylight Saving Time during the summer months, then you will need to add 1 hour to the local time given in this *Atlas*.

Finally, there is *Coordinated Universal Time*, or UTC, which is equivalent to *Greenwich Mean Time*. This is a fixed time system based on the Greenwich Meridian, and the time zone where you live will be measured in hours either ahead or behind UTC. So you may need to know how many hours ahead or behind of UTC you are to convert UTC to LT. Again, an Internet search should provide you with the information you need.

Radiants

Although there is a continuous influx of random meteors throughout the year—the so-called *sporadics*—meteors that belong to a specific shower appear to radiate from a particular patch of sky at a particular time of year: the *radiant* (Fig. 1.1). There are a number of things to note about radiants.

First, the radiant is an optical illusion. In the photograph of vehicles on a highway at night (Fig. 1.2), the headlights of each vehicle are obviously travelling along parallel paths but they appear to converge to a point on the horizon. Also note that within a specified time frame, the streaks of light are longer when they are closer to the observer. And so it is with shower meteors. Trace shower meteors backwards, and they should converge at the radiant. Meteors with the greatest length are those that are farthest from the radiant, and that dictates why you should *not* look directly at the radiant. You will see a far greater number of meteors if you look 20° to 45° away from the radiant, where the meteor paths appear longer.

Fig. 1.1 Taurid radiant
Shawn DeGroot's stunning picture of the Taurid meteor shower clearly shows the location of the radiant. (Credit: Shawn DeGroot, www.flickr.com/photos/88565330@N05)

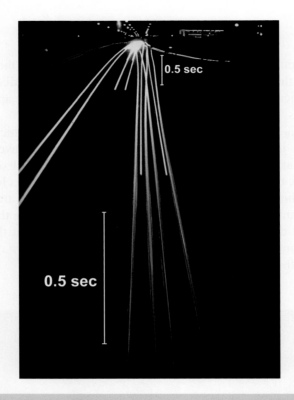

Fig. 1.2 Vehicles on a motorway at night
The radiant is an optical illusion, as shown by this image of vehicles on a highway at night. The lights on the vehicles are obviously a fixed distance from one another and therefore have parallel paths, but they seem to converge to a point on the horizon. This is analogous to a meteor shower's radiant. Note also that over a duration of half-a-second, the longest streaks of light are the farthest from the horizon

The second point is that when we talk about a radiant, what we actually mean is the *average radiant*. Within the average radiant there will be numerous *radiant points*, and a camera can often reveal these with far more precision than the human eye.

Radiants vary in size and there is often disagreement about how big a radiant is. A 'safe' figure is 8° in diameter—that's slightly less than the distance across your knuckles when you hold your fist at arm's length. The radiant of some showers appears to shrink at the time of maximum activity, indicating that the Earth is passing through a tight core of meteoroids, and some radiants are oval in shape. Radiants can also appear to be different sizes and shapes depending on the location of the observer. In Australia, when the Southern δ-Aquariid radiant is almost directly overhead, it appears round and small—some say less than 1° in diameter. However, from a little north of the Equator, with the radiant low on the horizon, it can appear large and diffuse, an artefact of a phenomenon known as *zenith attraction*.

Major Showers and Minor Showers

As a general rule of thumb, a major shower has a ZHR (see below) of 10 or more at maximum activity. A minor shower has a ZHR of less than 10.

Figure 1.3 shows the distribution of 94 radiants against the celestial background. The major showers are represented by the larger circles. What is immediately obvious is that most of the showers are in the Northern Hemisphere. However, this is likely to be due in large measure to the population distribution across the surface of the Earth. The Southern Hemisphere has a smaller population and fewer meteor observers, so many showers, and particularly minor showers, probably go undetected.

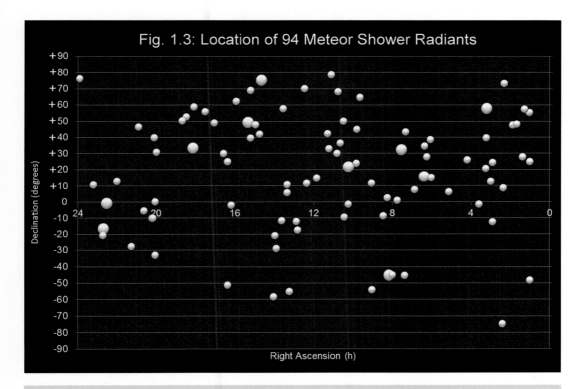

The location of meteor shower radiants, to some degree, reflects the distribution of observers

Meteoroid streams, the physical bands of dust in orbit around the Sun, produce meteor showers, the display of meteors in the Earth's atmosphere, when the Earth crosses the orbit of the stream. The term 'meteor stream', although often used, is nonsensical and should be avoided.

Shower Names

Showers are usually named after the constellation in which the radiant lies at maximum activity. So the *Perseids* are so named because the radiant is in Perseus, the *Orionids* after Orion, and the *Quadrantids*...well, they are named after a constellation that no longer exists—*Quadrans Muralis*—but that's the only shower that bears the name of a defunct constellation.

Many showers, particularly the minor showers, may be named after the star closest to the radiant, such as the *α (alpha) Capricornids*. It is commonplace to hyphenate the name of a shower, as in the *α-Capricornids,* but this is not strictly necessary. Some familiarity with the Greek alphabet helps, a copy of which can be found in *Appendix 1*.

A number of showers are named after the month in which they occur, such as the *April Lyrids* and the *June Boötids*, and daytime showers are indicated as, for example, the *Daytime κ (kappa) Aquariids*. Just for good measure, there are also showers named after both the month and daytime hours, e.g. the *Daytime April Piscids*. The meteor shower with the longest name is the *Daytime Capricornids-Sagittariids.*

Older texts will sometimes make reference to the *Bielids*, *Giacobinids, Pons-Winneckids* or the *Halleyids,* for example. These refer to the shower's parent comet: Biela's Comet (the shower is now called the *Andromedids*), Comet Giacobini-Zinner (*Draconids*), Comet Pons-Winnecke (*June Boötids*) and Halley's Comet, which produces two meteor showers: the *η (eta) Aquariids* and the *Orionids.*

Showers are also given a 3-letter code and a numeric code by the International Astronomical Union's Meteor Data Centre. So the *α-Capricornids'* 3-letter code is CAP and its number is 00001. The 3-letter code is a useful shorthand notation and is used throughout this *Atlas.*

The Zenithal Hourly Rate (ZHR)

Meteor shower activity is measured by a quantity called the *zenithal hourly rate*, or *ZHR*. Some people add meteors per hour or met/h after the number, e.g. *ZHR 35 met/h*, some people don't: it doesn't really make any difference.

The ZHR is the number of meteors an observer will see if the shower radiant is directly above their head and the faintest star visible, the *limiting magnitude*, is $m_v + 6.5$. At least in theory. In practice, there are all sorts of issues with the ZHR.

The height of the radiant above the observer's horizon will affect the number of meteors that can be seen, as shown in Fig. 1.4. Clearly, the higher the radiant, the more meteors the observer will see. This point is often missed on popular science writers and, as a result, they will tell people to 'go outside this evening' to watch the Perseids, when in fact, the shower doesn't really get started until after midnight because the radiant is too low in the sky. As a result, thousands of people are disappointed because they nip out for an hour at about 10 pm and see very little.

The Zenithal Hourly Rate (ZHR)

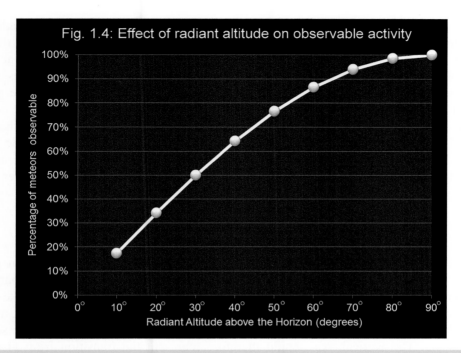

Fig. 1.4: Effect of radiant altitude on observable activity

The predicted ZHR of a shower should be adjusted to take into account the altitude of the radiant above the horizon. For example, if the radiant is 30° above the horizon, then the observer should see 50% of the observable meteors. In addition to radiant altitude, further adjustments must be made to take into account the limiting magnitude if it is less than $m_v + 6.5$ and how much of the sky is obscured, say by cloud, buildings or mountains. Table 1.1 is a ZHR calculator

Table 1.1: ZHR Calculator

	ZHR	5	10	20	30	40	50	60	70	80	90	100	110	120	130	140	150	160	170	180	190	200
Radiant Altitude	10°	1	2	3	5	7	9	10	12	14	16	17	19	21	23	24	26	28	30	31	33	35
	20°	2	3	7	10	14	17	21	24	27	31	34	38	41	44	48	51	55	58	62	65	68
	30°	3	5	10	15	20	25	30	35	40	45	50	55	60	65	70	75	80	85	90	95	100
	40°	3	6	13	19	26	32	39	45	51	58	64	71	77	84	90	96	103	109	116	122	129
	50°	4	8	15	23	31	38	46	54	61	69	77	84	92	100	107	115	123	130	138	146	153
	60°	4	9	17	26	35	43	52	61	69	78	87	95	104	113	121	130	139	147	156	165	173
	70°	5	9	19	28	38	47	56	66	75	85	94	103	113	122	132	141	150	160	169	179	188
	80°	5	10	20	30	39	49	59	69	79	89	98	108	118	128	138	148	158	167	177	187	197
	90°	5	10	20	30	40	50	60	70	80	90	100	110	120	130	140	150	160	170	180	190	200

The calculation of the ZHR is also problematic. Each observation carries a degree of error, and that error is different for each person. In addition, each observer's level of error changes as the night progresses because of a number of factors, such as fatigue, caffeine levels, time spent recording observations and loss of attention. Meteor shower analysts have gone to great pains to try to iron out these creases, and it is not uncommon for them to disagree depending on which errors they have taken into account and how they have mitigated them.

Activity predictions should always be taken with a large pinch of salt. Predictions are often based on past performance, but if the previous analyses contain errors, that can lead to a poor prediction. Another form of prediction depends on looking at when meteoroids have been ejected by a comet and undertaking orbital calculations to see when the ejected material will cross the Earth's path. Although researchers are getting better at this, there have still been a few red faces over the years.

With all this uncertainty about the accuracy of the ZHR, you may well ask what's the point of even quoting a ZHR figure? Well, if you look at the ZHR as a guide to the general level of expected activity, rather than a hard-and-fast figure, you will at least be able to decide whether you wish to make an effort to observe a particular shower.

Magnitude Distribution Graphs

There is a 'tradition' in meteor astronomy of producing magnitude distribution graphs that are back to front. Graphs of visual magnitude are usually shown as bar charts that have the brightest meteor closest to the vertical axis. This simply does not make sense. There is no upper limit to the visual magnitude of a meteor. There is, however, a lower limit, which is about $m_v + 6.5$. Any meteor fainter than $m_v + 6.5$ is too dim to be seen by most people. The vertical axis should therefore represent the transition from visible to invisible meteors.

We are going to break with this tradition. Continually repeating something that is wrong does not make it right! Figure 1.5 illustrates this further.

Magnitude distribution graphs are only useful to a certain level. Generally, the larger the meteoroid, the brighter the resulting meteor will be. In any stream, the number of meteoroids increases with decreasing size. That is to say that there are a lot more smaller meteoroids than larger meteoroids. In theory, then, observers should see more sixth magnitude meteors than fifth magnitude meteors, and more fifth than 4th. In practice, the number of observed meteors tends to peak around third magnitude and then falls off. This is not because the smaller meteoroids are missing, but because the human eye isn't very good at detecting fast-moving, faint meteors.

Despite the shortcomings of the human eye and the resulting magnitude distribution graphs, it is worth comparing your own observations to a larger dataset. If you are consistently over- or underestimating meteor magnitudes, then comparing your results should reveal this error. Observers who are new to meteor watching often struggle to accurately determine a meteor's magnitude—and let's face it, you have less than a second to do so!—but accuracy comes with practice.

Magnitude Distribution Graphs

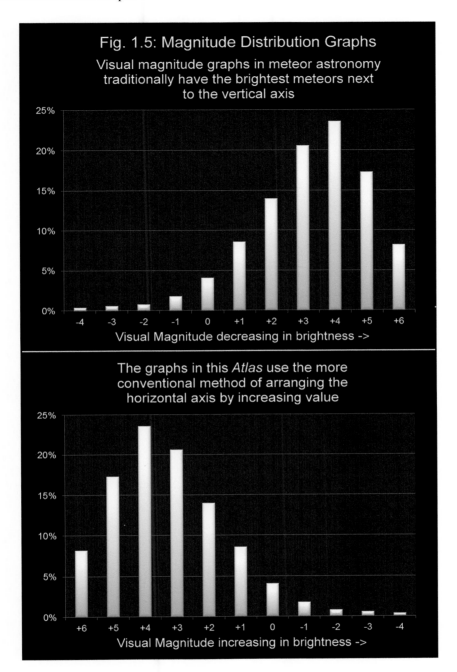

The magnitude distribution graphs in this Atlas follow the normal convention of placing the lowest value (magnitude $m_v + 6.5$) next to the vertical axis

Chapter 2

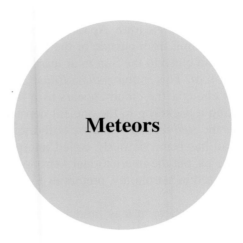

Meteors

The sudden, unannounced appearance of a meteor often takes people by surprise. They can appear in any part of the sky at any time during the night. Most are so swift they can be missed in the blink of an eye, flashing high above the observer's head for typically between 0.4 s and 0.8 s. They are usually faint to moderately bright; the brightest tend to last longer. They occur in a variety of colours, though most are white, some fragment and nearly all are silent. So why is there so much variety? And what is the science behind these brief streaks of light? (Fig. 2.1)

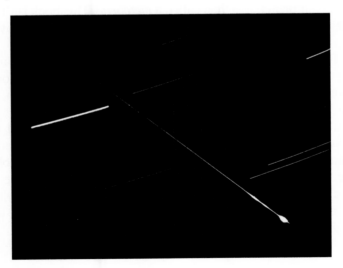

Fig. 2.1 Perseid meteor
Harold Ridley's classic 1958 image of a m_v -2.0 Perseid meteor showing terminal flaring
Credit: Harold B. Ridley

© The Author(s), under exclusive license to Springer Nature Switzerland AG 2021
P. M. Bagnall, *Atlas of Meteor Showers*, The Patrick Moore Practical Astronomy Series,
https://doi.org/10.1007/978-3-030-76643-6_2

Let's start with a few definitions.

We tend to imagine the space between the planets of the Solar System as being largely empty. Nothing could be further from the truth. Apart from being filled with radiation from the Sun, and cosmic rays originating from outside of the Solar System, there is also a vast cloud of dust that stretches from close to the Sun to at least as far as Jupiter, a distance of some 5.5 AU (816.6 million km). This is the *zodiacal dust cloud*. From those regions of Earth where the sky is dark enough, the cloud appears as a faint cone of light stretching along the ecliptic and creating a backdrop to the ancient constellations of the Zodiac. We refer to this ethereal glow as the *zodiacal light* and we will discuss it in more detail shortly.

The dust within this cloud has been variously called interplanetary dust, cosmic dust, micrometeoroids and meteoroids, but the International Astronomical Union (IAU), the body that defines the terminology used in astronomy, prefers us to simply call the dust particles, *dust* [1].

When the dust enters our atmosphere, it may produce a meteor. The term *meteor* refers to the streak of light you see in the night sky and also any accompanying phenomena such as heat, ionization, sound, etc.

If the meteor is brighter than the planet Venus (maximum visual magnitude m_v −4.4), then we refer to it as a *fireball,* a *bolide,* or *superbolide* if it is exceptionally bright. While most meteors are caused by dust from comets and tend to be quite friable, about 40% of fireballs and bolides have their origins in the asteroid belt and are caused by fairly large, solid rocks that have been ejected from the asteroid belt towards the Sun. Others seem to be related to the disintegration of Earth-crossing asteroids. These rocks are generally called *meteoroids*. If the meteoroid is not completely destroyed by its passage through the atmosphere and lands on our planet, it is then called a *meteorite*. This book isn't really concerned with meteorites but they will get a mention from time to time just to provide a more complete picture.

Thousands of fireball events take place every day, and some fireballs are bright enough to be seen during daylight hours. A regular meteor observer will witness on average one m_v −4 fireball for every 20 h of observing time. Increase the magnitude to m_v −6, and the fireball rate drops to 1 in every 200 h.

Some meteors, most fireballs and probably all bolides leave behind *meteoric dust* in their *train*. A very short-lived, almost instantaneous train is called a *wake*. Very bright fireballs and bolides sometimes leave behind a dense cloud of smoke particles—the smoke train (Fig. 2.3)—that can hang in the air for some time. These meteoric smoke particles (MSP) are tiny, typically just 100 nanometres across. High winds in the upper atmosphere can distort the meteoric smoke train in much the same way that aircraft vapour trails are distorted.

Meteorites

Larger, more solid and more robust meteoroids do not behave in the same way as the dust that produces the vast majority of meteors. Unlike friable dust particles that have their origins mainly in comets, meteoroids often come from the asteroid belt, Mars and the Moon. When they hit the atmosphere they do not immediately fragment. Instead, the meteoroid will stay intact for a short while and the column of air ahead of the meteoroid is compressed and rapidly heats up, in much the same way that a bicycle tyre pump gets hot as the air in the tube is squeezed.

The energy involved is mind-boggling. Even at the minimum entry velocity of 11 km/s, the kinetic energy is 60,000 joules per gram of the meteoroid's mass. By comparison, the chemical explosive TNT produces 'only' 4000 joules per gram. The heat generated by entry can be in excess of 2,000 °C and is sufficient to vaporize and ionize the surface of the meteoroid—a process known as *ablation*–producing a glowing plasma many times the diameter of the meteoroid and tens of kilometres in length. Molten drops are carried away from the surface of the meteoroid to eventually cool and settle in the atmosphere, but some are immediately sucked back and stick to the rear surface of the meteoroid as a momentary vacuum develops behind it. Despite the brilliance of the resulting fireball—sometimes so bright it can briefly turn night into day—only about 0.1–1% of the meteoroid's kinetic energy is actually converted to light: the rest is largely heat. Often the aerodynamic pressure will eventually exceed the crushing strength of the meteoroid, at which point it shatters. Any fissures in the surface of the meteoroid can result in the meteoroid fragmenting more quickly. Each fragment then produces its own smaller fireball.

If a meteoroid with a high crushing strength, $>10^6$ Newtons/m^2, has a mass of more than 100 g and an atmospheric entry velocity of less than 25 km/s, it stands a good chance of hitting the ground more or less intact, at which point it is termed a *meteorite*. Once the meteoroid loses its initial *cosmic velocity*—its original orbital velocity—it is no longer travelling fast enough to ionize and ablate, and the light phenomena cease at this so-called *point of retardation*. The object is then propelled forward many kilometres and falls only under the influence of the Earth's gravity, losing heat all the time to the cooler atmosphere. Any molten material on the surface cools to form a glassy coating called a *fusion crust*.

Meteorites found within minutes of landing are usually described as cold to warm, although a few seem to have been hot and one or two so hot as to cause fires. Many meteorites are stony, similar in some ways to most of the rocks found on the Earth's surface, and are poor conductors of heat, so the heat generated by entering the atmosphere does not penetrate the meteorite to any significant depth, often just a few millimetres (Fig. 2.2).

Although fireballs have been known since the dawn of civilization, their relationship to meteorite falls was only discovered a couple of hundred years ago. This is because those who witness a meteorite landing never actually see the fireball, which often peters out over the visible horizon. And those who do see the fireball are tens if not hundreds of kilometres from where the meteorite actually lands.

Fig. 2.2 The Middlesbrough meteorite
The Middlesbrough meteorite fell at 3:35 pm LT on 1881 March 14 at Pennyman's Railway Siding in North Yorkshire, England. Middlesbrough [2] is an oriented meteorite, meaning it did not tumble or spin as it passed through the atmosphere. As a result, it has been sculptured by air flow to form this beautiful cone shape, not unlike an Apollo Command Module

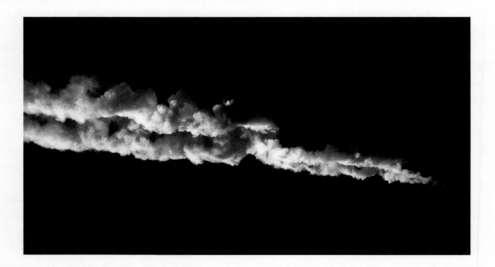

Fig. 2.3 Meteoric smoke train
The meteoric smoke train of the Chelyabinsk superbolide hangs in the daytime sky above Russia, 2013 February 15. This particular superbolide resulted in a meteorite impact. The original meteoroid was probably about 20 m in diameter with a mass of 12–13 Mg and hit the upper atmosphere at 19 km/s (68,400 km/h), exploding at a height of 29.7 km. The shock wave damaged more than 7000 buildings and injured more than 1000 people, though none seriously. The explosion showered the area with meteorites, the largest of which was a 60-cm fragment that weighed 654 kg. By the time the meteorite hit the Earth it had lost almost 98.8% of its entry velocity. The terminal impact velocity was around 810 km/h. Credit: Nikita Plekhanov / CC BY-SA 3.0

A Meteor Event

Meteors often first become visible at between 100 and 118 km altitude [3], but November's superfast Leonid meteors, which plunge into the atmosphere at 69.7 km/s, can become visible as high as 200 km.

Cosmic dust particles enter the upper atmosphere at between 11 and 72 km/s (39,600 to 259,200 km/h), which is a combination of the Earth's orbital velocity and that of the dust particle. Atmospheric entry at these speeds is a highly destructive process that rips the dust particle apart, preventing most cosmic dust from ever reaching the Earth's surface. At these incredibly high velocities, the Earth's atmosphere no longer behaves as most people would expect. Once the meteoroid meets air that is equivalent to its own mass, it rapidly decelerates. Instead of the air flowing smoothly around the dust particle, the atmospheric molecules tear into it and individual atoms start to break apart, electrons being unseated from their orbitals to create a stream of charged *plasma*. Plasmas in the atmosphere are highly unstable, however, and the individual components—the electrons and the atomic nuclei—quickly recombine.

It is during recombination that any excess energy is released as light to produce the meteor we see in the night sky. Velocity is an important factor in how bright a meteor shines. A 6-mm cosmic grain travelling at low speed will typically produce a m_v +5 meteor, just bright enough to be seen, while a high-speed object could brighten to zero magnitude or more, outshining the brightest stars in the night sky. In terms of power output, if a 0.008-g cosmic dust particle enters the atmosphere at the upper speed limit of 72 km/s, it will produce 200 Watts of light for a whole second. However, if the dust particle is small enough (less than 0.004 mm and 10^{-12} g), it will rapidly dissipate the heat caused by entry and temporarily settle, unscathed, in that part of the atmosphere called the mesosphere. Eventually it will find its way to the Earth's surface, sometimes inside a raindrop.

Any metal atoms left in the mesosphere by meteor and fireball events can potentially act as nuclei around which water vapour condenses to form a thin layer of cloud. The metal atoms react with CO_2 (carbon dioxide) which may then attract water molecules to form ice crystals. Normally, these clouds are too high and thin to be visible from the Earth's surface, but about a month either side of the summer solstice (around 21 June) at latitudes of between 50°N and 70°N, they can be seen as bright *noctilucent clouds* against the darkening evening sky because of the angle of the Sun (Fig. 2.4). In the Southern Hemisphere they can be seen a month either side of 21 December.

Fig. 2.4 Noctilucent cloud
Dust left behind by meteors can contribute to the formation of noctilucent clouds in the mesosphere at 75 to 85 km altitude. These clouds can only be seen at certain times of the year and at certain latitudes. Noctilucent is from the Latin for *night shining*. Credit: Natalia Eriksson, www.flickr.com/photos/natalia_erikkson

Meteors and fireballs can be very deceptive. They give the impression of being just above the observer's head, certainly no more than a few kilometres, but in reality most are well above 75 km altitude. Those who interview witnesses to a bright fireball are often told that the fireball 'landed just beyond those trees / hills / building' etc., but the researchers know that, in reality, 'just beyond' translates into tens and sometimes hundreds of kilometres!

Some particularly bright fireballs emit sound, often described as sonic booms, cracking or, in past centuries, like canon fire (sonic booms). Some of these sounds are caused by the meteoroid fragmenting, and they can reach the observer in the wrong order. Imagine a fireball heading towards you. As it enters the atmosphere, it produces a sonic boom. Twenty kilometres later, still heading towards you, it starts to fragment and emits a loud cracking sound. Finally, it makes a fizzing noise. Because you are physically closer to the fizz and the meteoroid is travelling a lot faster than the speed of sound, you will hear the fizzing first, then the crack and then the sonic boom as it eventually reaches you, having travelled a much greater distance. Someone watching the fireball travel away from them will hear the sounds in the correct order (Fig. 2.5).

The planets of the Solar System are thought to have formed by a gradual build-up of solid material—dust grains sticking together to form pebbles, pebbles sticking together to form boulders and so on. This process is known as *accretion* and was important in creating and shaping the early Solar System. The fact that the Earth still collects material from space demonstrates that accretion is not something that just happened in the past; it is still going on even today, 4.56 billion years after the birth of our planetary system.

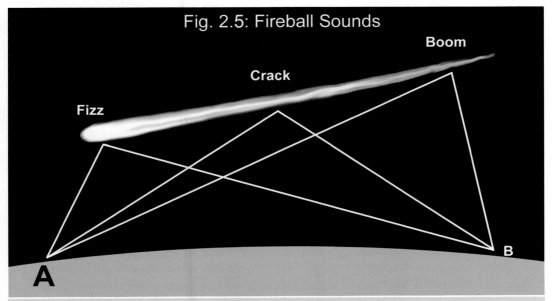

Fig. 2.5: Fireball Sounds

If a fireball is heading towards observer A, they will hear any sounds in the wrong order because the fireball is travelling faster than the speed of sound and the observer is closer to the 'fizz' than the 'boom.' For observer B, who is seeing the fireball travel away from them, they will hear the sounds in the right order

The Great Meteor Procession of 1913

A procession of fireballs is a very rare event. Perhaps the most famous is the so-called Cyrillids (or Chant) Great Meteor Procession of 1913 February 9. The procession was witnessed by hundreds and possibly thousands of people, all the way from Canada to ships off the coast of Brazil (Rosenfeld [9]).

The procession began at about 21^h EST (2^h on Feb 10 UT) with the appearance of a bright fireball on the north-western horizon. As the fireball crossed the sky, it soon became apparent that it actually consisted of several discrete objects. The fireballs moved slowly, taking 30–40 s to traverse the heavens, and were followed by other clusters on almost identical paths. In all, between 40 and 60 fireballs of various magnitudes, sizes and colours appeared over a 3 to 5 minute period. Observers also reported various sounds coming from the fireballs, described as cannon, thunder, swishing and hissing. Then, about $5^h\ 20^m$ later, a second procession occurred.

It has been estimated that hundreds of fireballs skimmed through the atmosphere at between 40 and 60 km altitude, travelling at about 8 km/s. Various theories have been put forward to explain the procession, but the one that looks most promising is that the meteoroids that caused the fireballs were temporarily captured orbiters (TCOs) that had gone into orbit around the Earth—effectively, minimoons.

Other processions include dozens of fireballs over Scandinavia on 1931 February 9 and possibly 40 fireballs over Prussian Saxony on 1830 December 12–13, which appear to have been Geminids.

The Zodiacal Dust Cloud

In spring or autumn, particularly in the Tropics, a faint conical glow can sometimes be seen stretching from the horizon along the ecliptic (Fig. 2.6), called the *zodiacal light*. City lights readily bleach out its ghostly presence but, if you are fortunate enough to be miles from civilization, you will not only notice the cone but also, as your eyes get used to the dark, an even fainter extension running right across the sky—the *zodiacal band*. And if you look carefully you may even notice a slightly brighter oval patch at the antihelion point, directly opposite the Sun, usually known by its German name, the *Gegenschein* (or *counterglow* in English). These three phenomena, the zodiacal light, band and Gegenschein, are evidence of the existence of the *zodiacal dust cloud (ZDC)* that pervades the inner Solar System out to at least as far as Jupiter.

The ZDC is not one huge homogenous mass of dust particles: it has considerable structure. Within the cloud are streams of meteoroids released by comets and some classes of asteroids called *active asteroids*, which are objects that appear to be a rocky asteroids but which sometimes display the appearance of a comet, releasing dust and volatiles into space. The meteoroid streams are sometimes very compact and dense, indicating they are young additions to the cloud. Older meteoroid streams are barely detectable; they spread

Fig. 2.6 The zodiacal light
The ghostly conical glow of the zodiacal light as seen from La Silla Observatory, Chile, in September 2009 and captured by Yuri Beletsky. The zodiacal light is caused by sunlight reflected from countless dust particles that lie along the plane of the Solar System. From La Silla's latitude at 29°S, the cone makes almost a right angle with the horizon at this time of year. Beautiful though it is, the zodiacal light can hinder deep-space studies of stars and nebulae. Credit: ESO and Y. Beletsky

out over millennia and will eventually blend completely into the background dust. The zodiacal dust cloud is losing dust particles continuously as some fall into the Sun and planets, while others are blown to the outer reaches of the Solar System by the solar wind and may even escape the pull of the Sun altogether to journey into interstellar space. The birth, ageing and dissipation of meteoroid streams help to continually replenish the cloud.

While the zodiacal light has probably been known since ancient times, the first published record of its appearance was not until 1660, when Joshua Childrey mentioned it in *Britannia Baconica*, a volume on natural history [4]. It was a further 23 years before the Franco-Italian astronomer Giovanni Cassini [5] undertook a study of the phenomenon. It was either he or his collaborator Nicolas Fatio de Duillier, a Swiss natural philosopher, who correctly reasoned that the glow was sunlight reflected from dust particles orbiting the Sun.

The ZDC is undoubtedly quite extensive, but it is not dense. The brightness of the zodiacal light suggests that if the ZDC were composed of dust particles 1 mm across and with *albedos*, or reflectivity, similar to that of the Moon (about 13%), then the spatial distribution of the individual dust particles would be about 8 km. Of course, as the sizes of the particles gets smaller, their number increases significantly. There is a lower limit of 10 μm (0.01 mm) for dust particles in the inner Solar System; anything smaller is blown away from the Sun by the solar wind. There is also a limit to how close dust particles can approach the Sun before they are vaporized—about 2.8 million kilometres. Given what we know about the zodiacal cloud, all of its dust particles could be condensed into a single body just 30 km in diameter with a density of 2.5 g/cm^3 and a total mass of around 35.4×10^{15} kg [6]. To put that into perspective, the zodiacal dust cloud has a total mass of just 0.7% of the Earth's atmosphere.

The cloud's origin has three components. Dust caused by the fragmentation of comets, particularly short period comets with orbital periods of <200 years, contribute about 70% of the mass of the cloud. Inter-asteroidal collisions and active asteroids account for a further 22% of the mass, with the remaining 8% being made up of interstellar dust particles that have drifted into the Solar System from the depths of the galaxy.

Over thousands of years, the ZDC can be regarded as dynamic. Most of its component dust particles will spiral in towards the Sun, a process known as the *Poynting-Robertson effect* [7, 8]. This is caused by the dust absorbing radiation from the Sun and then re-radiating it at a slightly different wavelength, producing a tangential drag on the particle. It mainly affects particles in the 1 μm to 1 mm range and is a very slow process. A dust particle released by, say, a main belt asteroid 2.7 AU from the Sun, will typically take about 10,000 years to spiral in to the position where it could intersect the Earth at 1 AU.

The View from Earth

As the Earth journeys around the Sun on its year-long orbit, it passes continuously through the various components of the zodiacal dust cloud. But from the point of view of a ground-based observer, what does this journey actually look like?

On a quiet day the Earth sweeps up about 5 Mg—five metric tonnes—of dust from the ZDC. At certain times of the year, however, the Earth will plunge into a compact stream of cosmic dust left behind by a disintegrating comet. When this happens, the total mass increases significantly, perhaps by as much as 270 Mg. Over the course of a year, the infall probably averages out at about 8 Mg a day.

With all this dust continually entering the Earth's atmosphere, we should see a never-ending drizzle of meteors. So why don't we? Well, there are numerous reasons, the most obvious being that half of the Earth is bathed in sunlight and the bright daytime skies bleach out just about every single meteor. A vast majority of meteors will burn up over the oceans and land masses that are sparsely populated, so even at night there is no one around to see them come to their fiery end. In more populous areas, light spill from our towns and cities has a similar effect as daylight, hiding meteor activity. And of course, people generally do not look skywards anyway.

Some dust particles are so small that they pass through our atmosphere unscathed, never producing a meteor to see. Many others result in meteors that are so swift and faint they are easily missed unless you make a deliberate effort to observe them. If you do, then you will notice that there is indeed a continuous rain of meteors, but the rate at which they appear varies throughout the day and year.

On a normal, run-of-the-mill day you will probably see between 3 and 10 meteors an hour in the night sky. The rate tends to be lowest at about 6 pm LT and highest 12 h later, at 6 am. This is because at 6 am, the Earth is heading into the ZDC, catching even the swiftest dust particles and gravitationally plucking them out of the cloud to burn up in the dawn skies. At 6 pm, the opposite is true. The Earth leaves behind the dust and only the fastest particles heading towards the Earth can catch up and plunge into the atmosphere (Fig. 2.7a, b). Because of their apparent randomness—they can appear anywhere in the sky at any time—these meteors are termed *sporadic*.

If you systematically observe the skies throughout the year, you will notice that there are somewhat more sporadic meteors in the late autumn in the Northern Hemisphere. In the south, the situation is reversed, though it is not exactly a mirror image, with sporadic rates peaking in June (Fig. 2.8). Although we call them 'sporadic' meteors, the truth is they are not as random as previously believed. Almost hidden among the sporadic haze are half a dozen areas that are the source of at least some of the so-called sporadics. These are known as the Helion and Antihelion (or, incorrectly, the Anthelion) sources, the North Apex and South Apex and the North and South Toroidal sources. The Helion and Antihelion lie in the plane of the ecliptic, 60° to 70° from the apex of the Earth's way—the apparent direction in which the Earth is travelling. The North and South Toroidal sources lie about 60° north and south of the ecliptic. The study of these sporadic sources can tell us much about the comets or comet families that led to their creation, but research is best carried out using, for example, radar equipment. Visual observers will be disappointed by the low hourly rates.

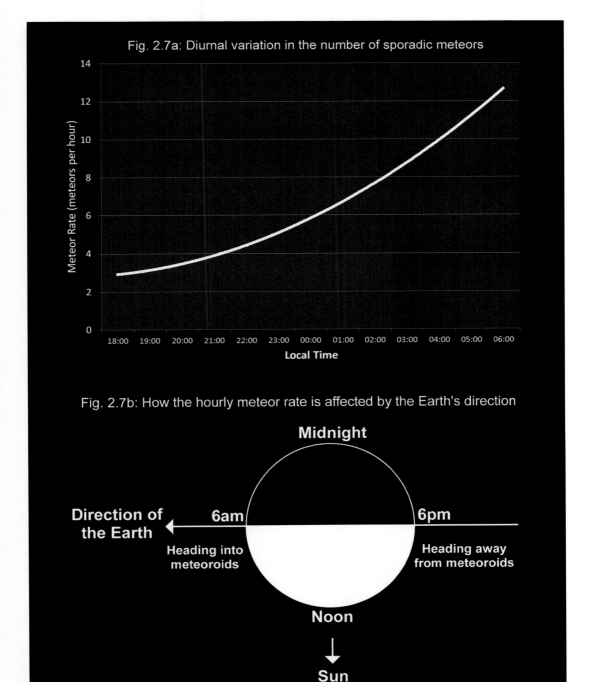

Fig. 2.7a Sporadic meteor rates vary according to the time of day, with most appearing at 6 am local time in latitudes where it is still dark at that hour

Fig. 2.7b At 6 am the Earth is heading into the cloud of zodiacal dust. At 6 pm it is leaving the dust behind. It is a similar situation to driving when it is snowing. The snow falls heavily on the front windscreen of your vehicle, whereas the back window appears to encounter very little snow

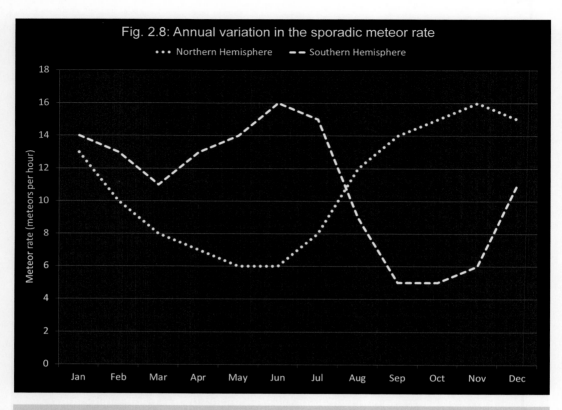

Sporadic meteor rates vary according to the time of year and hemisphere. In the Northern Hemisphere, maximum occurs in November, while south of the Equator maximum is in June

References

1. Retrieved from https://www.iau.org/static/science/scientific_bodies/commissions/f1/meteordefinitions_approved.pdf. Accessed 27 Aug 2020.
2. Grady, M. M. (2000). *Catalogue of meteorites* (5th ed.). Cambridge University Press.
3. Jacchia, L. G., Verniani, F., & Briggs, R. E. (1965). *An analysis of the atmospheric trajectories of 413 precisely reduced photographic meteors*. Smithsonian Institution Astrophysical Observatory.
4. Childrey, J. Britannia baconica
5. Jenniskens, P. (2006). *Meteor showers and their parent comets*. Cambridge University Press.
6. Pavlov, A. A. (1999). Irradiated interplanetary dust particles as a possible solution for the deuterium/hydrogen paradox of Earth's oceans. *Journal of Geophysical Research: Planets, 104*(E12).
7. Poynting, J. H. (1903). Radiation in the solar system: Its effect on temperature and its pressure on small bodies. *Monthly Notices of the Royal Astronomical Society, 64*(Appendix), 1a–5a.
8. Robertson, H. P. (1937). Dynamical effects of radiation in the solar system. *Monthly Notices of the Royal Astronomical Society, 97*, 423.
9. Rosenfeld, R. A. (2011). Gustav Hahn's graphic record of the great meteor procession of 1913 February 9. *Journal of the Royal Astronomical Society of Canada, 105*, 167.

Chapter 3

Meteoroid Steams and Meteor Showers

Against a backdrop of a constant hail of sporadic meteors, mainly from the zodiacal dust cloud, meteor activity can increase significantly at certain times of the year. Anyone who has even a passing interest in astronomy will know, of course, that these meteor showers are the debris left behind mainly by comets disintegrating as they approach the inner Solar System, and a few so-called active asteroids. But to leave it at that paints a simplistic picture that fails to place meteors in their true context.

In this chapter we'll look at how meteors, meteoroid streams and meteor showers are related to just about everything else in our planetary system.

The Structure of the Solar System

At the centre of our Solar System is, of course, the Sun, encircled by the four rocky terrestrial planets: Mercury (0.4 AU from the Sun), Venus (0.7 AU), Earth (1.0 AU) and Mars (1.5 AU). At about 2.1 AU is the start of a belt of mainly rocky objects, which most people refer to as *asteroids* but which are also called *minor planets*. These are likely to be the debris left behind when the Solar System was born 4.56 Gyr ago, and range in size from dust to mountainous chunks of rock and underdeveloped planets. Known collectively as the *asteroid belt*, the entire mass is somewhere between 3% and 4% of the mass of the Moon, or about 2.39 to 2.94×10^{21} kg.

The object (1) Ceres, with a diameter of 945 km (Fig. 3.1), accounts for about one-third of the entire mass of the asteroid belt. The next three minor planets in terms of size are (4) Vesta at 525 km diameter, (2) Pallas at 512 km diameter and (10) Hygiea at 434 km diameter. Together, they account for a further 17% of the belt's entire mass. The remaining 50% of the mass is distributed among millions of asteroids, of which about 1.7 million are larger than 1 km across. In popular culture, particularly science fiction films, the asteroid belt is depicted as a crowded and hazardous region of space to navigate. The truth is that between 2.1 to 3.1 AU from the Sun,

Fig. 3.1 Ceres
(1) Ceres was originally regarded as a planet when it was discovered on 1801 January 1 by Giuseppe Piazzi, then as an asteroid from the 1850s and later a minor planet. Since 2006, it has been classed as a dwarf planet along with the former planet Pluto, the trans-Neptunian objects (136199) Eris and (136108) Haumea, and the Kuiper belt object (136472) Makemake. It is also considered to be an active asteroid. Credit: NASA/JPL-Caltech/UCLA/MPS/DLR/IDA/Justin Cowart. CC BY 2.0

space is a big place and the asteroids are well spread out. Indeed, an alien residing on an asteroid may never see another within a typical human lifespan and so could be forgiven for thinking that they are living on the only asteroid in the entire Solar System.

Beyond the asteroid belt is the home of the gaseous giants: Jupiter at 5.2 AU from the Sun and 11 times bigger than the Earth, and Saturn at 9.5 AU, nine times as big as the Earth and 95 times more massive but with such a low density it could float on water—if you could find a pool big enough! Then there is Uranus at 19.2 AU and Neptune bringing up the rear at 30.1 AU.

From about the position of Neptune out to perhaps 55 AU is the *Kuiper Belt*, named after the Dutch-American astronomer Gerard Kuiper. It is also sometimes called the *Edgeworth–Kuiper Belt* in recognition of Kenneth Edgeworth, a British astronomer who first suggested the existence of the belt in a paper published in 1943. In some ways it is similar to the asteroid belt but much bigger by a factor of about 20 and perhaps containing as much as 200 times more mass. The main difference is that while the asteroid belt consists of mainly bare rock and rock fragments, the material in the Kuiper Belt is largely composed of icy objects that may have rocky or rubble-pile cores made up of several large fragments held together by ice. Some Kuiper Belt objects (KBOs) possibly do not have significant rock content. The least dense KBOs are about 0.4 g/cm^3, which is not far from the density of compacted snow here on Earth, 0.481 g/cm^3. The denser KBOs are in the order of 2.6 g/cm^3, which is about the same as terrestrial

Fig. 3.2 Asteroid (21) Lutitia
The minor planet (21) Lutitia is a typical asteroid whose orbit lies 2.037 to 2.833 AU from the Sun. It was discovered on 1852 November 15 by painter and astronomer Hermann Goldschmidt from the balcony of his Paris apartment. He named the object after the Latin word for Paris. This image was taken by the Rosetta probe in 2010 July and shows the 50-km asteroid to have a heavily cratered surface, typical of asteroids in this region. Credit: ESA/Rosetta

granite. Like the asteroid belt, the Kuiper Belt objects are probably leftovers from the formation of the planets. At more than 5 AU from the Sun, however, it is so cold that ices would have readily formed on their surfaces.

Until fairly recently, it was widely believed that the current structure of the Solar System had remained largely unchanged since the system's birth, and that planetary systems around other stars probably looked much the same as ours. However, improvements in the detection methods of exoplanets have shown that, far from being typical, our Solar System is rather unusual. Most other planetary systems seem to have giant gaseous planets close to the parent star and smaller rocky planets farther out—the exact opposite of the Solar System [1]. These observations, coupled with theoretical work showing that giant planet formation in the outer Solar System is difficult to achieve, have forced scientists to reassess their views on the birth and evolution of our planetary system. The current thinking is that the gaseous giants mostly formed closer to the Sun, but gravitational interaction, particularly with Jupiter, modified their orbits so that we have the arrangement that we find today.

Giant planets wandering through the Solar System can have a profound effect on the orbits of everything in our system, including the myriad of fragments that make up the asteroid and Kuiper belts. It is even possible that without this period of wandering, neither the asteroid belt nor the Kuiper Belt would have formed.

We can see evidence of how Jupiter still influences the asteroid belt today. The asteroid belt is not a band of evenly spaced chunks of rock. There are gaps in the belt at specific distances from the Sun, caused by orbital resonance with Jupiter. For example, an asteroid that strays into an orbit 2.50 AU from the Sun would have an orbital period of 3.95 years and would therefore go around the Sun three times during a single orbit of Jupiter, which has an orbital period of 11.86 years—a 3:1 *orbital resonance*. Each time the asteroid and Jupiter meet up, the giant planet gravitationally tugs the asteroid so that, over thousands of years, it pulls the asteroid and any others in that region into a new orbit, clearing a space within the belt. These spaces are called *Kirkwood Gaps* after the American astronomer and mathematician Daniel Kirkwood who first noticed the phenomenon in 1860. The problem with asteroids being deflected into new orbits is that they are more likely to cross the paths of other asteroids, resulting in a collision. Some of these collisions have sent fragments of asteroids hurtling towards the Sun and occasionally the Earth, where they produce brilliant fireballs and fall as meteorites.

Just as Jupiter affects the inner Solar System, the outermost gas giants, in particular Neptune, influence the outer Solar System, including the Kuiper Belt, which results in some KBOs being scattered into new orbits to form what is generally referred to as the *scattered disc*. Because of the massive gravitational pull of the gas giants, this is a highly unstable region and orbits evolve rapidly over just a few million years. *Scattered disc objects* (SDOs) typically go through a period during which their *perihelia*—their closest approach to the Sun—lie between Jupiter and Neptune. During this phase they are collectively called *Centaurs*, but it is only a transitional period, with some Centaurs being ejected from the Solar System altogether while others are sent into the inner Solar System where we detect them as comets. So although we used many different terms—KBOs, SDOs, Centaurs, comets (and others like trans-Neptunian objects or TNOs)—these are not different types of bodies, but often different phases in the life cycle of small Solar System objects.

Comets are usually classed as belonging to one of two categories: short and long period. *Short period comets* can be further subdivided into Jupiter-family comets (JFCs) and Halley-type comets (HTCs). Jupiter can substantially change the orbit of a comet, deflecting long period comets into short period orbits (Fig. 3.3). *Jupiter-family comets* are thought to have originated in the scattered disc. They have orbital periods of less than 20 years and lie within 30° of the plane of the ecliptic. Those with orbital periods of between 20 and 200 years and inclinations of up to 90° are *Halley-type comets* (named, of course, after the most famous of all comets). Any comet with a period in excess of 200 years is called a *long period comet*. The existence of long period comets suggests a reservoir of such objects lying at least 2000 AU from the Sun and extending to about 50,000 AU, although some scientists suggest 200,000 AU. This comet reservoir was originally the idea of the Estonian astronomer Ernst Öpik in 1932 and was revived in the 1950s by the Dutch astronomer Jan Oort [2]. Today it is known as the *Oort Cloud* or, more rarely, the *Öpik–Oort Cloud*.

The Structure of the Solar System

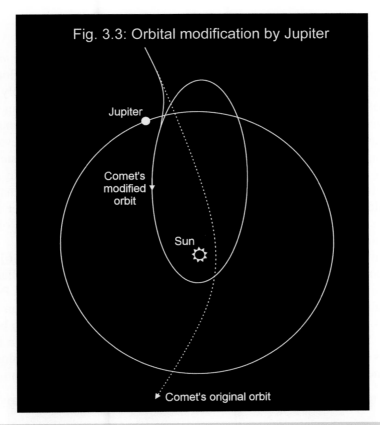

More than any other planet, Jupiter has shaped the Solar System over thousands of millions of years and continues to do so even today. This diagram shows how a long period comet can have its orbit radically modified to a much shorter period by the gravitational influence of Jupiter

The Oort Cloud is an unproven theory. The cloud is too far away to be detected with current technology, and no one is really sure whether it really exists and what shape it takes. A complete sphere, a partial sphere and an oblate shell have all been suggested at one point or another. The Oort Cloud, though bound to the Sun, can easily be affected by the gravitational pull of passing stars and large molecular interstellar clouds, which may dislodge some of the resident comets, sending them into the inner Solar System. Some will approach the Sun and travel outwards again on long, highly eccentric, elliptical orbits that have periods of anywhere from thousands to millions of years. Recently, there has been speculation that the cloud contains more interstellar objects than Solar System objects [3].

The Structure and Behaviour of Comets

There were two competing theories on the structure of comets. The *Whipple model*, devised by Fred Whipple at Harvard in the 1950s, was originally called the *icy* conglomerate model but soon became more widely known as the *dirty snowball* model, probably because it is easier to visualize [4]. Whipple believed that a comet consisted of a nucleus enveloped in ice and meteoric dust. As the comet approaches the Sun, the volatile ices would sublimate, turning from ice directly to gas without going through a liquid phase and evaporating into space, releasing the dust to form a meteoroid stream. Whipple also suggested that, as the comet aged, a layer or crust of non-volatile material would form that would insulate large parts of the comet's surface, inhibiting further sublimation. Where sublimation did take place, it would produce jets of outgassing that had a rocket effect, either increasing or decreasing the orbital energy of the comet and gradually changing the size of its orbit. He pointed to changes in the orbit of Encke's comet as proof [5].

The alternative *sandbank theory* was the work of Raymond Lyttleton [6]. He saw the formation of comets taking place as the Solar System passed through interstellar clouds, with the Sun focussing the dust and gas in its wake. There was no nucleus as such. Proof of which theory was right had to wait until the return of Halley's comet in 1986, when an armada of spacecraft was sent to investigate. It was Whipple's model that turned out to be correct—or at least partly. Some comets may not have a single rocky core but may be a rubble pile of several chunks of rock. While most cometary nuclei are quite small—a few kilometres across at most—Comet Hale-Bopp's (Fig. 3.4) was an exceptional 40–80 km in diameter.

Fig. 3.4 Comet Hale-Bopp, C/1995 O$_1$
Comet Hale-Bopp was a bright comet that was visible to the naked eye for nearly 19 months and could, for a time, be seen during daylight. This image was taken on 1997 March 14 and clearly shows a straight ion tail, which points directly away from the Sun, and a slightly curved dust tail that laid down a new meteoroid stream. The coma is the fuzzy head of the comet. Hale-Bopp, which was named after its discoverers, professional astronomer Alan Hale and amateur Thomas Bopp, has an orbital period of about 2500 years. Credit: ESO/E. Slawik

Most comets spend much of their time in the cold outer reaches of the Solar System and are effectively inert. If a comet is deflected from its original orbit and heads towards the Sun, then it will become active when the heat of the Sun penetrates the comet's surface. Once comets begin to warm up, they can take on a spectacular appearance. As the ice turns to gas, it cocoons the nucleus in a hazy cloud called a *coma* and the comet may brighten significantly and develop a tail, which always points away from the Sun because of the solar wind. The coma can be considerably larger than the nucleus; the Great Comet of 1811—Comet Flaugergues (C/1811 F_1)—had a coma estimated to be 1.4 times larger than the Sun! The tail may appear to split into two components: a straight ion tail that points directly away from the Sun due to its interaction with the solar wind, and a curved dust tail. The sunlight-reflecting dust tail is more noticeable, and some comets have more than one dust tail—De Cheseaux' Comet (C/1743 X_1) had at least six!

Comets and Meteoroid Streams

The sublimation of a comet's ices releases trapped dust along its orbit, forming a meteoroid stream. The stream begins almost immediately to decay, with the component dust particles being dispersed by both gravitational and non-gravitational forces, such as the solar wind and the Poynting-Robertson effect (Fig. 3.5). If the comet enters a short period orbit, then there is a good chance that the meteoroid stream will be replenished each time the comet returns to the Sun and releases a new dust trail. All meteoroid streams eventually disperse and feed the zodiacal dust cloud.

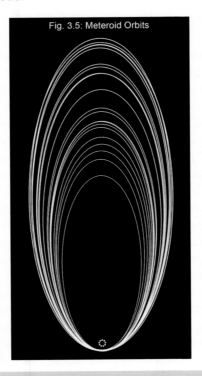

Meteoroids released by a comet, or an active asteroid, will eventually be dispersed by both gravitational and non-gravitational forces into a wide variety of orbits that may not resemble the original orbit of the parent body

Fig. 3.6 Comet Churyumov-Gerasimenko
This spectacular image of Comet 67P/Churyumov-Gerasimenko clearly shows the comet's dumbbell structure. Note the jets of gas and dust streaming out of parts of the comet's surface. Credit: ESA/Rosetta

About ten times a year, the Earth plunges into a dense stream of meteoroids and the hourly count of meteors can increase significantly. Unlike true sporadics, meteoroid streams produce showers of meteors that appear to radiate outwards from a small roughly circular area of sky known, unsurprisingly, as the *radiant* (see Chap. 1). The radiant is usually named after a nearby star, e.g. the κ-Cygnids (kappa Cygnids), or sometimes after an entire constellation; the Perseids for instance. In the past, meteor showers have also been named after comets, such as the Bielids [7], which are now known as the Andromedids. The Quadrantids are named after a constellation that no longer exists, Quadrans Muralis, abolished when the International Astronomical Union redrew the constellation boundaries in 1922.

Meteoroid streams are more complex than most people imagine. A stream can consist of several dust trails released each time the parent comet approaches perihelion. Individual dust trails are usually identified by the year the trail was created. Resonance with the planets, especially Jupiter, can gravitationally sculpt the meteoroid stream producing rope-like filaments of condensed material. The Perseid meteor shower has in the past revealed the presence of several filaments by sudden but short-lived increases in the number of meteors, some hours before the expected maximum. In 1991, one filament was three times more prolific than the main stream. Gaps are also found in meteoroid streams where the Earth or another planet has passed through and swept up all the individual dust particles. Similarly, there are clumps of material caused by sudden outbursts on the comet's surface or by the fracturing of the comet's nucleus. Where fracturing does occur, large chunks of material can join the dust trail and produce spectacular fireballs, and possibly meteorite falls, should they encounter the Earth.

Active Asteroids and Meteoroid Streams

While most meteoroid streams are the result of comets sublimating as they reach perihelion, and releasing dust trails, some asteroids can also produce meteoroid streams.

The idea that an asteroid could cause a meteoroid stream to form, and the resulting meteor shower, originally seemed unworkable. Asteroids do not just spontaneously fragment into dust. Some scientists suggested inter-asteroidal collisions as one possible solution, but the amount of dust in a meteoroid stream is so great that, unless at least one of the asteroids was completely destroyed, there simply would not be enough dust produced by most collisions. In 1983, views began to change. Two British astronomers, Simon Green and John Davies, spotted a new Apollo-type asteroid in data returned by the Infra-Red Astronomy Satellite (IRAS). Apollo-type asteroids have orbits that cross that of the Earth and are therefore potentially hazardous. A couple of weeks later, Fred Whipple noticed that the asteroid was embedded in the Geminid meteoroid stream. The object was later named (3200) Phaethon.

The presence of Phethon cause a bit of a stir. Some people postulated that it was just a coincidence: the asteroid and the meteoroid stream just happen to be in the same orbit. Others suggested that Phaethon was actually a dead comet, that it had outgassed all of its volatiles and just looked like an asteroid, and perhaps that was the fate of all comets. Then another asteroid turned up, 2003 EH_1, buried in the Quadrantid meteoroid stream. Perhaps Phaethon and 2003 EH_1 were once part of comets that had fragmented and they were the largest remnants? Anything was plausible. Since these early discoveries, asteroid surveys have gone digital and a number of what are now called *active asteroids* have been discovered.

The discovery of asteroids in meteoroid streams opened up new avenues of research aimed at explaining how such bodies could lose mass. Trapped water was one possibility, but impacts, landslides caused by rapid rotation, and thermal shock as the asteroid approaches the Sun, have all been cited as possibilities. In 2019, Quanzhi Ye predicted that a new meteor shower would be found in the constellation Sculptor, caused by the disintegration of the active asteroid (101955) Bennu. Tim Cooper and four members of the Astronomical Society of South Africa subsequently observed the shower in September of that year (see Chap. 7).

Types of Meteor Showers

Meteor showers come in two basic flavours: major showers and minor showers. *Major showers* are the most active and have a *Zenithal Hourly Rate* (ZHR) of 10 or more. There are currently 10 major showers if we count the Northern and Southern Taurids as a single shower (the meteors from the two components are so difficult to separate that this is actually a reasonable approach).

Minor showers are considerably more numerous. The IAU Meteor Data Center lists more than a thousand, although only about 100 have sufficient data to elevate them to the status of 'established' [8]. Certain showers show variable activity. The June Boötids and the

Phoenicids, two minor showers, usually have low ZHRs but can suddenly become highly active (see Chap. 7). Outbursts such as these are usually rare but are sometimes predictable if there is sufficient data available on the shower's parent body.

Meteor showers often have their own characteristics. The Perseids have a well-defined and reliable maximum; they rarely disappoint. The Leonids are superfast, tearing into the atmosphere at 69.7 km/s (250,920 km/h)—see Fig. 3.7—while the Taurids arrive at a more leisurely 26.1 km/s (93,960 km/h). Quadrantid fireballs are said to be yellow to bluish, whereas α-Capricornid meteors are often described as electric blue. Some showers last just a few days, others can run for weeks. Of the major showers, the Ursids have the dimmest average visual magnitude at $m_v + 3.2$, while the Perseids appear to be the brightest at $m_v + 1.7$ (Fig. 3.8).

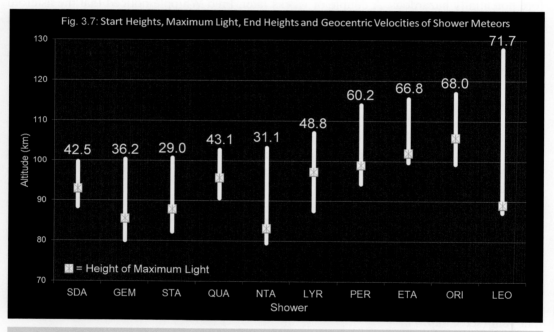

In the 1960s Luigi Jacchia, Franco Verniani and Robert Briggs [9] undertook a study of 413 meteors captured by the Super-Schmidt Camera Network in New Mexico and found that shower meteors had their own characteristics. The Lyrids (LYR) reached maximum light halfway through their journey (square box), while the Southern δ-Aquariids (SDA) and the Quadrantids (QUA) were at their most brilliant two-thirds of the way, and the η-Aquariids (ETA) and Leonids (LEO) were brightest at the end of their passage. The Leonids also became visible at a much greater height than other shower meteors, and plunged deeper into the atmosphere. Although the velocity of a meteoroid (shown at the top of each data bar) clearly has a role to play in when meteors first become visible, it is not the only factor and other features, such as the friability of the meteoroid, must also be taken into account. Velocities quoted are from the original study and have been refined since.

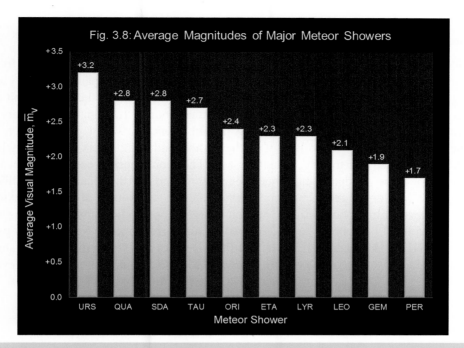

The Perseids have the brightest average visual magnitude at $\bar{m}_v$ +1.7, followed closely by the Geminids, $\bar{m}_v$ + 1.9. At $\bar{m}_v$ + 3.2 the Ursids are the dimmest of all the major showers. Note that the symbol for the average magnitude has a bar over the m (see Glossary for other symbols)

The meteor year is back-loaded, with half of the major meteor showers occurring during the last 3 months. February and March are completely devoid of major showers, as is September. Even so, it pays to go on a meteor watch whenever the opportunity arises. First, it helps to hone your skills in plotting and magnitude estimation. Second, there's always the chance that you will stumble across a new shower or witness an unexpected outburst.

Summary

The never-ending rain of meteors is caused by dust particles that have their origins mainly in comets, but also to a much lesser degree, in the disintegration of active asteroids. Earth-crossing debris from the asteroid belt is by comparison relatively rare, but can be sizeable and sometimes results in stunning fireballs with the object occasionally surviving its plunge to land on Earth as a meteorite. The zodiacal cloud provides a background display of sporadic meteors, the cloud being replenished by ageing meteoroid streams that disperse over thousands of years. The Earth encounters a number of meteoroid streams, which leads to meteor showers at certain times of the year.

Although scientists have identified many classes of Solar System objects—comets, asteroids, Centaurs, KBOs, SDOs, etc.—all of these objects are related and sometimes interchangeable, and they tell us much about the origin and subsequent history of the Solar System.

> **Nemesis: The Death Star**
>
> In 1984, Dr. Richard Muller [10], Professor of Physics at the University of California at Berkeley, suggested that, like so many other stars, the Sun was part of a binary system. In his theory, the Sun's companion was a red dwarf in a highly eccentric orbit that came to within 32,000 AU of the Sun and stretched out to 1.5 light years. Every 26 million years the red dwarf, unofficially named *Nemesis*, would pass through the Kuiper Belt, disrupting the orbits of perhaps thousands of comets, many of which would be perturbed into the inner Solar System. The orbital period was chosen because it fits with some evidence that mass extinctions on Earth occur in 26-million-year cycles. Muller suggested his theory could identify the source of the mass extinctions. However, a binary companion in such a large, elongated orbit would be unstable in itself and would risk being tugged out of the Sun's gravitational well by other passing stars and molecular clouds. The 26-million-year cycle is also disputed by some, and the impact histories of the Moon, Mars and Mercury do not support such a cycle.

References

1. Nesvold, E. (2016). *Dynamics of exoplanet systems*. Retrieved from https://www.planetary.org/articles/1031-dynamics-of-exoplanet-systems. Accessed 10 Aug 2020.
2. Oort, J. J. (1950). The structure of the cloud of comets surrounding the solar system, and a hypothesis concerning its origin. *Bulletin of the Astronomical Institutes of the Netherlands, 11*, 91.
3. Amir, S. & Abraham, L. (2020). *Interstellar objects outnumber solar system objects in the oort cloud*. (Draft). eprint arXiv:2011.14900.
4. Whipple, F. L. (1949). A comet model. I. The acceleration of Comet Encke. *Astronomical Journal, 54*, 179.
5. Whipple, F. L. (1950). Comets, meteors and the interplanetary complex. *Astrophysical Journal, 111*, 375.
6. Lyttleton, R. A. (1953). *The comets and their origin*. Cambridge University Press.
7. Olivier, C. P. (1925). *Meteors*. Williams and Wilkins.
8. IAU Meteor Data Center. Retrieved from https://www.ta3.sk/IAUC22DB/MDC2007/. Accessed 1 Aug 2020.
9. Jacchia, L. G., Verniani, F., & Briggs, R. E. (1965). *An analysis of the atmospheric trajectories of 413 precisely reduced photographic meteors*. Cambridge, MA.
10. Muller, R. (1988). *Nemesis: The death star*. Wm. Heinemann.

Chapter 4

Observing Meteor Showers

Since the turn of this century, there has been a significant change in the way meteor showers have been observed and recorded. For much of the nineteenth and twentieth centuries, most observations relied on astronomers—often amateur astronomers—who used nothing more technical than their eyes, a pen and paper. However, the advent of cheap, reliable technology now means that it is now much easier and cost-effective to use digital cameras, video and radio detection equipment to monitor meteor activity. From a purely scientific point of view, this change has resulted in a significant increase in high quality raw data that, thanks to powerful desktop and mainframe computers, has allowed astronomers to probe the nature of meteoroid streams as never before. But what the technology cannot do is make the observer's heart jump with excitement when a piece of cosmic debris comes to a spectacular end in the night sky. Meteor astronomy is a dynamic subject that, along with the aurora and a total eclipse of the Sun, is one of the few branches of astronomy that can surprise and amaze the observer in real time.

All methods of observation have their limitations, which is why meteor showers should be monitored by every available means. Technology and the way it is used change at a breath-taking pace, and any advice or recommendations can become obsolete in a very short time. This chapter is therefore intended primarily for the visual observation of meteor showers. Anyone wishing to complement their visual observations with a digital camera or use radio observations will find more current information on the Internet. Better still, join an astronomical society where you'll find others that have been there, done that and bought the T-shirt. Astronomy clubs are a great source of help!

Supplementary Information The online version of this chapter (https://doi.org/10.1007/978-3-030-76643-6_4) contains supplementary material, which is available to authorized users.

Before You Start

If you are planning on observing meteors, there are a few things that you need to know before starting your meteor watch, the first of which is of course whether a meteor shower is active.

The major meteor showers are listed in Table 4.1 and in Appendix J. A list of the minor showers mentioned in this *Atlas* can be found in Appendix I.

Table 4.1: Major Meteor Showers

IAU No.	IAU Code	Shower Name	Duration (date)[1] (Solar longitude $\lambda_\odot$)	Maximum (date)[2] (Solar longitude $\lambda_\odot$)	ZHR[3]	RA[4] α	Decl[5] δ
0010	QUA	Quadrantids	Dec 28 / 274° — Jan 12 / 274°	Jan 4/5 / 291°	91	15^h 18^m / 229°	+49.5°
0006	LYR	April Lyrids[6]	Apr 14 / 24° — Apr 30 / 40°	22 Apr / 32.32°	18	18^h 07^m / 272°	+33.1°
0031	ETA	η-Aquarids	Apr 19 / 29° — May 28 / 68°	May 6/7 / 46.2°	83	22^h 32^m / 338°	-0.8°
0005	SDA	Southern δ-Aquarids	Jul 12 / 109° — Aug 19 / 147°	Jul 29/30 / 126.5°	28	22^h 44^m / 341°	-16°
0007	PER	Perseids[7]	Jul 18 / 115° — Aug 25 / 153°	Aug 12/13 / 140°	130	03^h 11^m / 48°	+58°
0002	STA	Southern Taurids	Sep 10 / 168° — Nov 20 / 238°	Nov 5/6 / 223°	5	03^h 33^m / 53°	+12.9°
0008	ORI	Orionids	Oct 3 / 190° — Nov 7 / 220°	Oct 22/23 / 209°	26	06^h 24^m / 96°	+15.7°
0017	NTA	Northern Taurids	Oct 20 / 207° — Dec 10 / 258°	Nov 12/13 / 230°	5	03^h 57^m / 59°	+22.3°
0013	LEO	Leonids	Nov 6 / 224° — Nov 30 / 248°	Nov 18 / 236°	17	10^h 17^m / 154°	+21.4°
0004	GEM	Geminids	Dec 4 / 252° — Dec 17 / 265°	Dec 14 / 262.2°	100	07^h 27^m / 112°	+32.3°
0015	URS	Ursids	Dec 17 / 265° — Dec 26 / 274°	Dec 23 / 271°	15	14^h 28^m / 217°	+75.4°

(1), (2) Approximate dates. Use the Solar Longitude value for greater accuracy.
(3) Zenithal Hourly Rate (2010-2020 average)
(4) Right Ascension in hours and minutes with degrees underneath.
(5) Declination.
(6) Often just called the Lyrids.
(7) Sometimes called the August Perseids.

The Meteor Shower Calendar (Fig. 4.1) shows you what the meteor year actually looks like. You will notice that no two meteor showers are the same. The Quadrantids (QUA) last just a few days and climb rapidly to maximum before fading even more quickly. The April Lyrids (LYR) have a gradual build-up and an equally slow decline, while the Geminids (GEM) climb to maximum over about 10 days but then fizzle out just a few days later. The η-Aquariids (ETA) have a broad maximum lasting several days. Once you know which shower you will be observing, you should make a copy of the radiant chart from Chap. 5 or 6 and, if you intend to plot meteors, use the relevant gnomonic chart from Appendix H.

If it is your first time observing meteors, then it is probably best to plan your meteor watch around the time of maximum activity. If you attempt to observe a shower at any other time, then you may be disappointed by the low number of meteors you see. Which brings us nicely to…

Before You Start

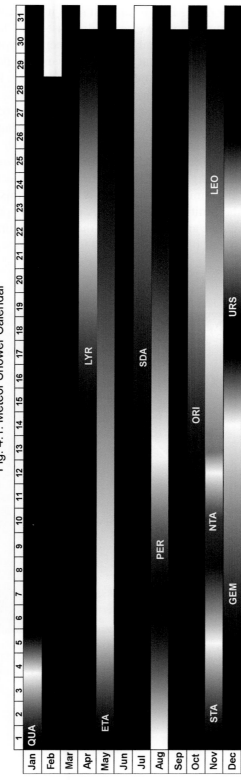

Fig. 4.1: Meteor Shower Calendar

This is a graphical representation of what the meteor year actually looks like for the major meteor showers. Note the wide variation in length of activity.

…the Zenithal Hourly Rate or ZHR. This is a measure of how active the meteor shower is likely to be, but there are a number of caveats, as discussed in Chap. 1. The ZHR rate varies from one year to the next, so the actual number of meteors you will see will likely be less than the quoted ZHR. Having said that, meteor showers are notoriously unpredictable and you could well be surprised by a sudden burst in activity. Figure 4.2 shows the effect of radiant altitude on the number of meteors you are likely to see. For example, if the radiant is 40° above the horizon then you are likely to see about 65% of the observable meteors. At 50° the likelihood rises to about 78%. Bear in mind that the ZHR is usually quoted for maximum activity and meteors will be fewer in number either side of maximum. A couple of other factors to keep in mind are the *limiting magnitude*—the faintest star visible—and any obscuration. Again, ZHRs are quoted for perfect seeing with a limiting magnitude of m_v +6.5. If the limiting magnitude is less than this, then you won't see meteors that are fainter than the limiting magnitude. A clear sky is preferred but you may have to put up with passing cloud and other things that can obscure your view such as mountains, trees and buildings, all of which may reduce the number of meteors you are likely to see.

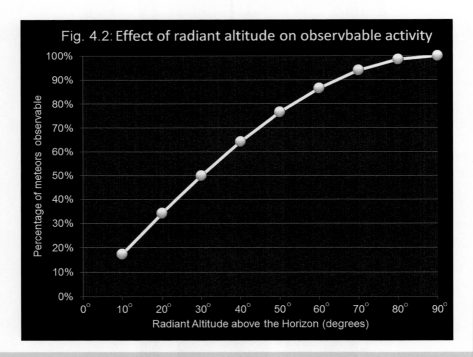

Locate the radiant's altitude above the horizon on the horizontal axis, and the curved line directly above will reveal the percentage of meteors you are likely to see under perfect conditions where the limiting magnitude is m_v +6.5

Next you need to check the Moon. The best time to observe meteors is when the Moon is absent from the sky altogether. As the Moon grows in size and brightness, its light can bleach out the fainter meteors, making them invisible. A full Moon can ruin a meteor watch completely. For any particular meteor shower, moonlight can seriously interfere one year in every three on average. So the Moon has to be between about third quarter and first quarter full, or between 21 and 7 days old. Search for 'Moon Phases (year)' on the Internet. Alternatively, check out the American Meteor Society's website at www.amsmeteors.org.

You also need somewhere to observe from. Light-flooded towns and cities are just as bad as a full Moon, so you should head out to the countryside and as far away from population centres as you can. Wherever you go, always bear in mind your personal safety. Check out your proposed observing site during the day to avoid slipping down a grassy bank, putting your foot in a hole, falling into a river or walking off the edge of a cliff. You should cordon off any hazardous areas. If you are organizing a meteor watch for several people you may be under a legal duty of care to do this, depending on which country you are living in. And beware of wild animals or, worse still, wild farmers! Don't trespass and follow the countryside code (e.g. close gates, do not leave litter behind, etc.).

Check the times of sunset and sunrise. You need the Sun to be well below the horizon—at least 12°—with twilight just about faded away completely. Generally, the best chances of seeing a good meteor display is in the pre-dawn hours rather than after sunset. This is because at 6 am local time, the Earth in plunging headlong into the meteoroids, whereas at 6 pm local time, it is leaving the meteoroids behind (see Figs. 2.6a and 2.6b in Chap. 2). However, there is a caveat in terms of when a particular meteor shower peaks. If maximum activity is expected at 10 pm local time, and the shower has a short, sharp maximum that lasts for only a couple of hours, then 10 pm is your best chance of seeing the display.

Finally, check the weather forecast. You need little or no cloud. And don't forget to check the temperature. It can get quite chilly in the middle of the night, even in midsummer in some parts of the world. When checking the weather forecast, bear in mind that the temperatures quoted are just *air temperature*. If there is a wind blowing, then it may seem considerably colder. You can get some idea of how cold it will feel by using the Wind Chill Chart in Fig. 4.3 or Appendix E.

Regardless of what the weather forecast is, you should always take warm and waterproof clothing with you just in case the weather unexpectedly turns nasty. As a general rule of thumb, never believe anyone who claims to be able to predict the future—especially meteorologists!

Fig 4.3 Wind Chill Diagram

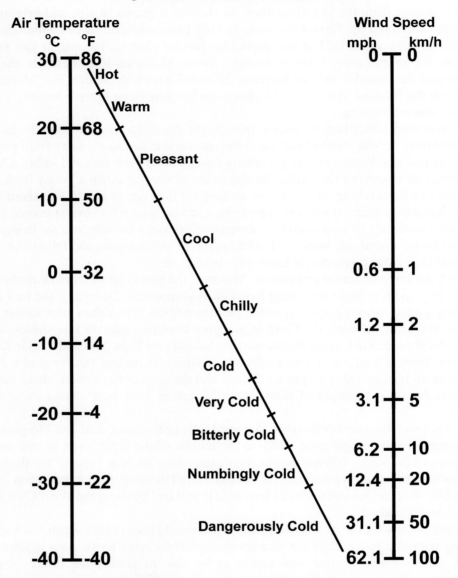

Take a ruler and line it up with the Air Temperature on the left axis and the Wind Speed on the right axis, both of which are usually given in weather forecasts. Where the ruler meets the diagonal line is an indication of how cold or warm it will feel

So, your pre-watch checklist is:

- Activity
- Moon
- Location
- Hours of darkness
- Weather
- Radiant Chart
- Gnomonic Chart

And try to get some sleep before you start your watch.

Essentials

If you stand in the middle of a field looking up towards the sky, then it won't be long before you get a crick in your neck. You need something to sit in or lie on, such as a reclining lawn chair.

You need to keep warm and dry, so think about wearing several layers of clothing and a shower-proof jacket and trousers. Some observers use a sleeping bag. Rain isn't a problem, of course, but dew can be. If the weather is likely to be cold, take a hot drink in a flask. Wear gloves if necessary. If you wear something on your head it will keep you warm in the winter and keep the insects out of your hair during the summer. Insect repellent will also help. Shoes or boots with thick rubber soles will insulate you from the cold ground. Some observers have even been known to take a hot water bottle on a cold night, while others have been known to fall asleep!

Obviously, if it's likely to be a warm night then you may prefer a cold drink, but avoid alcohol as it may hinder your ability to accurately estimate magnitude, angular distance and time for those meteors that display long-duration trains. Consider taking something to eat. If it looks like it's going to be a long night then you should take regular breaks, say 5 to 10 minutes every hour, to give your eyes a rest and reduce the effects of fatigue. Walk around and have a stretch. Some nourishment will help.

You will require a dim torch with a red lens or filter (or a bright torch with a thick red lens or filter). When you first go outside to observe, any white light you have been exposed to will bleach your eyes, making them less sensitive to low light levels. As time passes, your eyes will adapt to the darkness and you will see an increasing number of the fainter stars and meteors. It takes at least 20 min to become adapted to dark conditions—*dark adaption*—and around an hour to become fully adapted. The problem of course is that you need some light to work by, e.g. to record the meteors you see, and the moment you are exposed to white light, you ruin your dark adaption, which is why a dim red torch is necessary. Red light, providing it isn't too bright, has minimal effect on dark adaption.

You will need a watch you can read in the dark. Make sure you have set your watch to Coordinated Universal Time (UTC). This is effectively the same as Greenwich Mean Time (GMT). You can find out what your UTC is by searching the Internet for 'Universal Time Now'. Alternatively, you can keep your watch on local time and convert the time to UTC later.

There are various methods to record the meteors you see. A digital recorder is great, provided you are not at risk of disturbing sleeping neighbours, otherwise you may find something flying towards you that is not a meteor. Some observers prefer to use pen and paper and record the meteors they see directly onto an observing sheet fixed to a clipboard (a generic form can be found in Appendix F, which is also available as a download. If you are a member of an astronomical society or meteor organization, then they may use a slightly different form). Other observers like to plot meteors on a star chart. *Norton's Start Atlas* is brilliant for general astronomy—most amateur astronomers seem to own a copy—but the projection system it uses means that some meteors will appear as curves. To avoid this, meteor observers use gnomonic star charts, a set of which is available in Appendix H and as downloads (https://doi.org/10.1007/978-3-030-76643-6_4).

The idea behind plotting meteors is to see whether their paths intersect the radiant. If they do, then they likely belong to that particular meteor shower. But you need to be flexible. The radiant diameter is based on an average, and radiant sizes can vary from year to year, day to day, even hour to hour, so if your plotted meteor just misses the radiant, then chances are it may well be a shower member. The following day when you have time to analyse the plots, do not be surprised to find that the radiant is larger, and sometimes smaller, than the quoted value.

The issue with recording meteors using a pen or pencil is that you may miss a meteor while looking at your observing sheet or star chart. That can appear to slightly reduce the ZHR, which is often not critical, but you risk missing a spectacular meteor or fireball. If you are observing with a friend, then they may well call out when a bright meteor flashes across the sky, but by the time you look up and locate the part of the sky they are observing, the meteor will have disappeared from view.

The Observing Session

The basic idea behind a visual observing session is that you watch an area of sky and record any meteors that may appear.

You should not observe the radiant directly. Meteors appearing near the radiant tend to be short in both length and duration (see Figs. 1.1 and 1.2 in Chap. 1). Instead, you should choose the darkest area of sky, 20°–45° from the radiant and 45° to 75° above the horizon. Unless they are exceptionally large or small, you can use your hands to roughly estimate angular distance (Fig. 4.4). If you hold your hand out at arm's length and splay out your fingers, the distance from the tip of your small finger to the tip of your thumb is about 22°. Your fist is about 9° from the first to the last knuckle, and your thumb is about 2.5° across (Fig. 4.5). You can get a more accurate estimate by holding out your splayed hand or fist and rotating a full 360°, then divide 360° by the number of times it took for your hand to cover the distance.

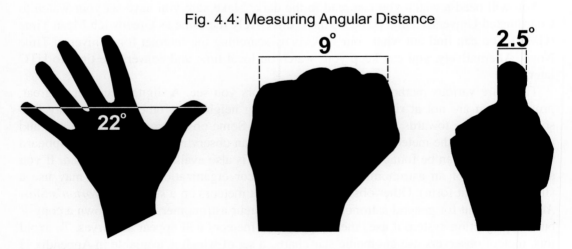

Fig. 4.4: Measuring Angular Distance

Holding your hand out at arm's length will enable you to measure angular distance. The average hand splayed wide is about 22°, across the knuckles is about 9°, and a thumb's width is about 2.5°

It is best to observe the same region of sky throughout the entire watch. This can take a little self-discipline—you may find your focus wandering to other parts of the sky—and may not always be possible, e.g. if cloud starts to drift in. As the night progresses, your area of sky will of course move and you will therefore have to reposition your lawn chair when you take a break. Concentrating on a particular constellation is common among meteor observers.

When you see a meteor, freeze! Try to capture the following information:

How bright was the meteor? You can estimate its magnitude by comparing it to the stars it passes close by. If you are new to estimating magnitude, then Appendix G contains a number of magnitude comparison charts that can be used specifically for this purpose.

Did the meteor appear to come from a shower radiant or was it a member of the sporadic background? You should know where the radiant is and, in your mind's eye, you should backtrack the meteor to see if it started within or close to the radiant. This becomes more difficult with increasing distance, which is why it is best to avoid looking more than 45° from the radiant. If you do not know how big the radiant is, assume it is about 8° in diameter. In reality, radiants vary in size and shape. Young meteoroid streams in which the dust has not spread out produce small, compact, well-defined radiants less than 3° across. Older streams have much larger, diffuse radiants. While most are circular, some are elongated or oval. Meteors that are long and close to the radiant are probably not members of that particular shower.

Was there anything noteworthy about the meteor? Did it fragment into several pieces? Was there a terminal flare? Could you detect any colour? Did it have a particularly long path? Did it have a persistent train? If so, for how long did it persist? A pair of binoculars is handy for studying persistent trains. Was it very fast or very slow compared to other meteors you have witnessed during the watch? Sometimes a sketch helps. Many of the details you record will not be detectable by technology.

If you are really lucky, you may see a meteor heading directly towards you. They are more common than most people realize, appearing as a momentary flash and on photographs as a new star (Fig. 4.5).

Finally, you should record the time of the meteor's appearance to the nearest minute in UTC.

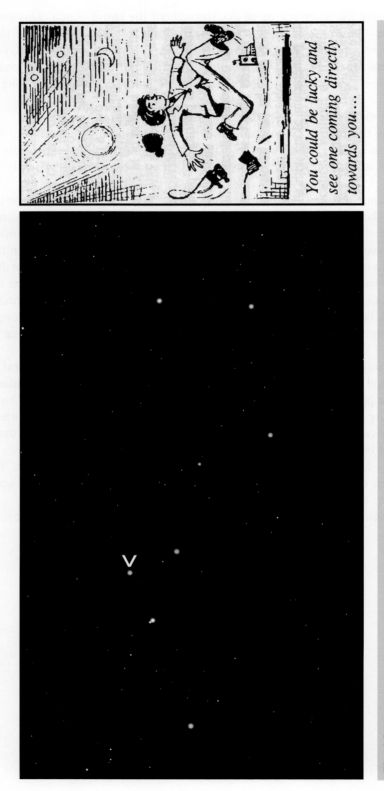

Fig. 4.5 Head-on meteor
This new star in Ursa Major was actually a meteor heading directly towards the camera. On the right is the late Cyril Blount's unique take on the event. (Credit: Cyril Blount, courtesy Impact! magazine).

The Observing Session

Appendix F provides a generic record sheet for a meteor watch. In addition to the details of each meteor, you should also record the limiting magnitude, the sky conditions and the 'age 'of the Moon, i.e. the number of days since the last New Moon. The limiting magnitude tends to vary throughout the night, especially if there is a rising or setting Moon. Sky conditions can also change, particularly at coastal or mountain sites as mist and cloud roll in. Cloud cover should be given as the percentage of sky obscured in the direction you are looking. Figure 4.6 shows a completed record sheet.

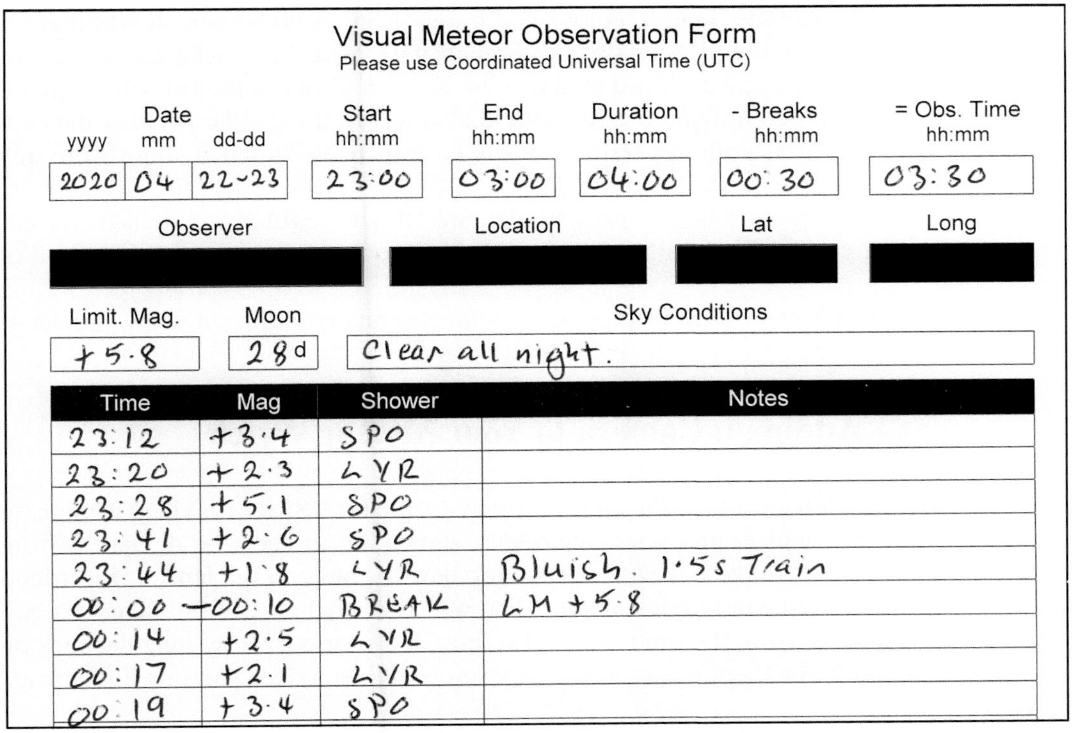

Visual Meteor Observation Form
Please use Coordinated Universal Time (UTC)

Date yyyy mm dd-dd	Start hh:mm	End hh:mm	Duration hh:mm	- Breaks hh:mm	= Obs. Time hh:mm
2020 04 22-23	23:00	03:00	04:00	00:30	03:30

Observer	Location	Lat	Long
▓▓▓	▓▓▓	▓▓▓	▓▓▓

Limit. Mag.	Moon	Sky Conditions
+5.8	28 d	Clear all night.

Time	Mag	Shower	Notes
23:12	+3.4	SPO	
23:20	+2.3	LYR	
23:28	+5.1	SPO	
23:41	+2.6	SPO	
23:44	+1.8	LYR	Bluish. 1.5s Train
00:00 – 00:10		BREAK	LM +5.8
00:14	+2.5	LYR	
00:17	+2.1	LYR	
00:19	+3.4	SPO	

Fig. 4.6 Meteor watch record sheet
Part of a completed record sheet for a meteor watch

If you are exceptionally lucky, you may witness a sudden but short-lived outburst, or even a more prolonged and unexpected storm, when meteors fall like rapid snowflakes. How do you record all the details when so much is happening? The fact is, you don't. You should just count how many meteors you see during 5-min time intervals.

It is always best to submit your observations to a meteor organization to be added to the body of knowledge. The International Meteor Organization (IMO) is by nature the largest and has an online reporting facility. In the UK, the British Astronomical Association and the Society for Popular Astronomy both have active and welcoming Meteor Sections. In the United States, the American Meteor Society is the principle body and works closely with the IMO. Different organizations may have different methods for recording meteors and sky conditions.

Fireballs and Bolides

Fireballs are meteors that are brighter than Venus, m_v −4.4. *Bolides* are generally regarded as exceptionally bright fireballs that explode. The scientific community is particularly interested in reports of fireballs because they may indicate the arrival of a meteorite. Consequently, detailed information about the fireball's flight path can aid any subsequent search for meteorites and can also lead to the calculation of an orbit, indicating the origin of the body.

In addition to magnitude, length, colour and duration in seconds, you should make a note of the beginning and end points of the fireball with regard to background stars, or compass direction, and angular elevation above the horizon. Plotting the fireball's path on a gnomonic star chart certainly helps. You should also record things like whether the fireball left a long-duration luminous train or a smoke trail behind, and if there were any sounds.

Fireballs are uncommon but not particularly rare. If you regularly undertake meteor observations then you are likely to see a m_v −4, fireball, on average once in every 20 h of observation.

The International Meteor Organization accepts fireball reports online at www.imo.net.

Adding a Camera to Your Meteor Watch

It is relatively straightforward to add a digital camera, either a DSLR or a video camera, to a meteor watch. You will need a wide-angle lens, something to keep the dampness from seeping into the camera body, a dew heater to stop dew forming on the lens, a firm tripod and a shutter release cable or wireless device. You will also need several fully charged batteries as you will be leaving the shutter open for quite long periods. How long depends on how much light pollution there is. Try a 30-s exposure and then reduce the time if the image is washed out.

There are numerous articles scattered around the Internet on astrophotography, and with technology improving all the time, it is worth revisiting sites regularly for new tips. However, the best advice is to join your local astronomical society. There will undoubtedly be members who are a mine of invaluable information.

Analysing Your Observations

If you submit your observations to the International Meteor Organization, they will be added to all the others received and, at some point, a report may be produced to show how the meteor shower performed. However, there is no guarantee that this will happen, especially if there are too few observations to produce a meaningful report.

You can of course analyse your own observations and an Excel spreadsheet can be downloaded from Springer's website (https://doi.org/10.1007/978-3-030-76643-6_4) for this purpose. Part of the spreadsheet is shown in Fig.4.7.

	A	B	C	D	E	F	G	H	I	J	K	L	M	N
1														
2													r	
3													2.3	
4		Start	End	Time	Altitude of the Radiant (degrees)	Limiting Magnitude LM		Cloud Cover f	Field Obscuration F	Number of Meteors and Shower or Sporadic				
										Shower	Sporadics	Total	$r^{6.5-LM}$	ZHR
5		23:00	00:00	1.00	36	5.8		0%	1.00	2	3	5	1.79	6.1
6		00:10	01:00	0.83	45	5.8		0%	1.00	4	4	8	1.79	12.2
7		01:10	02:00	0.83	54	5.8		0%	1.00	7	5	12	1.79	18.6
8		02:10	03:00	0.83	61	5.8		0%	1.00	5	5	10	1.79	12.3
9		00:00	00:00	0.00	0	0.0		0%	1.00	0	0	0	224.51	#DIV/0!
10		00:00	00:00	0.00	0	0.0		0%	1.00	0	0	0	224.51	#DIV/0!
11		00:00	00:00	0.00	0	0.0		0%	1.00	0	0	0	224.51	#DIV/0!
12		00:00	00:00	0.00	0	0.0		0%	1.00	0	0	0	224.51	#DIV/0!
13		00:00	00:00	0.00	0	0.0		0%	1.00	0	0	0	224.51	#DIV/0!
14		00:00	00:00	0.00	0	0.0		0%	1.00	0	0	0	224.51	#DIV/0!
15		00:00	00:00	0.00	0	0.0		0%	1.00	0	0	0	224.51	#DIV/0!
16		Totals								18	17	35		

Fig. 4.7 Spreadsheet for analysing meteor observations

This spreadsheet can be downloaded from springer.com (https://doi.org/10.1007/978-3-030-76643-6_4)

The spreadsheet requires eight inputs and produces a ZHR output. The inputs (dark grey columns) are:

(1 & 2) Start and End Times (columns B and C): You may have to input midnight as 24:00 instead of 00:00. In this example, times are 1-h or 50-min slots. If there are rapidly changing sky conditions, it may be best to use a shorter time slot, such as 20 or 30 min.

The spreadsheet will convert the observation time to a decimal (column D).

(3) Altitude of the Radiant (column E): Radiant altitudes for various latitudes and times are given in Chaps. 5 and 6 for all the major showers and some of the more noteworthy minor showers. Alternatively, you may estimate the altitude yourself or use an Internet calculator (search 'Altitude of a star').

(4) LM (column G): Is the Limiting Magnitude. In this case, the limiting magnitude remained the same for the duration of the watch, m_v +5.8.

(5) Cloud Cover (column H): Estimate how much of the sky *in the direction you are observing* is covered with cloud or is otherwise obscured, e.g. by hills, buildings, etc.

The spreadsheet will return an obscuration factor (column I).

(6 & 7) Number of Meteors (columns J and K): Enter the number of meteors as *Shower* meteors or *Sporadics*.

The spreadsheet will total the number of meteors (column L).

(8) r: The Population Index: The *Population Index* is the ratio between meteors of different magnitudes. Most showers have an r-value of between 2 and 3.5. For the major showers, the value can be found in Chaps. 5 and 6. The lower the r-value, the brighter the meteors overall, so r = 2 means the meteors are bright while r = 3.5 means they are faint. If in doubt, use r = 2.4.

Input the r-value in Cell M5.
The spreadsheet will output a new r-value dependent on the Limiting Magnitude.
The ZHR will be shown in the last column, N.

Mathematical Approach

If you do not have access to an Excel spreadsheet, then the ZHR can be worked out mathematically as in Table 4.2.

Mathematical Approach

Table 4.2: Manual Calculation of the ZHR

Step	Process	Example	Formula and Notes
(a)	Enter start time	01:10	Break up your watch with 10-minute rest periods every hour.
(b)	Enter end time	02:00	
(c)	Calculate time decimal	0.83	Formula: (b) – (a) /24, e.g. 01:10 – 02.00 = 50 minutes, then 50 minutes / 60 minutes = 0.83
(d)	Enter Altitude of the Radiant	45°	Use the Radiant Altitude Charts in this *Atlas* for your latitude for midway through your observation hour.
(e)	Enter Limiting Magnitude (LM)	5.8	The limiting magnitude can change throughout the night, especially shortly after dusk or before dawn, or if there is a rising or setting Moon.
(f)	Enter percentage or proportion of sky obscured by cloud, hills, buildings, etc.	5% or 0.05	Limit obscuration to the part of the sky you are observing.
(g)	Calculate the Field Obscuration Factor (F)	1.05	Formula: $1/[1-(f)]$ e.g. $1 / [1-0.05] = 1 / [0.95] = 1.05$
(h)	Enter the number of shower meteors	4	
(i)	Enter the number of sporadic meteors	4	*Steps (i) and (j) are not required for the ZHR calculation but are included for completeness.*
(j)	Add the number of shower meteors to the number of sporadic meteors	8	
(k)	Enter the Population Index, r	2.3	The r-value is given for all the major showers in this *Atlas*. If no value is available, then try r = 2.4
(l)	Calculate $r^{(6.5-LM)}$	1.79	Formula: 6.5 – (e) e.g. 6.5 – 5.8 = 0.7 Then $(k)^{0.7}$ e.g. $2.3^{0.7} = 1.79$ The sequence on a calculator is typically: $\boxed{2.3}$ $\boxed{x^y}$ $\boxed{0.7}$
(m)	Calculate the ZHR	12.8	Formula: (g) x (h / c) x [1 / sin(d) x (l)] Where: g = 1.05 h = 4 c = 0.83 d = 45° (and sine d = 0.707) l = 1.79 So: 1.05 x (4 / 0.83) = 1.05 x 4.819 = 5.060 1 / 0.707 = 1.414 x 1.79 = 2.531 Therefore: ZHR = 5.060 x 2.531 = 12.8

The spreadsheet and the manual method are both rough-and-ready ZHR calculations. In practice, and providing there are sufficient observations, analysts go to great pains to refine this method by taking into account other factors such as observational bias (some observers overestimate magnitude, some underestimate, some tend to count sporadics as shower members and vice versa). Even so, it is always worth estimating your personal ZHR and making comparisons with published reports. If your ZHR is consistently and significantly higher or lower than the published reports, then you may need to hone your skills, particularly when assigning meteors to a shower.

And Why Not Try…

…setting up a local group of meteor observers in your area? This is relatively easy to do, especially if you are a member of an astronomical society, and with social media platforms like Facebook, Twitter and Instagram, getting people together and distributing your results is so easy.

As a reminder, the following charts are available for download from Springer's website (https://doi.org/10.1007/978-3-030-76643-6_4):

D1: Meteor Observation Record Sheet
D2: Magnitude Comparison Charts
D3: Gnomonic Plotting Charts
D4: Major Meteor Shower Radiant Charts
D5: Minor Meteor Shower Radiant Charts
D6: ZHR Calculation (Excel Spreadsheet)

Chapter 5

The Major Meteor Showers: January to August

Introduction

This chapter, along with Chapter 6, covers the ten major meteor showers that appear annually—or eleven, if you split the Taurids into a northern and a southern component.

There is nothing written in stone, but a *major shower* is generally regarded as one that has a maximum ZHR of 10 or more. Some minor showers occasionally have outbursts that exceed 10 meteors/hour but, because the outburst is not an annual event, they do not qualify as major showers.

Each of the ten major showers has its own section and each ends with a series of charts and diagrams. These introductory notes will help you get the most out of this chapter.

Activity and Additional Data Tables

Solar Longitude, $\lambda_\odot$: As noted in Chap. 1, calendar anomalies cause the dates of meteor shower activity to drift. Solar longitude avoids this problem and should be used when planning a meteor watch. For convenience, the approximate civil calendar dates are also given.

Right Ascension, α, and Declination, δ: The position of the radiant is given for the expected time of maximum activity. Some star atlases indicate right ascension (RA) in degrees, others in hours and minutes. Both are given here.

Mean daily motion of radiant, Δ: Radiants drift eastwards by a certain amount, which varies on a daily basis. The mean, or average, daily motion is given as the change in right ascension, $\Delta\alpha$, and declination, $\Delta\delta$. For short duration showers, adding the change in motion to the quoted position of the radiant will give its location after maximum activity, while deducting the change will give the radiant's position before maximum to a reasonable degree of accuracy. This method is not sufficiently accurate for long duration showers.

Time of Transit: is the local time (LT) when the radiant crosses the observer's meridian. The time should be adjusted if Daylight Saving is in force.

Zenithal Hourly Rate, ZHR: The ZHR is the theoretical number of meteors an observer will see if the shower radiant is directly above their head and the faintest star visible, the limiting magnitude, is m_v +6.5. The ZHR varies with time as the Earth passes through different segments of the meteoroid stream. Do not be surprised to find disagreement between different publications and websites as the information some use is long out of date and has not been checked. Predictions for the Quadrantid meteor shower nearly always quote a ZHR of 120 met/h. While that figure was a true average between 1998 and 2005, activity over the past decade has been somewhat lower with an average ZHR of 98. The ZHRs quoted in this *Atlas* are the averages for the period 2010–2020.

Radiant Diameter, Ø: Each shower has a number of radiant points, and the radiant diameter is a circle that encloses most of these points. The radiant can therefore be regarded in effect as the average radiant. Whilst some radiants are well defined and compact, others are large and diffuse, and it is not uncommon to find a range of radiant diameters quoted. Some radiants are said to be oval, and some shrink as maximum activity approaches. It is useful to note the radiant diameter as it is undoubtedly true that it will change in size, and perhaps shape, over time.

IAU Abbreviation and Code: The International Astronomical Union's Meteor Data Center has assigned a number and three-letter code to uniquely identify each meteor shower. The three-letter code is used throughout this *Atlas*.

Geocentric velocity, V_g: is given in both km/s and km/h.

Population Index, *r*: is the ratio between meteors of different magnitudes. Most showers have an r-value of between 2 and 3.5, with 2 being the brightest meteors and 3.5 being the dimmest.

Mean magnitude, $\overline{m}_v$: is the average magnitude of the meteors from a particular shower.

Parent body: The name or IAU designation of the body responsible for the formation of the meteoroid stream.

Best visibility: Indicates when and where the shower is best placed for observation.

Maxima 2020–2030: A list of dates of when the shower will peak together with the age of the Moon. A Moon about 9 or 10 days either side of full can potentially ruin a meteor watch, depending on the rising and setting times of the Moon and the radiant. On average, moonlight causes problems one year in every three.

Other Active Meteor Showers: A list of other showers active during the period of the major shower. Observers should note the location of the showers listed, as some may be mistaken for major shower meteors (contamination). Most of the showers will not cause contamination and some will not be visible from the same hemisphere as the major shower.

Orbital Data: The *orbital elements* or the meteoroid stream's parent body.

Charts, Graphs and Diagrams

Sky View: A night-time view of the shower's radiant position.

Radiant Chart: Essentially, a negative image of Sky View with additional key information such as the radiant's position in right ascension and declination for a specific date (given as solar longitude, $\lambda_\odot$), the probable diameter of the radiant and the average ZHR. Radiants drift eastwards across the sky by about 1° per day, a consequence of the Earth's passage through the meteoroid stream. The charts in this *Atlas* show radiant drift for all of the major showers.

Activity Profile: A graph showing how the shower's ZHR develops over a specific period of time, with maximum indicated by a vertical white line. Much of this data has been compiled by the IMO with additional observations helping to fill any gaps and refine the results.

ZHR: The estimated ZHR for the shower over a specific number of years, taken from a variety of sources.

Average ZHR: The ZHR average for the periods 1990–2020 and 2010–2020.

25-year Activity Levels: How the shower has performed over the past 25 years in terms of Low, Medium, High and Very High levels of activity. The levels have been set by the author but other researchers may choose different limits.

Magnitude Distribution: The visual magnitude profile of the shower's meteors.

Geocentric Velocity: A speed dial of how fast the shower's meteors are.

Popularity: The popularity of certain meteor showers is taken from the International Meteor Organization's published data relating to the number of people who have submitted observations. Popularity will vary from one year to the next depending on the Moon's phase—few observers will take time to view a shower when there is a full Moon—but the important thing is the trend. Some showers, like the Quadrantids, have become less popular over the past few decades while others, such as the Southern δ-Aquariids, are gradually becoming more popular.

Orbit of the Parent Body: Shown in relation to the Earth. Three-dimensional images are always difficult to depict on two-dimensional paper and no single image truly captures the shape and size of the orbit. Fortunately, there are a couple of websites that show orbits in three-dimensions and allow the image to be rotated. The IAU Minor Planet Center includes a wealth of additional information and can be found at https://minorplanetcenter.net/db_search.

NASA's Jet Propulsion Laboratory's offering looks as though it has been designed on a 1984 Sinclair Spectrum+ but at least has the advantage of an animated depiction of the parent body orbiting the Sun. It can be found at https://ssd.jpl.nasa.gov/sbdb.cgi.

Comets, asteroids and everything else in the Solar System orbit the Sun on elliptical paths, as shown in Fig. 5.1.

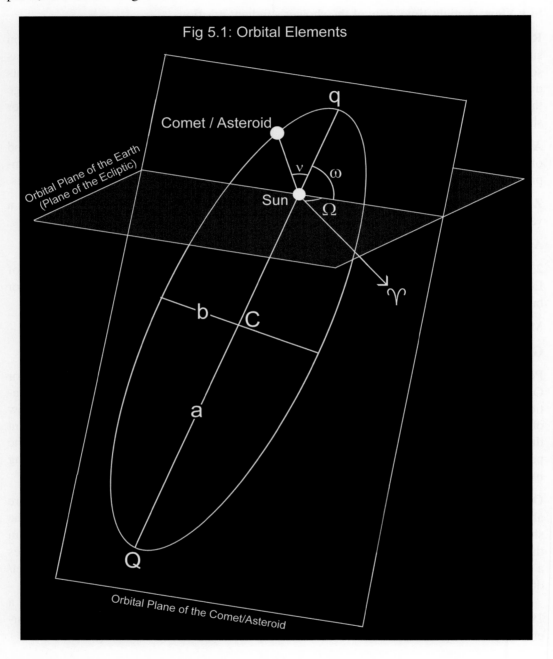

Charts, Graphs and Diagrams

If **C** is the centre of the ellipse, then the various elements are:

1. **a** is the Semi-Axis Major (or Semi-Major Axis), in AU, which is half the length of the orbit.
2. **b** is the Semi-Axis Minor (or Semi-Minor Axis), in AU, which is half the width of the orbit.
3. **q** is the Perihelion Distance, in AU, which is the closest approach to the Sun.
4. **Q** is the Aphelion Distance, in AU, which is the farthest distance from the Sun.
5. ♈ is the First Point of Aries, a fixed reference point used to orientate the orbit in space.
6. **Ω** is the Longitude of the Ascending Node, in degrees, measured in relation to the First Point of Aries and where the plane of the Earth crosses the plane of the comet or asteroid.
7. **ω** is the Argument of the Perihelion, in degrees, and indicates the angle between the plane of the Earth's orbit to the line connecting the Sun to the Perihelion Distance, q.
8. **ν** is the True Anomaly, in degrees, which indicates the position of the comet or asteroid in its orbit at a specific point in time.

In addition, there is also **e**, the eccentricity, which is how much the ellipse varies from a perfect circle, and **P**, the orbital period, in years.

The relationship between these elements is relatively straightforward:

- The perihelion distance: $q = a(1-e)$
- The aphelion distance: $Q = a(1+e)$
- The semi-axis major: $a = q/(1-e)$
- The semi-axis minor: $b = a\sqrt{1-e^2}$
- The length of the major axis is $2a$
- The length of the minor axis is $2b$
- The eccentricity: $e = 1-(q/a)$
- The orbital period: $P = a^{3/2}$

The orbital elements that are commonly given for any orbit are q, e, a, ω, Ω, i and P.

Radiant Altitude Charts: Just like the stars, a radiant will rise in the east, reach a high point and then set in the west. Ideally, the observer will want the radiant to be as high in the sky as possible, as this maximizes the number of meteors that can be seen. The radiant altitude depends on the observer's latitude and the time of night, and its visibility is dependent of course on sunset and sunrise. Each of the major showers contains a set of radiant altitude charts, so an observer can determine a radiant's altitude regardless of where they are located. Major towns and cities close to the various latitudes are identified. Local time is given but this may need to be adjusted for countries that use Daylight Saving.

The Quadrantids

The Quadrantids[1] are one of the most prolific showers of the year, but this northern midwinter event is often shunned by all but the hardiest of observers.

Observing Notes

The meteor year starts with a strong but short-lived meteor shower: the Quadrantids. Named after the now obsolete constellation of Quadrans Muralis, activity begins towards the end of December through to about 7 January, with a short, sharp peak around January 3 or 4 that lasts for just a few hours. In the 2010–2020 period, the ZHR averaged just over 90 but the shower has surprised observers in the past with bursts in the 130–140 range and occasionally higher.

Observation of the 'Quads' can be hampered by a number of factors. Obviously, the time of year is a major issue. Cloudless skies are needed for good quality meteor observation but these are few and far between in January. When cloud is absent, the temperature can plummet to well below zero, especially in the more northerly latitudes, which can put a lot of people off; observations of the shower have gradually declined over the past 30 years. As with many of the major showers, the Quadrantids are often at their best in the pre-dawn hours. That's great if you don't have to go to work or college the next day, but many people do and are reluctant to take a day's holiday, having just taken a Christmas, Hanukkah or a New Year break (Tables 5.1a and 5.1b).

Maximum lasts only a few hours at best, so some observers will miss the maximum completely while others on another continent may witness a fine display. Following maximum, activity rapidly declines to about half the peak value within four hours. In addition, moonlight spoils about one year in every three on average. However, if you are willing to brave the cold night air, and you can get just the right combination of maximum activity with the radiant high in the sky on a moonless night, then you will often be rewarded with a strong display. The Quadrantids reach maximum at solar longitude $\lambda_\odot$ 283.2°, which falls on January 3 or 4 over the next decade. Table 5.2 gives further details. Ideally the Moon should be younger than 7 days or older than 23 days to minimize interference.

The Quadrantids are best seen from the Northern Hemisphere with the radiant being circumpolar for those living north of 40°N. It is a challenge for those south of the Equator where the radiant never rises much above the horizon, if at all, for most observers. The radiant transits at 8.5^h LT. reaching its lowest point at 20^h LT. The radiant itself is difficult to pinpoint—there are no bright stars anywhere near—but it lies roughly on a line from γ UMa (Phad) through ζ UMa (Mizar). For most of the duration of the shower, the radiant appears to be about 8° in diameter shrinking to less than 1° at the time of maximum. In reality, the area contains multiple radiants. In the early 1950s J.P.M. Prentice and G.E.D. Alcock found 13 active radiants, and photographic work has subsequently revealed more than 60 active components [1] between right ascension α 226° and 237° ($15^h 04^m$ and $15^h 48^m$) and declination $\delta + 47.5°$ and $+ 51.8°$. Not all radiants are active each year. This multitude of radiants indicates considerable structure within the meteoroid stream.

---

[1] Is it 'Quadrantid meteors' or 'Quadrantids meteors'? 'Lyrid shower' or 'Lyrids shower'? It doesn't really matter. There are no hard and fast rules and it is not uncommon to see both terms used together.

The Quadrantids

Table 5.1a: Quadrantid Activity Details				
		Start	Maximum	End
Dates (approx.)		Dec 28	Jan 3/4	Jan 12
Solar longitude:	$\lambda_\odot$	274°	283.2°	291°
Right Ascension at max:	α		$15^h 18^m$ (229°)	
Declination at max:	δ		+49.5°	
Time of transit :			8.5^h L.T.	
ZHR (2010-2020 average):			91	
Radiant diameter:	$\varnothing$	8°	<1°	8°

Table 5.1b: Quadrantid Additional Data		
IAU Abbreviation: QUA IAU Code: 010 AKA: Quadrantis Muralids, Boötids		
Mean daily motion of radiant:	Δ	$\Delta\alpha$ = +0.56° $\Delta\delta$ = -0.25°
Geocentric velocity:	V_g	41.2 km/s (148,320 km/h)
Population index:	r	2.1
Mean magnitude:	$\overline{m}_v$	+2.8
Parent body: 2003 EH$_1$		
Best visibility: Pre-dawn skies in the Northern Hemisphere		
The radiant is circumpolar from 40°N, i.e. north of Madrid, Spain; Naples, Italy; Baku, Azerbaijan; Thessaloniki, Greece; Ankara, Turkey; Salt Lake City, Omaha, Peoria, Sapporo, Chicago, Cleveland, Pittsburgh, Philadelphia and New York, USA.		

Table 5.2 : Quadrantid Maxima 2020-2030				
	Maximum ($\lambda_\odot$ 283.2°)			Moon
year	month	day	hour (UT)	Age (d)
2020	01	04	09:27	9
2021	01	03	15:40	3
2022	01	03	21:52	1
2023	01	04	04:02	12
2024	01	04	10:09	23
2025	01	03	16:15	4
2026	01	03	22:26	14
2027	01	04	04:36	11
2028	01	04	10:41	8
2029	01	03	16:49	25
2030	01	03	22:59	14

About 40% of Quadrantids are of the first or second magnitude, with about 25% at zero magnitude or brighter. The average magnitude is usually around $\overline{m}_v$ +2.9, with the brighter meteors tending to leave persistent trains. Fireballs are rare but not unknown, and are often described as yellow or bluish. The meteors appear medium-swift, entering the Earth's atmosphere at about 41.2 km/s (148,320 km/h), but when close to the horizon they can appear much slower.

The maximum ZHR is often quoted as 120, but this is a myth [2, 3]. There was an eight-year period between 1998 and 2005 when the ZHR did indeed average 120, but over the much longer timescale of 1990–2020, the average ZHR was just 104, and in the past decade it has fallen to a little over 90. Don't let this put you off though: the Quadrantids can surprise, attaining ZHRs of around 130 in 1994, 1999 and 2001; 140 in 1987, 145 in 1992, 153 in 1970, 156 in 2009; 190 in 1965 and 202 in 1909. There is some suggestion that this pattern indicates a 5.6-year period for the shower—the parent body has an orbital period of 5.52 years—but no one can say for sure when the next strong display will occur.

Finally, note that there are a number of minor showers active during the Quadrantids (Table 5.3) though none is likely to contaminate the shower.

Discovery

The Quadrantids were probably first recorded in 1825 by the Italian astronomer Antonio Brucalassi, who commented that the heavens were "traversed by a multitude of the luminous bodies known by the name of falling stars," clearly suggesting a strong display [4]. Various other people witnessed the shower during subsequent years, but it was Adolphe Quetelet [5], the Director of Brussels Observatory, and the Yale University librarian Edward Claudius Herrick [6] who in 1839 independently suggested the shower appeared annually, emanating from the now defunct constellation of Quadrans Muralis. However, it wasn't until 24 years later that Stillman Masterman in Maine, USA, managed to determine the radiant as being at α 238° (15^h 52^m), δ +46° 26' [7]. In 1918, the British meteor enthusiast William F. Denning and Mrs. Fiametta Wilson were surprised to find that the radiant appeared to have shifted about 8° north of its expected position [8]. The normal radiant produced only modest activity that year. This was the first hint of complexity within the Quadrantid meteoroid stream (Table 5.4).

Origins

The origin of the Quadrantids was long regarded as a mystery, as the stream did not appear to be associated with any known comet. In the early 1960s, researchers Salah E. Hamid and Mary N. Youssef looked at six Quadrantid meteors photographed from two stations, which allowed their orbits to be calculated, and speculated that the Quadrantids' parent comet was captured by Jupiter 4000 years ago [9]. Hamid later teamed up with Fred Whipple, the long-time doyen of meteor science, and pointed out that the shower may have had a common origin with the δ-Aquariids, which are active in July-August [10]. Not only did the Quadrantid and the δ-Aquariid meteors have very similar light curves, but their orbital planes and perihelion distances were almost identical 1300–1400 years ago.

In 1986, Donald Machholz of California discovered a new periodic comet using 130-mm binoculars [11]. Named 96P/Machholz in his honour, this Jupiter-family comet comes to within 0.13 AU (19.5 million km) of the Sun before journeying out to 5.94 AU on its 5.29 year orbit. The comet caught the attention of Bruce McIntosh, who noticed that 1500 years ago the Quadrantids and Machholz had similar orbits [12].

Earlier, in 1979, Ichiro Hasegawa proposed that the bright comet $C/1490\ Y_1$, observed from China, Korea and Japan in the fifteenth century, was the progenitor of the Quadrantid meteoroid stream, based on orbital calculations [13]. Subsequent research by Peter Jenniskens uncovered an Amor-type asteroid, 2003 EH_1, imbedded in the Quadrantid meteoroid stream [14]. This is not the first asteroid to be found in a meteoroid stream: the Geminids also have an asteroidal parent body—(3200) Phaethon. They were once believed to be dead comet nuclei, but current thinking in many quarters is that 2003 EH_1 and (3200) Phaethon may be fragments of what was originally a large comet, and that Phaethon may also be an active asteroid, which can shed mass by various processes such as the sublimation of volatiles when close to the Sun, ejection of material due to orbital spin and the thermal degradation of its surfaces, causing rock to fracture, fragment and escape the low-gravity environment (Table 5.5).

It also seems possible, based on orbital data, that 2003 EH_1 and $C/1490\ Y_1$ are related objects, observed at different stages in their evolution. There are some interesting comparisons. 2003 EH_1 has a diameter of about 2.9 km, but estimations of the diameter of the comet's nucleus suggest it was twice the size at 5.8 km and had an estimated mass of 5×10^{13} kg. This is somewhat different from the mass of the Quadrantid meteoroid stream: 1.3×10^{15} kg according to David Hughes and Neil McBride [15]. A close encounter with Jupiter in 1650 radically changed the orbit of $C/1490\ Y_1$, pushing the aphelion to more than 20 AU and making the comet invisible from Earth. If the relationship between the asteroid and the comet is real, then rather than the Quadrantids being an ancient stream, they quite likely were created by the breakup of $C/1490\ Y_1$ and are therefore no more than 500 years old.

The Machholz Complex

The Quadrantids may have had a common origin with a number of bodies that collectively are often called the *Machholz Complex*. At least eleven comets, two asteroids and four meteor showers have been nominated as members of the complex. The list includes:

Meteor showers

- Quadrantids (QUA)
- Southern δ-Aquariids (SDA)
- Northern δ-Aquariids (NDA)
- Daytime Arietids (ARI)

Asteroids

- 2003 EH_1
- 5496 (1973 NA)

Periodic comets

- 8P/Tuttle
- 12P/Pons-Brooks23P/Brorsen-Metcalf
- 38P/Stephan-Oterma
- 96P/Machholz 1
- 206P/Barnard-Boattini (1892 T_1)
- 226P/Pigott-LINEAR-Kowalski (1783 W_1)

Non-periodic comets

- C/1490 Y_1
- C/1860 D_1 (Liais),
- C/1939 B_1 (Kozik-Peltier)

Lost comet

- 5D/Brorsen

The evidence for some memberships is stronger than others.

Comet 5D/Brorsen had a 5.5-year orbital period and was observed on five occasions between its discovery in 1846 and 1879, after which it seemed to disappear completely and has not been seen since.

Table 5.3: Other Active Meteor Showers During the Quadrantids
Calendar dates are approximate. Use solar longitude $\lambda_\odot$

Shower	IAU Code	Start $\lambda_\odot$	Max $\lambda_\odot$	End $\lambda_\odot$	At Max. α	At Max. δ	V_g (km/s)	ZHR
December Leonis Minorids	DLM	Dec 6 254°	Dec 20 268°	Feb 4 315°	$10^h 40^m$ 160°	+30°	63	5
σ-Serpentids	SSE	Dec 7 255°	Dec 27-28 275°-276°	Jan 12 292°	$16^h 16^m$ 244°	-1.7°	45.5	<1
α-Hydrids	AHY	Dec 17 266°	Jan 1 280°	Jan 17 296°	$08^h 19.5^m$ 124.9°	-7.0°	43.2	1
January Leonids	JLE	Dec 31 279°	Jan 4 283°	Jan 8 287°	$09^h 51^m$ 147.7°	+24.1°	55.2	<1
λ-Boötids	LBO	Dec 31 279°	Jan 16 295°	Jan 17 296°	$14^h 40^m$ 220°	+43°	41	<1
γ-Ursae Minorids	GUM	Jan 9 288°	Jan 19 298°	Jan 22 301°	$15^h 12^m$ 228°	+67°	30	3
o-Leonids	OLE	Jan 1 281°	Jan 8 288°	Jan 28 308°	$09^h 36^m$ 144°	+7°	41.5	<1
κ-Cancrids	KCA	Jan 8 287°	Jan 10 289°	Jan 15 294°	$09^h 12^m$ 138°	+9°	47.3	<1

Table 5.4: Quadrantid Timeline	
1825	Antonio Brucalassi, Italy, makes first known observations of the meteor shower on January 2.
1839	Adolphe Quetelet, Brussels Observatory, and E.C. Herrick, Connecticut, independently suggest annual activity from the shower.
1863	Stillman Masterman, Maine, USA, determines the position of the radiant.
1864	Alexander Herschel observes 60 meteors per hour emanating from the radiant. Almost a century later, J.P.M. Prentice calculates the ZHR at 131.
1909	Exceptionally strong activity with an estimated ZHR of 202.
1918	W.F. Denning and Mrs F. Wilson find that a radiant 8° north of the expected radiant exists and is more active.
1952	Dual observations of the Quadrantids by J.P.M. Prentice and G.E.D. Alcock reveal a cluster of 13 radiants.
1953	J.P.M. Prentice shows that the shower consistently peaks at $\delta_\odot$ 282.9° (1950.0 Epoch).
1963	S.E. Hamid and M.N. Youssef undertake a study of Jupiter's gravitational effect on the meteoroid stream. They suggest that the Quadrantids' parent comet was originally captured by Jupiter some 4,000 years ago.
1963	S.E. Hamid and Fred Whipple suggest that the Quadrantids may be associated with the δ-Aquariids.
1965	Shower puts on its second most prolific display with a ZHR of 190.
1970	ZHR reaches 153.
1971	Keith Hindley, British Astronomical Association, determines that rates are higher than half the maximum for just 16 hours (since amended to 14 hours).
1989	David W. Hughes and Neil McBride estimate the mass of the Quadrantid meteoroid stream to be 1.3×10^{15} g
1990	B. McIntosh suggests Comet 96P/Machholz as the parent body.
1999	Popularity of the shower peaks, with more than 1,000 observations. ZHR 103
2003	Object 2003 EH$_1$ discovered by LONEOS in March. Peter Jenniskens announces in December that it is the parent body of the Quadrantids.
2009	ZHR 156

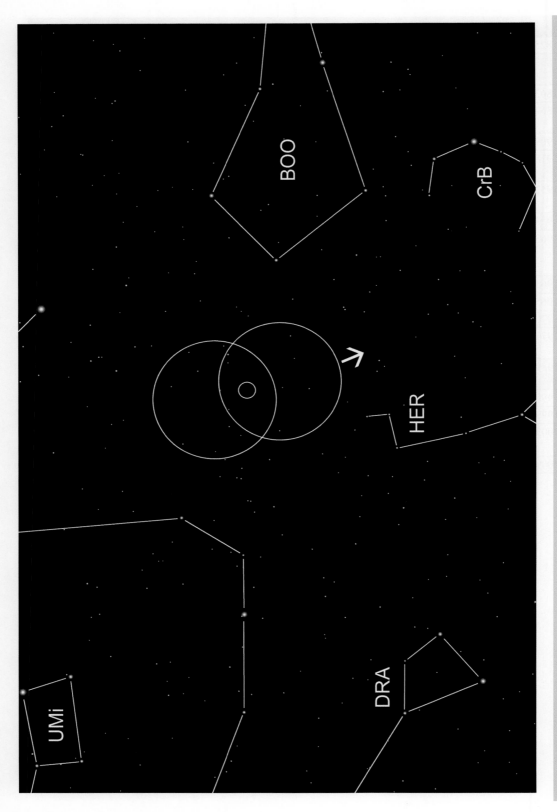

Fig. 5.2 Quadrantids sky view

The Quadrantids

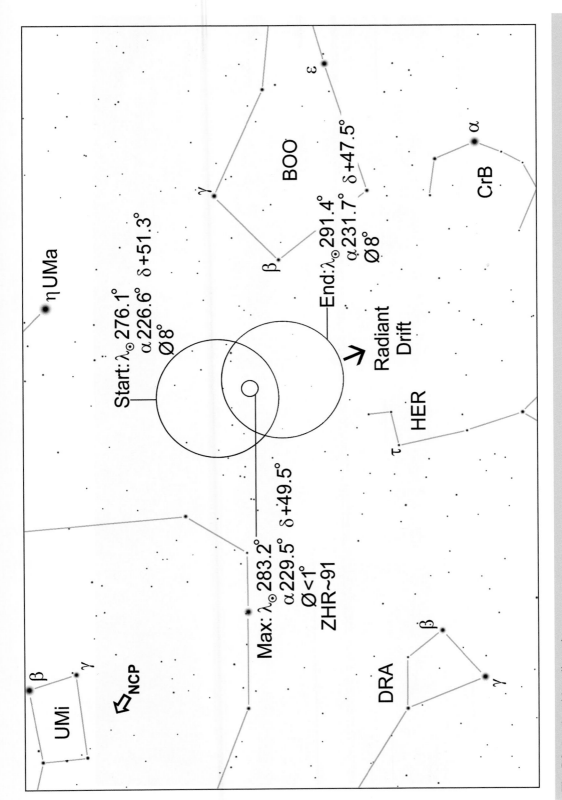

Fig. 5.3 Quadrantids radiant chart

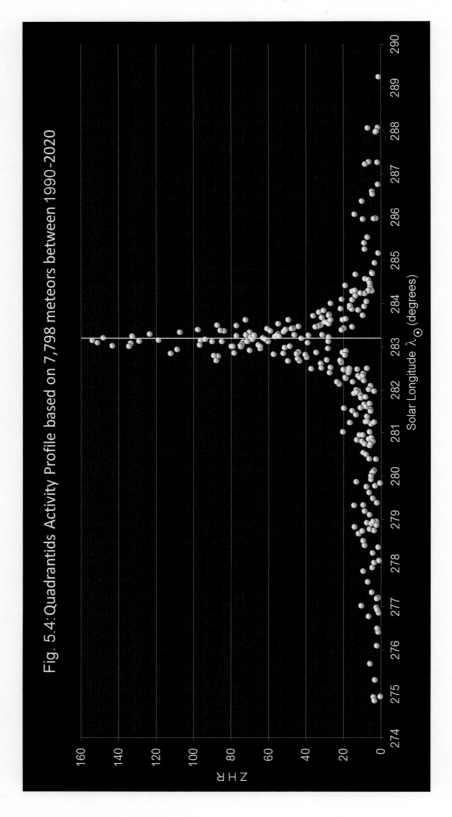

Fig. 5.4: Quadrantids Activity Profile based on 7,798 meteors between 1990-2020

The Quadrantids 67

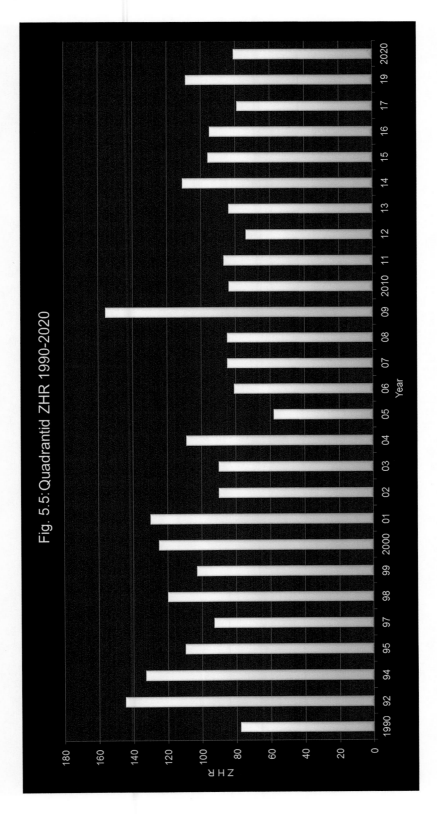
Fig. 5.5: Quadrantid ZHR 1990-2020

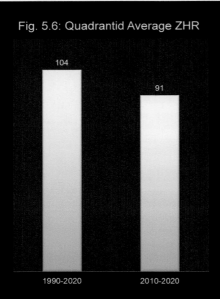

Fig. 5.6: Quadrantid Average ZHR

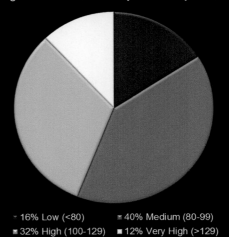

Fig. 5.7: Quadrantid 25-year Activity Levels

- 16% Low (<80)
- 40% Medium (80-99)
- 32% High (100-129)
- 12% Very High (>129)

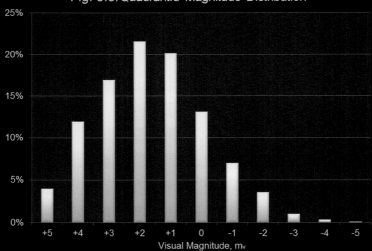

Fig. 5.8: Quadrantid Magnitude Distribution

Fig. 5.9: Quadrantid Geocentric Velocity

The Quadrantids

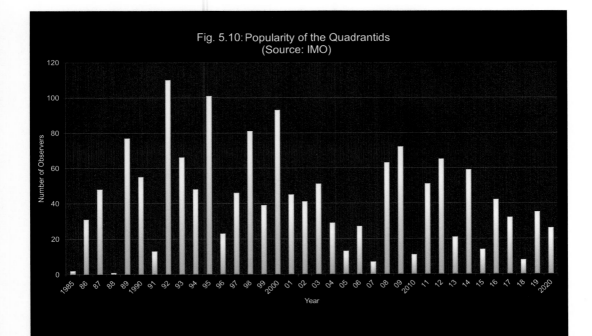

Fig. 5.10: Popularity of the Quadrantids (Source: IMO)

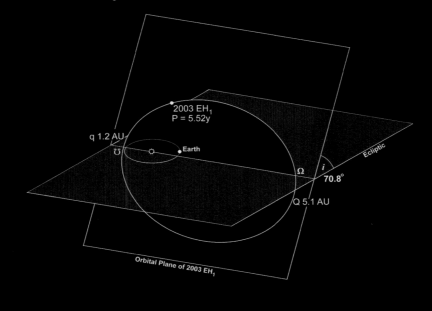

Fig. 5.11: Orbit of the Quadrantid Parent Body

Table 5.5: Orbital Data for the Quadrantid Parent Body*								
2003 EH$_1$								
q	e	a	ω	Ω	i	P	Q	b
AU		AU	°	°	°	y	AU	AU
1.1904	0.6189	3.137	171.3396	282.9816	70.8402	5.52	5.0570	2.4536

*For Epoch 2020-05-31.0

70 5 The Major Meteor Showers: January to August

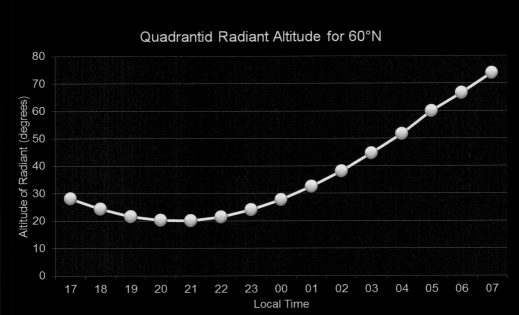

Lerwick, Scotland; Oslo, Norway; Stockholm, Sweden;
Helsinki, Finland; St. Petersburg, Russia; Nanontalik, Greenland.

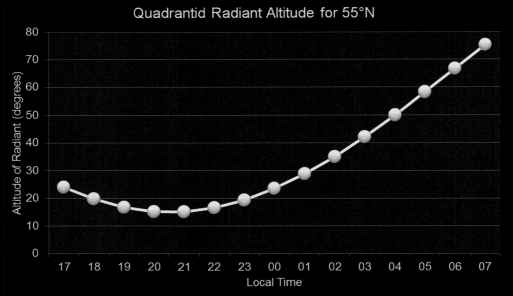

Newcastle upon Tyne, England; Copenhagen Denmark; Malmö, Sweden;
Vilnius, Lithuania; Moscow, Russia; Chelyabinsk, Russia; Edmonton, Canada.

The Quadrantids

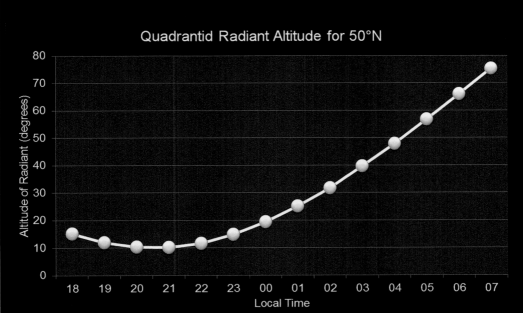

London, England; Frankfurt, Germany; Bern, Switzerland; Kiev, Ukraine; Qaraghandy, Kazakhstan, Qiqihar, China, Vancouver and Winnipeg, Canada.

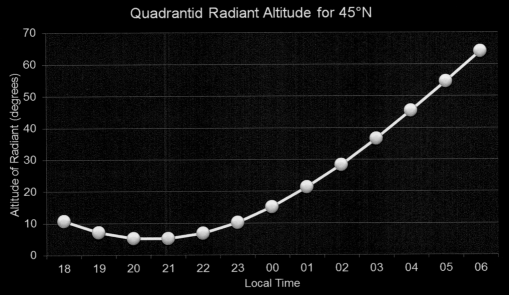

Bordeaux, France; Zagreb, Croatia; Ulaanbaatar, Mongolia; Harbin, China; Portland and Minneapolis, USA; Toronto and Montreal, Canada.

72   5 The Major Meteor Showers: January to August

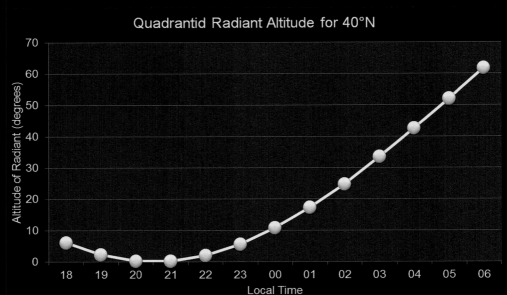

Porto, Portugal; Rome, Italy; Istanbul, Turkey; Toshkent, Uzbekistan; Beijing, China; Sapporo, Japan; Sacramento, Salt Lake City, Denver, Kansas City, Indianapolis, Pittsburgh and Philadelphia, USA.

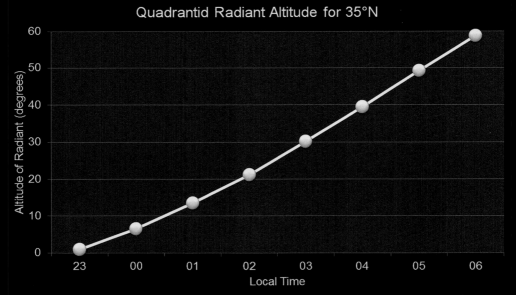

Casablanca, Morocco; Malaga, Spain; Athens, Greece; Tehran, Iran; Jinan, China; Soul, South Korea; Los Angeles, Memphis and Charlotte, USA.

The Quadrantids

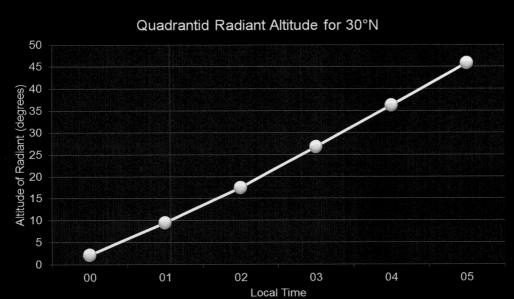

Madeira, Portugal; Tripoli, Libya; Cairo, Egypt; Basra, Iraq; Multan, Pakistan; Lahore, India; Shanghai, China; Tokyo, Japan; San Diego, Austin and Jacksonville, USA.

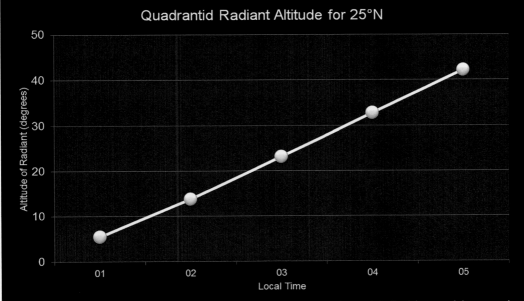

Las Palmas, Canary Islands; Aswan, Egypt; Karachi, Pakistan; Varanasi, India; Kunming, China; T'aipei, Taiwan; Torreon, Mexico; Miami, USA.

74 5 The Major Meteor Showers: January to August

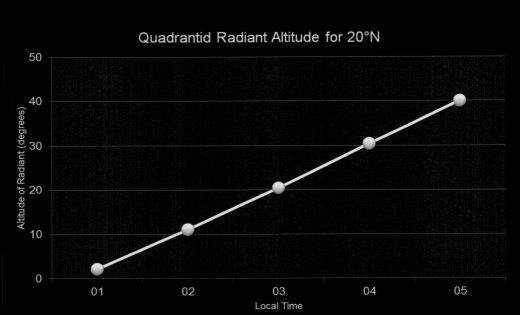

Jeddah, Saudi Arabia; Mumbai, India; Haikou, China; Campeche, Mexico; Santiago de Cuba, Cuba.

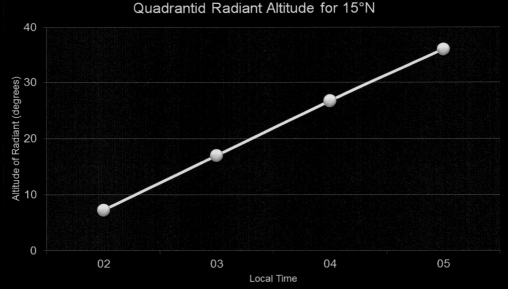

Dakar, Senegal; Khartoum, Sudan; Hyderabad, India; Yangon, Myanmar; Manila, Philippines; Guatemala City, Guatemala

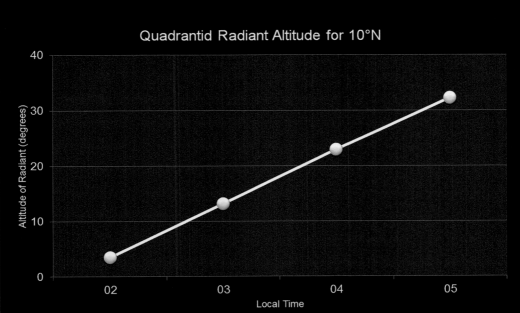

Conakry, Guinea; Kaduna, Nigeria; Addis Ababa, Ethiopia; Ho Chi Minh City, Vietnam; Manila, Philippines; San Jose, Costa Rica; Caracas, Venezuela.

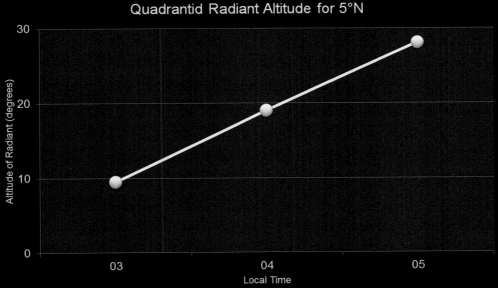

Ghana; Colombo, Sri Lanka; George Town, Malaysia; Bogota, Colombia; Cayenne, French Guiana.

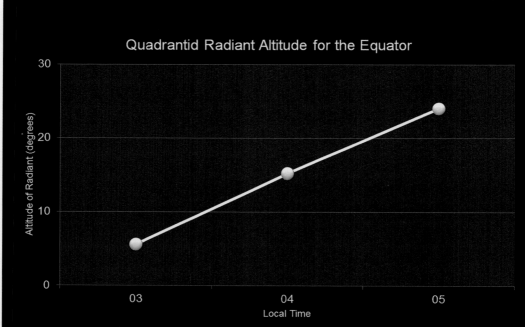

Libreville, Gabon; Kampala, Uganda; Singapore; Quito, Ecuuador; Belem, Brazil.

The April Lyrids

After the northern winter hiatus, meteor activity picks up in the spring with the April Lyrids, the earliest observed meteor shower.

Observing Notes

With the bitterly cold nights of a northern winter often just a distant memory for many, it's a real pleasure to get back to hunting for meteors without having to dress like a polar explorer. The April Lyrids' popularity has had its ups and downs—down mainly when moonlight has spoilt observation—but observers are just as likely to make an effort to record the shower today as they were 40 years ago.

The name of the shower is a bit of a misnomer. The shower's radiant spends much of its time in Hercules, only crossing the border into Lyra after maximum. That particular portion of Hercules is devoid of bright stars however, so the shower derives its name from the proximity to α Lyrae, which shines like a beacon, making the radiant easy to locate (Tables 5.6a and 5.6b).

Activity runs from about April 14 ($\lambda_\odot$ 24°) to April 30 ($\lambda_\odot$ 40°) with maximum on April 22 or 23 ($\lambda_\odot$ 32.32°). Rates tend to be low, just a couple of meteors per hour, until about two days before maximum, when you can expect a ZHR of 5, rising swiftly to a sharp peak of about 18 before falling off just as rapidly. Activity stays above half the maximum value for typically about 6 hours, though it can last longer when rates are low. Some years are better than others. There have been half a dozen times in the past few decades when the ZHR has been in the low 20s, though in 2001 it reached 26. There is a 36% chance that the rates will be high (between 20 and 23) and a 20% chance of the ZHR exceeding 23 if the past 25 years are anything to go by (Fig. 5.18). When rates are low, peak activity may not be so sharp but may spread out for a while. High rates may result in the radiant apparently shrinking, similar to the Quadrantids but by not so much: 8° down to 5° in diameter [16].).

There have been a number of outbursts, most notably in 1803 when the ZHR was estimated to be about 650, as witnessed by Virginians in the USA, as well as in Greece in 1922 (ZHR = 430), Japan 1945 (112 and 110 the following year) and an impressive 253 in 1982. Researchers have suggested various periodic cycles for enhanced activity, including 9.7, 12, 16, 20, 27, 29.7 and 47 years, all with varying degrees of merit, though Peter Jenniskens of the SETI Institute and NASA Ames Research Center, favours a 60-year cycle. That ties in with 1922 and 1982, so if he is right, then the next good display will be in 2040/41 [17]. Only time will tell.

The shower is most favourable to observers in the Northern Hemisphere. For mid-northern sky watchers, the radiant ascends above the horizon at a little before 21:00 LT and reaches about 40° altitude by midnight, so late evening observation is possible for those who live in a dark sky area. For most though, the shower is best after midnight. At 4:00 LT, the radiant is virtually at the zenith for observers at latitudes between 30°N and 35°N. This makes the Lyrids a good test for the ZHR estimate. With the radiant almost directly overhead, it is ideally placed to compare the actual ZHR with the predicted value.

The April Lyrids

Table 5.6a: April Lyrid Activity Details				
		Start	Maximum	End
Dates (approx.)		Apr 14	Apr 22/23	Apr 30
Solar longitude:	λ☉	24°	32.32°	40°
Right Ascension at max:	α		$18^h 07^m$ (272°)	
Declination at max:	δ		+33.1°	
Time of transit :			4.1^h L.T.	
ZHR (2010-2020 average):			18	
Radiant diameter:	∅	8°	5°	8°

Table 5.6b: April Lyrid Additional Data		
IAU Abbreviation: LYR	**IAU Code:** 006	**AKA:** Lyrids
Mean daily motion of radiant:	Δ	Δα = +0.84° Δδ = -0.34°
Geocentric velocity:	V_g	47.3 km/s (170,280 km/h)
Population index:	r	2.1 average (2.9 at maximum activity)
Mean magnitude:	$\bar{m}_v$	+2.3
Parent body: C/1861 G1 Thatcher		
Best visibility: Post-midnight in the Northern Hemisphere		
The radiant is circumpolar from 57°N, i.e. north of Aberdeen, Scotland; Gothenburg, Sweden; Riga, Latvia; Sverdlovsk, Russia; Port Nelson, Canada		

Table 5.7: April Lyrid Maxima 2020-2030				
	Maximum (λ☉ 32.32°)			Moon
year	month	day	hour (UT)	Age (d)
2020	04	22	06:30	28
2021	04	22	12:40	10
2022	04	22	18:53	21
2023	04	23	01:02	3
2024	04	22	07:14	14
2025	04	22	13:32	24
2026	04	22	19:37	5
2027	04	23	01:41	17
2028	04	22	07:52	28
2029	04	22	14:01	16
2030	04	22	20:10	20

Astronomers in the Southern Hemisphere generally get a raw deal with meteor showers, but the Lyrids can put on a reasonable display all the way down to about 25°S. That covers most of South America, Africa and Northern Australia. Obviously, the farther south you are, the closer the radiant will be to the horizon and the fewer meteors you will see.

April Lyrid meteors are not particularly fragile; the Harvard Meteor Survey showed they are similar in density to most sporadics [18] – 0.41 ± 0.12 g/cm^3 – and they have an average magnitude of $\overline{m}_v$ + 2.3, with about 16% being brighter than zero. The shower produces the occasional fireball. There is an increase in fainter meteors at the time of maximum, which lowers the average magnitude. Consequently, the population index, r, which is usually around 2.0–2.1, increases to about 2.3–2.9 at maximum.

Table 5.8 gives details of other meteor showers that are active during the April Lyrids.

Discovery

The Lyrids are the first recorded meteor shower. In 687 March 23 BCE, Chinese royal astrologers noted that after the middle of the night, 'stars fell like rain. They were 10 to 20 degrees long. This phenomenon was repeated continually. Before arriving at the Earth they were extinguished.' [19] Although March 23 may seem too early to be associated with the April Lyrids, it actually equates to April 19 once calendar changes have been taking into account. The Japanese and Koreans were also keen on recording meteor activity. Charles P. Olivier, the founder of the American Meteor Society, in his classic work *Meteors* (1925), republished records collated by Jean-Baptiste Biot, Michel Chasles and Yale University librarian Edward Claudius Herrick, shown in the Table 5.9. These catalogued Lyrid activity over several centuries but particularly in the 11th and 12th centuries [20].

On 1803 April 19–20, a storm of about 670 meteors per hour was witnessed by those in the eastern USA from New Hampshire to North Carolina, but despite being such a rare phenomenon, little attention was paid during subsequent years. It was not until the great Leonid storm of 1833 that astronomers began to take a serious interest in meteor showers. Consequently, it was left to François Arago[2] to speculate in 1835 that the Lyrids were an annual phenomenon with a return date of April 22.

Between 1838 and 1839, the German physicist and astronomer Johann Benzenberg, founder of the Bilk Observatory at Düsseldorf, could only detect limited activity from the shower, as did Herrick in Connecticut. But Herrick was able to estimate the position of the radiant, which he put at α 273° (18^h 12^m), δ +45° on 1839 April 18, about 12° farther north in declination from the current average position [21]. A further five years went by before Alexander Herschel was able to refine the position of the radiant, which he placed at α 277° (18^h 28^m), δ +35° [22]. Henry Corder, the English naturalist, archaeologist and astronomer, and Director of the BAA Meteor Section 1892–99, estimated the radiant to be at α 275° (18^h 20^m), δ +36° from his observations between 1876 and 1879 [23], but in 1882 the radiant appeared at α 268° (17^h 52^m), δ +37°. In 1885 between April 18–20, another English amateur astronomer, William F. Denning, noted that the radiant drifted from one night to the next—a

---

[2] Arago had an eventful life. He was once imprisoned as a spy, incarcerated in a windmill, and later succeeded in abolishing slavery in the French colonies.

phenomenon that had only previously been seen in the Perseids [24]. Incidentally, Denning succeeded Corder as Meteor Section Director and stayed in post from just 1899 to 1900.

Meteoroid streams that have high inclination orbits tend to suffer relatively little from the gravitational effects of the planets. In 1969, Keith Hindley, another Director of the BAA's Meteor Section, concluded that the stream's orbital nodes had shown little or no movement in more than 2600 years [25]. By comparison, streams that lie in the plane of the ecliptic gyrate as they are tugged by the planets, which can lead to significant changes in their nodes over a relatively short period of time.

In 1992, Bertil A. Lindblad and Vladimír Porubčan discovered that intense bursts of activity historically occurred about 6 hours before the normal maximum, and that these were due to a dense filament of meteoroids, though they were at a loss to explain its longevity [26]. Later work by Porubčan and Kornoš concluded that there were two filaments, one with a period of 40 years and another with a much greater orbital stability of about 600 years [27].

Origins

The relationship between comets and meteor showers was not made until 1866 by Giovanni Schiaparelli, whose two-year study of the Perseids revealed orbital similarities with Comet Swift-Tuttle (1862 III) [28]. The following year, the Director of the Vienna Observatory, Edmond Weiss, noticed that Comet Thatcher (1861 I) came within 0.002 AU (300,000 km) of the Earth's orbit on April 20, close to the date of Lyrid maxima [29]. His calculations were confirmed by Johann Galle, who also managed to determine an orbit for the meteoroid stream that was very similar to that of the comet. He also traced the history of the Lyrid shower back to 687 BCE. In fact, the close proximity of Comet Thatcher's orbit to that of the Earth had previously been discovered in 1861 by C.F. Pape and independently by Theodor von Oppolzer (1841–1881), though neither had made the connection with the Lyrids.

The Lyrids are one of only two showers that are associated with long period comets (P > 200 y); the other is Comet Kiess and the Aurigids. The Earth encounters the Lyrid meteoroid stream at its descending node (Table 5.11).

Table 5.8: Other Active Meteor Showers During the April Lyrids
Calendar dates are approximate. Use solar longitude $\lambda_\odot$

Shower	IAU Code	Start $\lambda_\odot$	Max $\lambda_\odot$	End $\lambda_\odot$	At Max. α	At Max. δ	V_g (km/s)	ZHR
α-Virginids	AVB	Mar 10 350°	Apr 7-18 17°- 28°	May 6 46°	$13^h 36^m$ 204°	-11°	17.7	5-10
σ-Leonids	SLE	Apr 8 18°	Apr 15 25°	April 28 38°	$13^h 28^m$ 202°	+3°	19	1
April ρ-Cygnids	ARC	Apr 11 21°	Apr 22? 32°?	May 8 48°	$21^h 24^m$ 321°	+45.5°	41.4	<1
π-Puppids	PPU	Apr 15 25°	Apr 23 33.5°	Apr 28 38°	$7^h 20^m$ 110°	-45°	15	Var.
η-Aquariids	ETA	Apr 19 29°	May 6 46.2°	May 28 68°	$22^h 32^m$ 338°	-0.8°	65.7	2-5
h-Virginids	HVI	Apr 20 30°	Apr 28-30 38°- 40°	May 4 44°	$13^h 40^m$ 205°	-11°	18.4	<1

Table 5.9: Historical Records of April Lyrid Activity

Date	Equivalent to (Epoch 1850)	Author
BCE 687 March 23.7	CE April 19.9	Biot
15 March 27	April 19.6	Biot
CE 582 March 31	April 18.1	Chasles
590 April 4?	April 22.1?	Chasles
1093 April 9.6	April 20.7	Chasles
1094 April 10	April 20.8	Chasles
1095 April 9.6	April 20.2	Herrick
1096 April 10	April 21.3	Herrick
1122 April 10.6	April 20.2	Herrick
1123 April 11	April 20.4	Chasles
1803 April 19.6	April 19.9	Herrick

Table 5.10: April Lyrid Timeline

687 BCE	First record of April Lyrid activity (then March 23).
1122	Storm.
1803	Storm on April 19-20, estimated at 670 meteors per hour.
1835	François Arago suggests that the Lyrids are an annual occurrence.
1839	E.C. Herrick locates the shower's radiant.
1849	High level of activity.
1850	High level of activity.
1861	The Lyrids' parent comet discovered on April 5 by amateur astronomers A.E. Thatcher in New York, and independently by Carl Wilhelm Baeker of Nauen, Germany.
1864	Alexander Herschel refines Herrick's radiant position.
1867	Edmond Weiss notices that Comet Thatcher comes to within 0.002 AU of the Earth's orbit on April 20. Johann Galle confirms the calculations and traces the shower's history to 687 BCE.
1885	W.F. Denning notes the radiant drifts eastward by about 1° per day.
1922	Strong display. H.N. Russell, of Hertzsprung-Russell Diagram fame, sees 96 meteors/h from Nauplia, Greece on April 21. ZHR 430 (other estimates range from 180 to 600).
1923	Denning produces first ephemeris from April 8 to 30.
1934	ZHR 55-80
1945	ZHR 110, Japan
1946	ZHR 80-110
1982	ZHR 253, North America. Peak lasts 1 hour. Europeans report slow rates.
1983	Porubčan and Štohl discover the average magnitude falls at time of maximum.
1992	Lindblad and Porubčan discover historic pre-maxima outbursts due to a filament of material in the stream.
1997	Arter and Williams estimate the age of the Lyrids to be 1.5 million years.
2008	Porubčan and Kornoš identify two filaments.

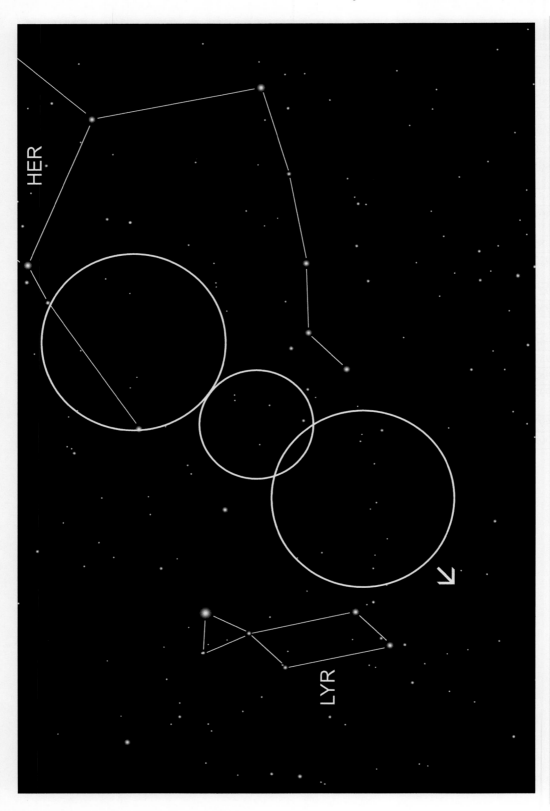

Fig. 5.13 April Lyrids sky view

The April Lyrids

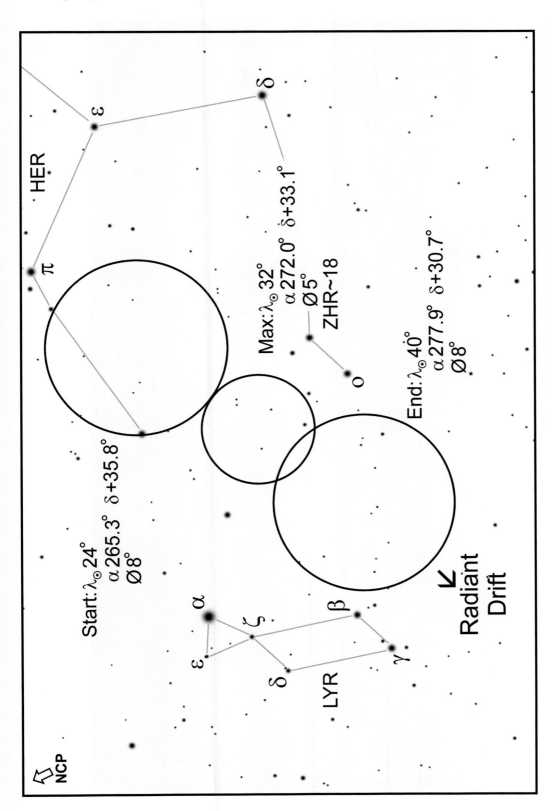

Fig. 5.14 The April Lyrid radiant

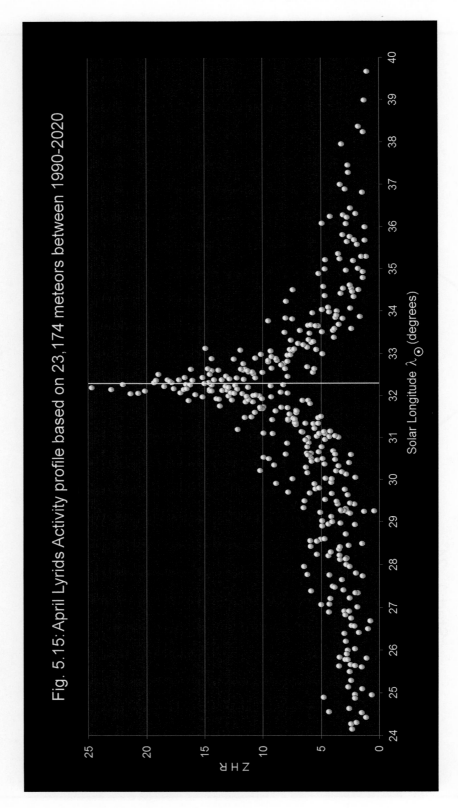

Fig. 5.15: April Lyrids Activity profile based on 23,174 meteors between 1990-2020

The April Lyrids

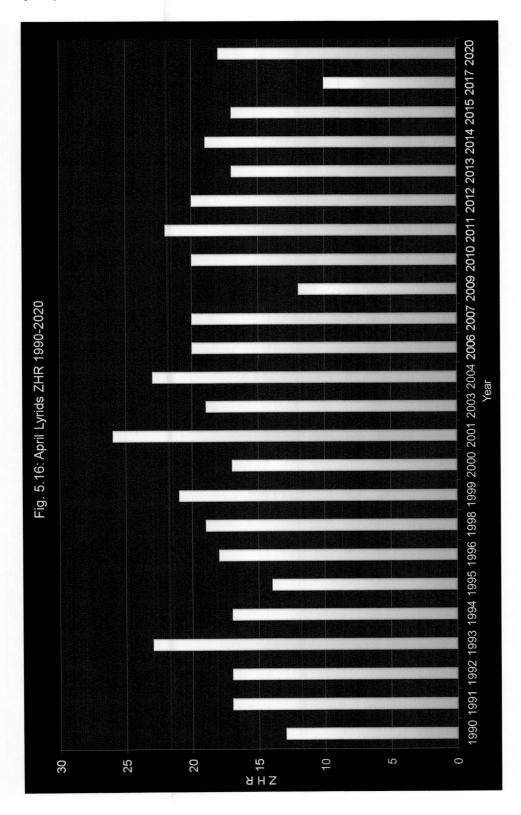

Fig. 5.16: April Lyrids ZHR 1990-2020

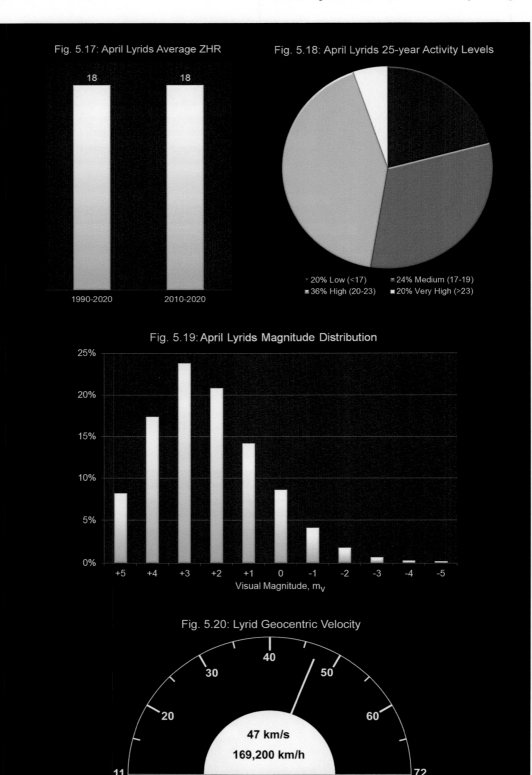

The April Lyrids

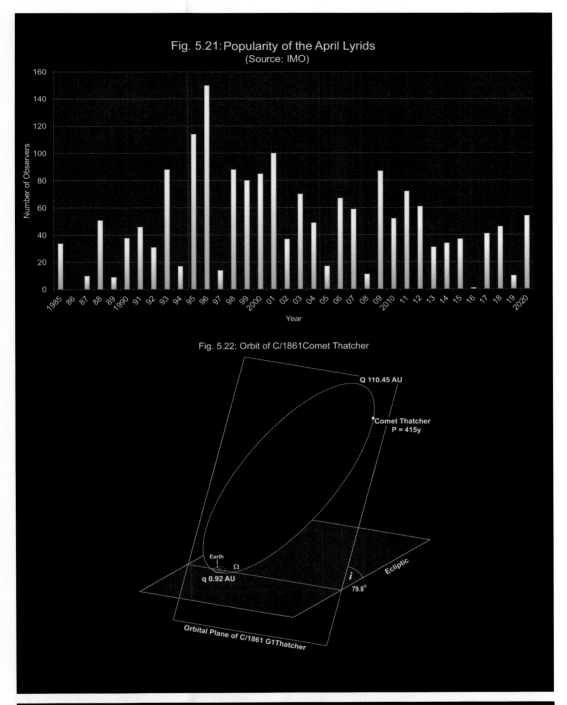

Fig. 5.21: Popularity of the April Lyrids (Source: IMO)

Fig. 5.22: Orbit of C/1861 Comet Thatcher

Table 5.11: Orbital Data for the April Lyrids Parent Body*									
Comet Thatcher (C/1861 G$_1$)									
q	e	a	ω	Ω	i	P	Q	b
AU		AU	°	°	°	y	AU	AU
0.9207	0.9835	55.6819	213.4496	31.8674	79.7733	415	110.4431	10.0733

*For Epoch 1860-05-25.0

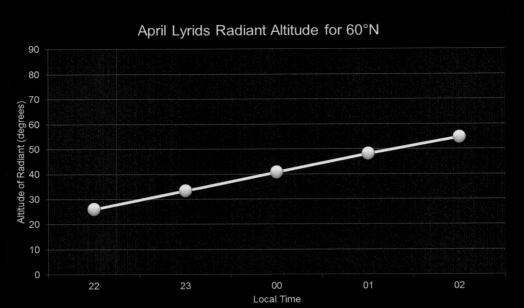

Lerwick, Scotland; Oslo, Norway; Stockholm, Sweden; Helsinki, Finland; St. Petersburg, Russia; Nanontalik, Greenland.

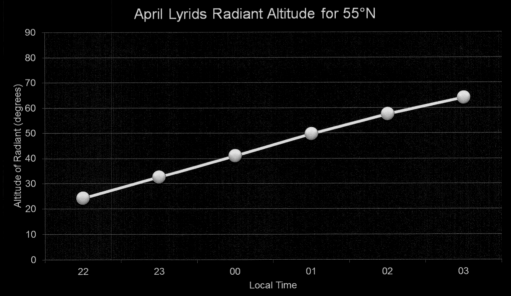

Newcastle upon Tyne, England; Copenhagen Denmark; Malmö, Sweden; Vilnius, Lithuania; Moscow, Russia; Chelyabinsk, Russia; Edmonton, Canada.

The April Lyrids

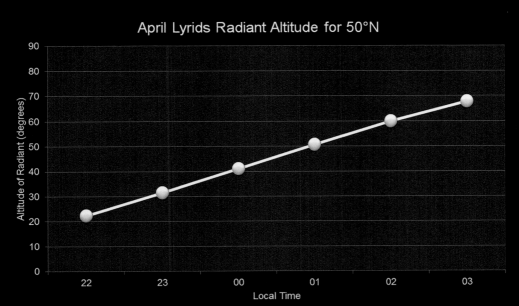

London, England; Frankfurt, Germany; Bern, Switzerland; Kiev, Ukraine; Qaraghandy, Kazakhstan, Qiqihar, China, Vancouver and Winnipeg, Canada.

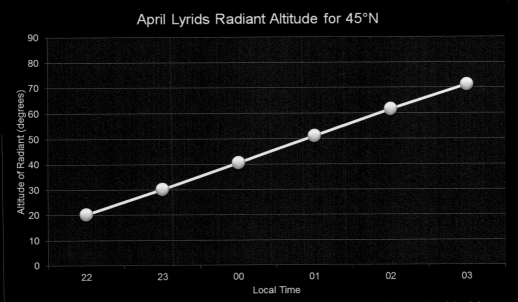

Bordeaux, France; Zagreb, Croatia; Ulaanbaatar, Mongolia; Harbin, China; Portland and Minneapolis, USA; Toronto and Montreal, Canada.

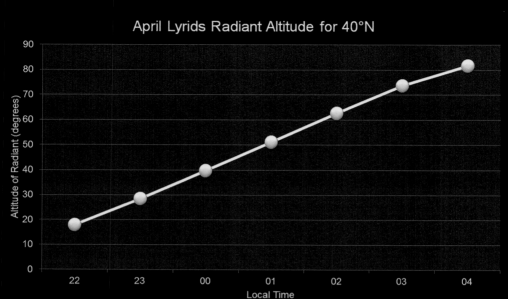

Porto, Portugal; Rome, Italy; Istanbul, Turkey; Toshkent, Uzbekistan; Beijing, China; Sapporo, Japan; Sacramento, Salt Lake City, Denver, Kansas City, Indianapolis, Pittsburgh and Philadelphia, USA.

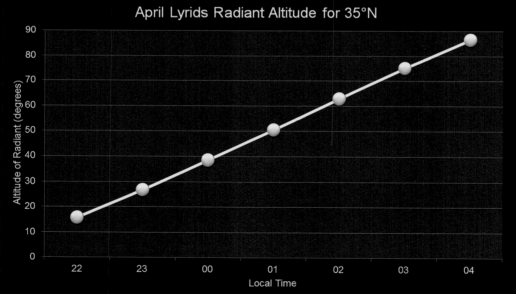

Casablanca, Morocco; Malaga, Spain; Athens, Greece; Tehran, Iran; Jinan, China; Soul, South Korea; Los Angeles, Memphis and Charlotte, USA.

The April Lyrids

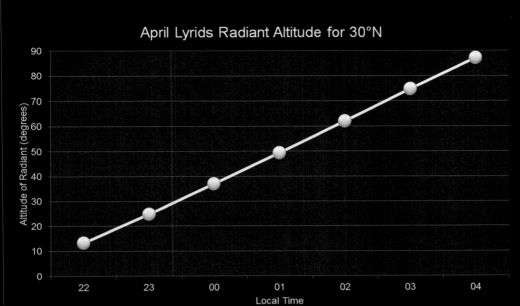

Madeira, Portugal; Tripoli, Libya; Cairo, Egypt; Basra, Iraq; Multan, Pakistan; Lahore, India; Shanghai, China; Tokyo, Japan; San Diego, Austin and Jacksonville, USA.

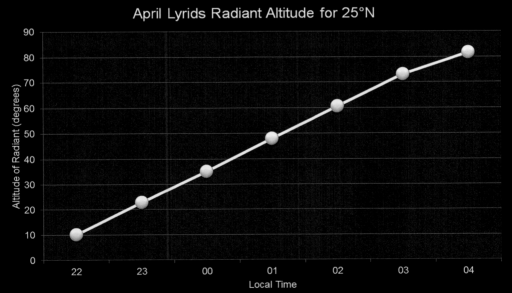

Las Palmas, Canary Islands; Aswan, Egypt; Karachi, Pakistan; Varanasi, India; Kunming, China; T'aipei, Taiwan; Torreon, Mexico; Miami, USA.

94   5 The Major Meteor Showers: January to August

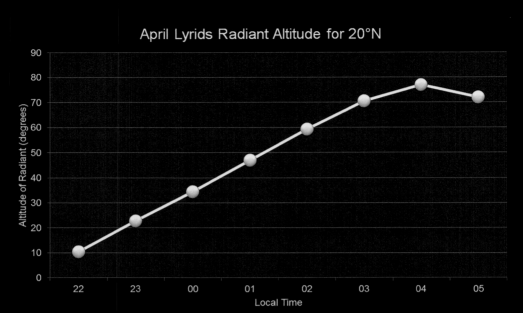

Jeddah, Saudi Arabia; Mumbai, India; Haikou, China; Campeche, Mexico; Santiago de Cuba, Cuba.

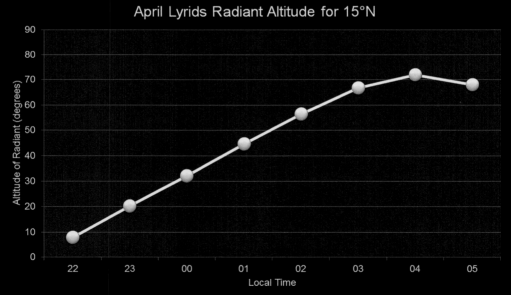

Dakar, Senegal; Khartoum, Sudan; Hyderabad, India; Yangon, Myanmar; Manila, Philippines; Guatemala City, Guatemala

The April Lyrids

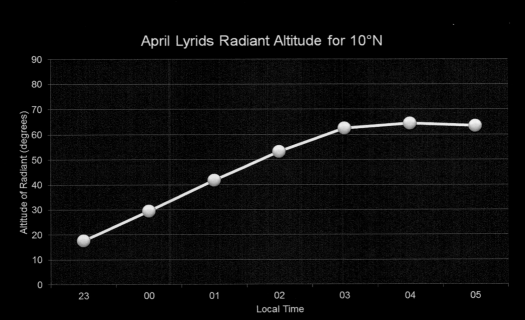

Conakry, Guinea; Kaduna, Nigeria; Addis Ababa, Ethiopia; Ho Chi Minh City, Vietnam; Manila, Philippines; San Jose, Costa Rica; Caracas, Venezuela.

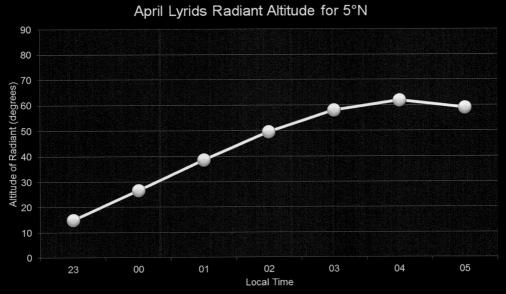

Accra, Ghana; Colombo, Sri Lanka; George Town, Malaysia; Bogota, Colombia; Cayenne, French Guiana.

96 5 The Major Meteor Showers: January to August

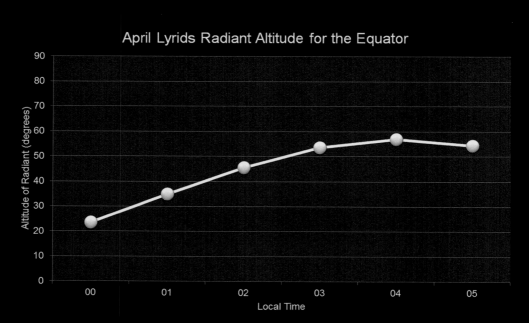

Libreville, Gabon; Kampala, Uganda; Singapore; Quito, Ecuuador; Belem, Brazil.

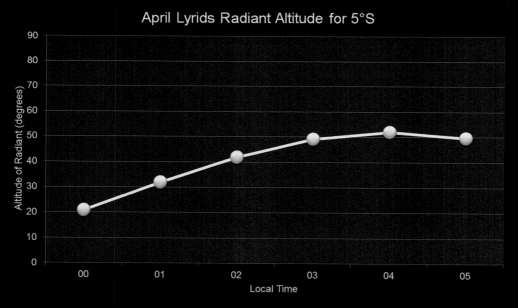

Kinshasa, D.R.Congo; Jakarta, Indonesia; Piura, Peru; Teresina, Brazil.

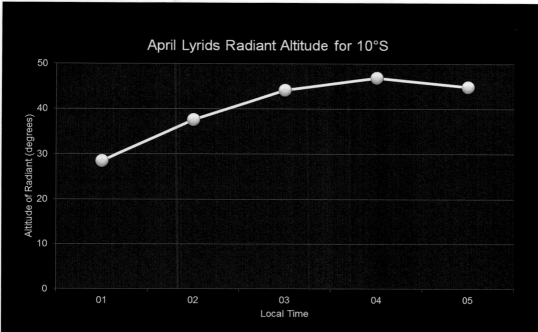

Luanda, Angola; Kolwetzi, D.R.Congo; Kupang, Indonesia, Maceio, Brazil.

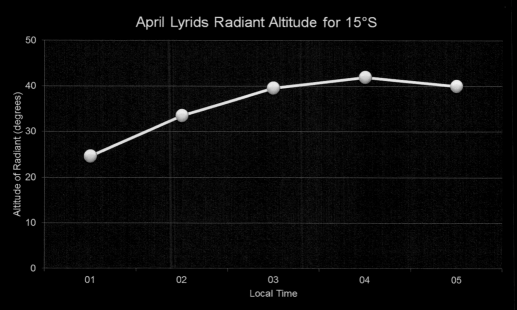

Lusaka, Zambia; Blantyre, Malawi; Coronation Island, Western Australia.

98   5 The Major Meteor Showers: January to August

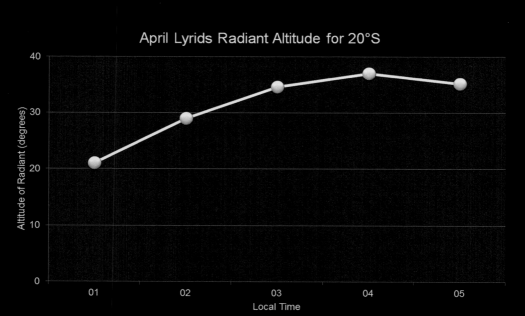

Bulawayo, Zimbabwe; Beira, Mozambique; Port Hedland, Western Australia; Bowen, Queensland, Australia; Iquique, Chile; Cariacica, Brazil.

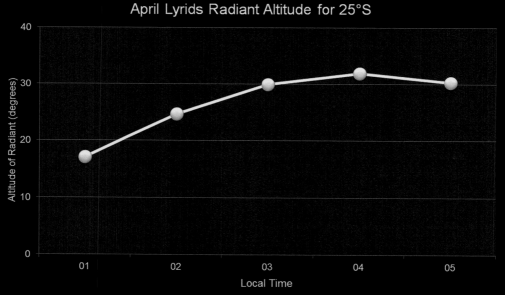

Gibeon, Namibia; Pretoria, South Africa; Carnarvon, Western Australia; Bundaberg Central, Queensland, Australia; Taltal, Chile; Asuncion, Paraguay.

The η-Aquariids

The eta-Aquariids

May sees the arrival of the η-Aquariids, the most active meteor shower in the Southern Hemisphere.

Observing Notes

At first glance, it doesn't make sense. At maximum activity, the radiant of the η-Aquariids sits less than 1° south of the Celestial Equator, yet it is virtually impossible to observe from most of the Northern Hemisphere. How come? The secret is the 23.4° tilt of the Earth's axis. At the beginning of May, the Northern Hemisphere is halfway through spring and the Sun is rising earlier each day. In the Southern Hemisphere, it is mid-autumn. The Sun rises later and later and the nights are getting longer and longer. For someone in Rome, Italy, at 40°N, the radiant reaches 27° altitude at 4 am LT but the first rays of dawn are only about half an hour away. Meanwhile, for an observer at 40°S in Wanganui, New Zealand, the radiant is also at 27° altitude at 4 am LT but the Sun doesn't rise until after 6 am, giving them up to two hours' observing advantage over their northern counterpart. Just before dawn, the radiant will be at 50° altitude and southern observers can expect to see around 80% of all observable meteors before the sky starts to lighten, or about 65 meteors an hour in an average year.

The first η-Aquariids meteors begin to make an appearance around 19 April with the radiant close to β Aquarii, *Sandalsuud*. There are not many at first—one or two an hour—but they gradually pick up to about 7 to 10 by the last week of April and then follow a steady, rapid rise to maximum in the first week of May. Unlike the Quadrantids or the April Lyrids, there is no sharp peak but a nice, rounded maximum that can last up to about 36 hours as the radiant sweeps past γ and ζ Aquarii to rendezvous with the star from which it takes its identity, η Aquarii, about May 6. As the radiant continues its journey, crossing the border into Pisces, heading for γ Piscium, activity starts to tail off, but more slowly than its rise to maximum. A few stragglers can be seen until May 28.

From 1990 to 2020 the shower maximum averaged 71 meteors per hour, but rates have been higher than usual in the past decade, so the ZHR now averages 83. There is the possibility of a 12-year cycle. 2004 and 2005 saw ZHRs of 112 and 108 respectively, while 12 years later, in 2016 and 2017, the rates were about 115 after a decade of fairly modest activity. Except for 2013, when rates shot up to 135 due to the Earth encountering trails of dust ejected by the stream's parent comet, Comet Halley, in 1197 and 910 CE.

The radiant seems to be a steady 5° in diameter throughout the entire period of activity, although some observers put it slightly larger at 7°. Average magnitude is about $\bar{m}_v + 2.5$ but around 6% of all η-Aquariids are brighter than zero magnitude. On a moonless night, when the shower reaches maximum, there is a 28% chance that activity will be high to very high if the past couple of decades are anything to go by (Fig. 5.30).

Entering the atmosphere at a breath-taking 65.7 km/s (236,520 km/h), the η-Aquariids are among the swiftest meteors of the year, beaten only by the Leonids. They often leave behind persistent trains that typically last for couple of seconds. In a good year, as many as 2 out of 3 η-Aquariids produce persistent trains, but typically it is around 40%.

The η-Aquariids

Table 5.12a: η-Aquariid Activity Details				
		Start	**Maximum**	**End**
Dates (approx.)		Apr 19	May 6/7	May 28
Solar longitude:	λ☉	29°	46.2°	68°
Right Ascension at max:	α		22ʰ 32ᵐ (338°)	
Declination at max:	δ		-0.8°	
Time of transit :			7.7ʰ L.T.	
ZHR (2010-2020 average):			83	
Radiant diameter:	⌀	5°	5°	5°

Table 5.12b: η-Aquariid Other Data		
IAU Abbreviation: ETA	**IAU Code:** 031	**AKA:** April Halleyids
Mean daily motion of radiant:	Δ	Δα = +0.92° Δδ = +0.37°
Geocentric velocity:	V_g	65.7 km/s (236,520 km/h)
Population index:	*r*	2.4
Mean magnitude:	$\overline{m}_v$	+2.3
Parent body: Comet 1P/Halley		
Best visibility: Pre-dawn in the Southern Hemisphere		
The radiant is circumpolar from the South Pole only.		

Table 5.13: η-Aquariid Maxima 2020-2030				
	Maximum (λ☉ 46.2°)			Moon
year	month	day	hour (UT)	Age (d)
2020	05	06	13:23	14
2021	05	06	19:39	24
2022	05	07	01:39	7
2023	05	07	07:55	16
2024	05	06	14:08	27
2025	05	06	20:19	8
2026	05	07	02:34	19
2027	05	07	08:32	1
2028	05	06	14:43	13
2029	05	06	20:59	22
2030	05	07	02:57	5

Observers in the Northern Hemisphere, especially north of the 40th parallel, struggle to observe the shower with it being so close to the horizon and rising only a little before dawn. However, when meteors are observed from this shower they are often very long; several tens of degrees. These are the so-called *Earthgrazers*: meteoroids that just skim the outermost layers of the atmosphere, sometimes even skipping back out into space. Northern observers shouldn't be too disappointed at not being able to see the η-Aquariids though, for they put in a second appearance in October as the Orionids.

Discovery

Perhaps not surprisingly, it was the Royal Court Astrologers of China who first recorded the η-Aquariids. Hubert Newton, a mathematics professor at Yale, discovered Chinese records dating back 14 centuries to 401 CE [30]. In 1863, Newton suggested the existence of an annual shower towards the end of April but six years passed before any concrete evidence came to light. Johann Schmidt of the Athens Observatory in Greece encouraged a British military officer, Lieutenant-Colonel George L. Tupman, that he should make an effort to look for meteors in late April while he was serving aboard *HMS Prince Consort* in the Mediterranean [31]. Not only did Tupman rediscover the η-Aquariid meteor shower in 1869, but he also observed a total of about 300 meteors and was one of the first to calculate their velocity.[3] Some years later, William F. Denning discovered that the members of the Italian Meteoric Society had accidentally observed 45 η-Aquariids between 1868 and 1870, predating Tupman's observations by a year, but not realizing at the time that they belonged to a particular annual shower [32].

Almost a decade after Tupman's observations, in 1878, Alexander Herschel discovered that the η-Aquariids radiant drifts eastwards from one night to the next due to the Earth's passage through the stream. Although the geometry of radiant drift was fairly easy to demonstrate, it later led to a bitter disagreement between Charles Olivier and William Denning, with Denning insisting that some radiants, at least, did not drift but remained stationary. Olivier devoted a whole chapter in his 1925 book *Meteors* to demonstrating that Denning and his followers were wrong.

The relatively low population in the Southern Hemisphere meant that good-quality observations were few and far between and much about the η-Aquariids remained a mystery until the 1920s. Then Ronald A. McIntosh came along and, from his base in New Zealand, undertook a series of observations that led to a detailed report published in the *Monthly Notices of the Royal Astronomical Society* in 1929 [33]. A few years later, he determined the radiant drift to be $\Delta\alpha$ +0.96° and $\Delta\delta$ +0.37° per day (the currently accepted average values are $\Delta\alpha$ +0.92° and $\Delta\delta$ +0.37°).

Radio observations of the η-Aquariids began in 1947, when Bernard Lovell and his team at Jodrell Bank, England, turned their attention to the shower. But perhaps the most significant finding had to wait until 1973, when Anton Hajduk announced the discovery of a double maximum, which he related to a filamentary structure within the stream. Hajduk also worked out that the meteor shower begins when the Earth is within 0.065 AU (9.75 million km) of the meteoroid stream. A decade later, Hajduk and Bruce McIntosh showed that the

---

[3] Tupman is most famous for his work with the British Expeditionary Force on the 1874 and 1882 transits of Venus.

The η-Aquariids

meteors currently seen were actually released during previous apparitions of Comet Halley many centuries earlier [34]. The current orbital configuration of 1P/Halley takes the comet farther away from the Earth, with no prospect of encountering the dust trails that it now leaves behind. As a result, the 1985–86 apparition did not result in enhanced meteor activity from either the η-Aquariids or the Orionids. An outburst did occur in 2013 but this was due to dust ejected in 1197 and 910 [35].

During the 1970s and 1980s, the Western Australia Meteor Society, led by Jeff Wood, was instrumental in popularizing several meteor showers including the η-Aquariids. Data published by the International Meteor Organization shows that the shower's popularity has varied over the past 30 years but remains a favourite among those fortunate enough to be able to observe it.

Origins

The progenitor of the η-Aquariids is that most famous of comets, Halley, which also produces a sister shower, the Orionids, in October.

In 1868, inspired by Newton's work on the history of the shower, the German astronomer Rudolf Falb successfully worked out that the shower was associated with both Halley's Comet and the Orionids [36]. The comet association is often wrongly attributed to Alexander Herschel in 1876, and the link with the Orionids is usually credited to Charles Olivier in 1911. Figures 5.24 and 5.34 show the orbital configuration of the stream in relation to the Earth's orbit. The stream comes from below the Earth's orbit and at the intersection, the *ascending node*, it produces the Orionids in the Northern Hemisphere. The stream then swings around the Sun and on its descent encounters the Earth six months later, in mid-April, at the *descending node* to produce the η-Aquariids in the Southern Hemisphere. The showers are collectively known as the *Halleyids*.

The stream is rich in small particles with masses of less than 10^{-5} g. In 1989, University of Sheffield researchers David Hughes and Neil McBride [15] estimated the total mass of the meteoroid stream to be 3.3×10^{16} g.

Table 5.14: Other Active Meteor Showers During the η-Aquariids								
Calendar dates are approximate. Use solar longitude $\lambda_\odot$								
Shower	IAU Code	Start $\lambda_\odot$	Max $\lambda_\odot$	End $\lambda_\odot$	At Max. α	At Max. δ	V_g (km/s)	ZHR
α-Virginids	AVB	Mar 10 350°	Apr 7-18 17°- 28°	May 6 46°	$13^h\,36^m$ 204°	-11°	17.7	5-10
April ρ-Cygnids	ARC	Apr 11 21°	Apr 22? 32°?	May 8 48°	$21^h\,24^m$ 321°	+45.5°	41.4	<1
April Lyrids	LYR	Apr 14 24°	Apr 22-23 32.32°	Apr 30 40°	$18^h\,07^m$ 272°	+33.1°	47.3	18
π-Puppids	PPU	Apr 18 28°	Apr 23 33.5°	Apr 25 35°	$7^h\,28^m$ 112°	-43°	15	Var.
h-Virginids	HVI	Apr 20 30°	Apr 28-30 38°- 40°	May 4 44°	$13^h\,40^m$ 205°	-11°	18.4	<1
η-Lyrids	ELY	May 3 43°	May 8 48°	May 14 54°	$19^h\,08^m$ 287°	+44°	43	3
ε-Aquilids	EAU	May 4 44°	May 21-22 61°	May 27 67°	$19^h\,18^m$ 289.5°	+18°	31.2	<1

Table 5.15: η-Aquariid Timeline

1863	H.A. Newton proposes the existence of the η-Aquariid meteor shower based on ancient Chinese observations from CE 401, 839, 927, 934 and 1009 between April 28-30.
1868-1870	45 η-Aquariids accidentally recorded by members of the Italian Meteoric Society between April 29 and May 5 and later uncovered by William F. Denning. The Italian astronomers did not realize their discovery.
1868	R.Falb first to suggest link between Halley's Comet, the η-Aquariids and the Orionids.
1869 Apr 29	Lieutenant-Colonel George L. Tupman in the Mediterranean makes the first planned observations of the η-Aquariids following a suggestion by J.F.J. Schmidt of Athens Observatory, Greece, based on Newton's prediction (see 1863).
1876	Prof. Alexander S. Herschel also associates Halley's comet with the η-Aquariids.
1878	Prof. Herschel discovers that the η-Aquariids radiant drifts eastward.
1911	Charles P. Olivier also notes similarities between the η-Aquariids and the Orionids.
1929 Nov	Ronald A. McIntosh publishes first detailed report on the η-Aquariids in the *Monthly Notices of the Royal Astronomical Society*.
1935	McIntosh determines the radiant drift to be $\Delta\alpha$ +0.96° and $\Delta\delta$ +0.37° per day.
1947	First radio observations of the shower by Bernard Lovell and his team at Jodrell Bank.
1973	A. Hajduk identifies double maxima in radar studies and relates this to filaments within the stream. He states the shower occurs when Earth is 0.065 AU from the core of the meteoroid stream
1983	B.A. McIntosh and A. Hajduk recognize that the meteoroids that are currently encountered were released by Halley's Comet many revolutions previously.
1985-86	The stream's parent comet, 1P/Halley, returns to perihelion but there is no noticeable increase in meteor rates.
1989	David W. Hughes and Neil McBride estimate the mass of the η-Aquariid and Orionid meteoroid stream to be 3.3×10^{16} g.
2006	Peter Jenniskens shows that the shower's dust particles have not separated by mass.
2013	M. Sato and J.Watanabe successfully predict an outburst due to material ejected by Halley in 1197 and 910.
2020	A. Egal, P. Wiegert, P. G. Brown, M. Campbell-Brown and D. Vida find a possible 10.7 periodicity for the η-Aquariids (and a stronger 11.8 years for the associated Orionids).

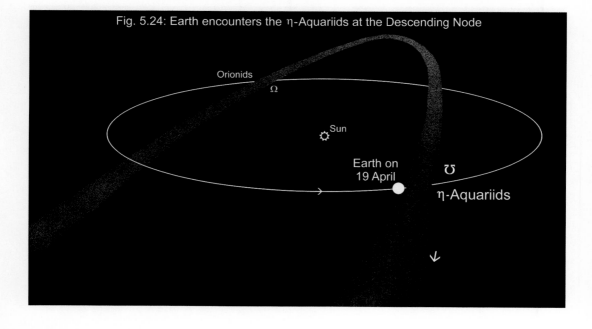

Fig. 5.24: Earth encounters the η-Aquariids at the Descending Node

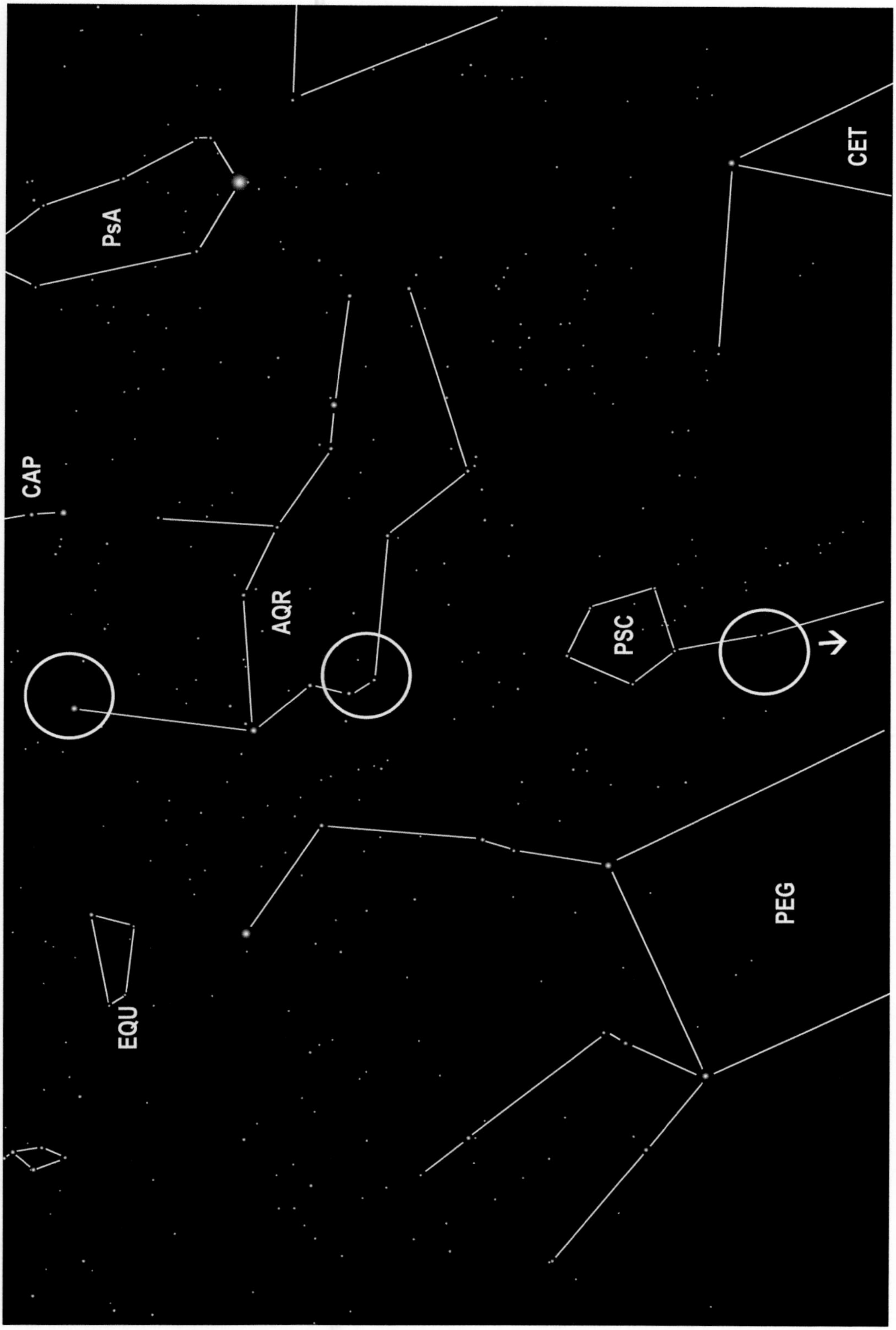

The η-Aquariids

Fig. 5.25 η-Aquariids sky view

106　　　5　The Major Meteor Showers: January to August

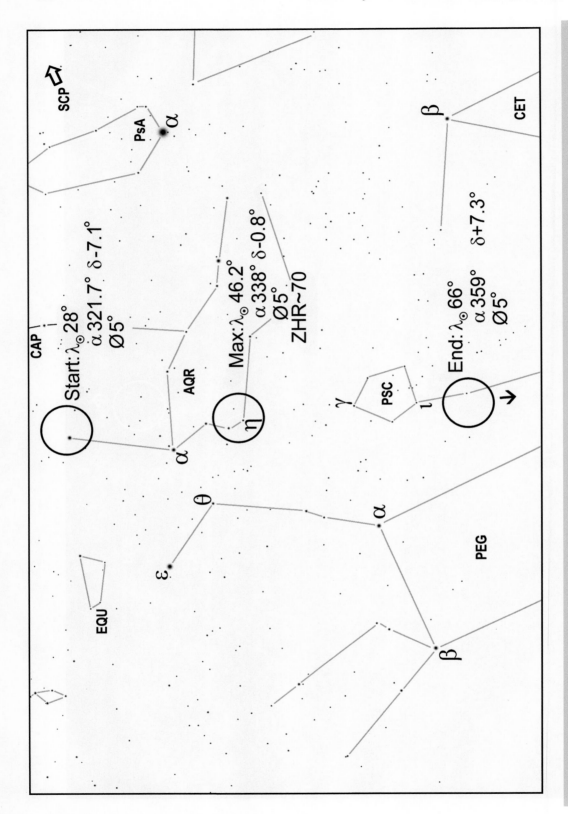

Fig. 5.26 η-Aquariids radiant

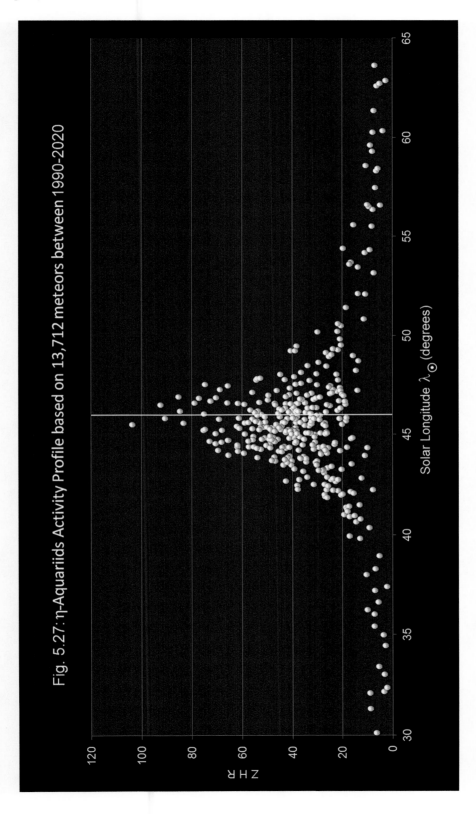

Fig. 5.27: η-Aquariids Activity Profile based on 13,712 meteors between 1990-2020

108 5 The Major Meteor Showers: January to August

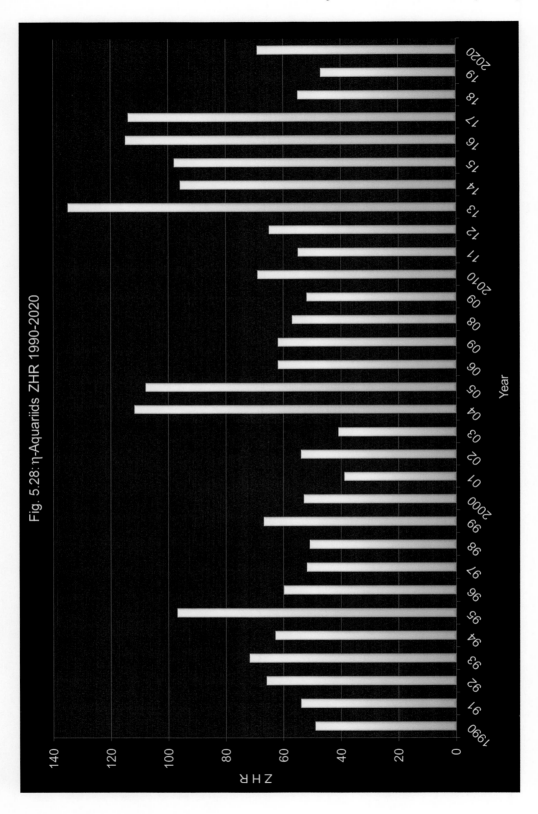
Fig. 5.28: η-Aquariids ZHR 1990–2020

The η-Aquariids

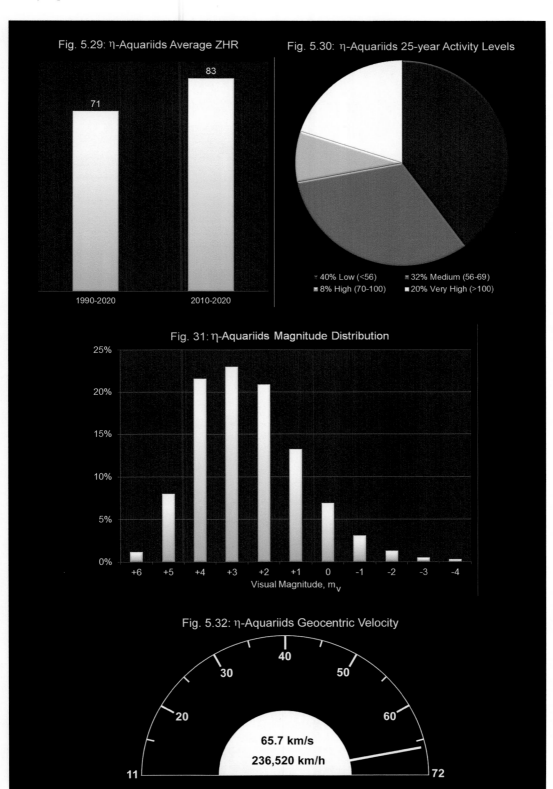

Fig. 5.29: η-Aquariids Average ZHR

Fig. 5.30: η-Aquariids 25-year Activity Levels

- 40% Low (<56)
- 32% Medium (56-69)
- 8% High (70-100)
- 20% Very High (>100)

Fig. 31: η-Aquariids Magnitude Distribution

Fig. 5.32: η-Aquariids Geocentric Velocity

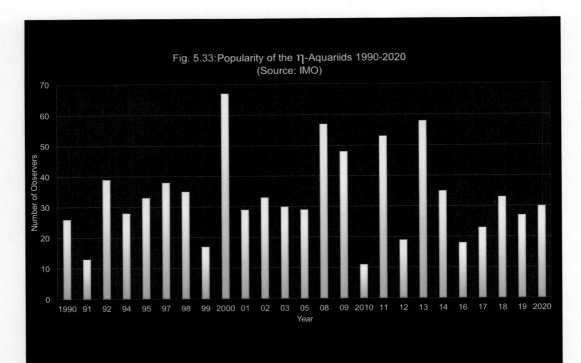

Fig. 5.33: Popularity of the η-Aquariids 1990-2020 (Source: IMO)

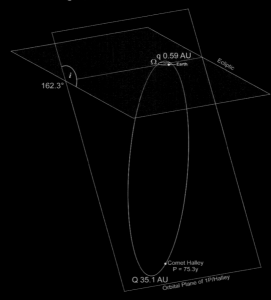

Fig. 5.34: Orbit of 1P/Comet Halley

Table 5.16: Orbital Data for the η-Aquariid Parent Body*									
Comet 1P/Halley									
q	e	a	ω	Ω	i	P	Q	b	
AU		AU	°	°	°	y	AU	AU	
0.5860	0.9671	17.8341	111.3325	58.4201	162.2627	75.32	35.0823	4.5340	

*For Epoch 1994-02-17.0

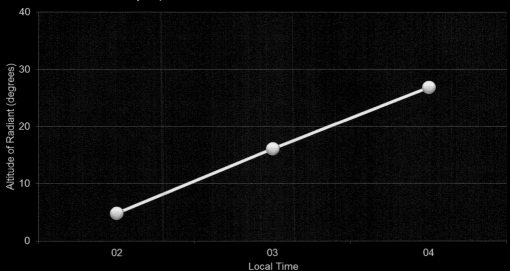

Porto, Portugal; Rome, Italy; Istanbul, Turkey; Toshkent, Uzbekistan; Beijing, China; Sapporo, Japan; Sacramento, Salt Lake City, Denver, Kansas City, Indianapolis, Pittsburgh and Philadelphia, USA.

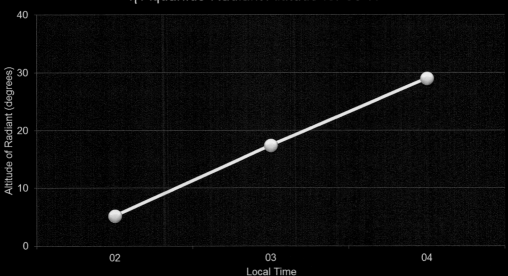

Casablanca, Morocco; Malaga, Spain; Athens, Greece; Tehran, Iran; Jinan, China; Soul, South Korea; Los Angeles, Memphis and Charlotte, USA.

Adjust local time for daylight saving if applicable

112　　5　The Major Meteor Showers: January to August

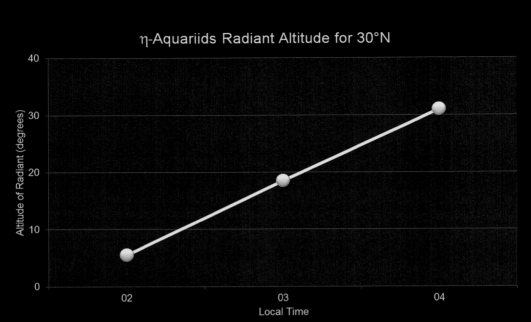

Madeira, Portugal; Tripoli, Libya; Cairo, Egypt; Basra, Iraq; Multan, Pakistan; Lahore, India; Shanghai, China; Tokyo, Japan; San Diego, Austin and Jacksonville, USA.

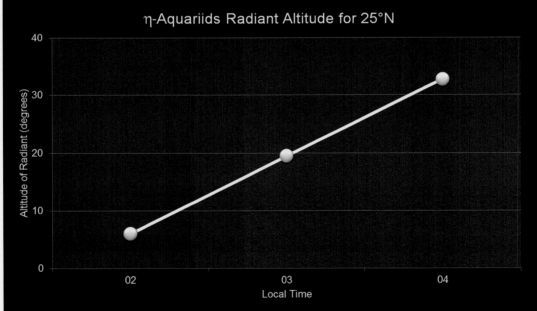

Las Palmas, Canary Islands; Aswan, Egypt; Karachi, Pakistan; Varanasi, India; Kunming, China; T'aipei, Taiwan; Torreon, Mexico; Miami, USA.

The η-Aquariids

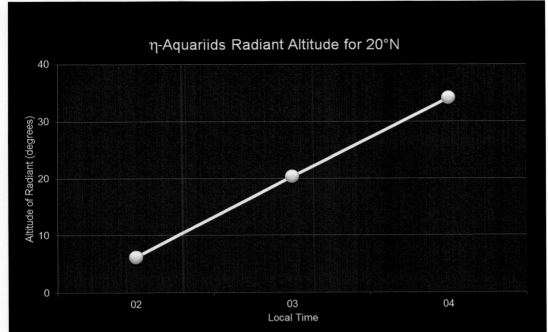

Jeddah, Saudi Arabia; Mumbai, India; Haikou, China; Campeche, Mexico; Santiago de Cuba, Cuba.

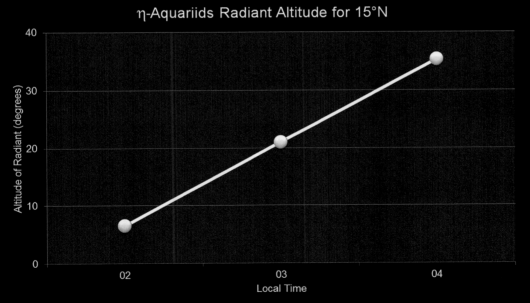

Dakar, Senegal; Khartoum, Sudan; Hyderabad, India; Yangon, Myanmar; Manila, Philippines; Guatemala City, Guatemala

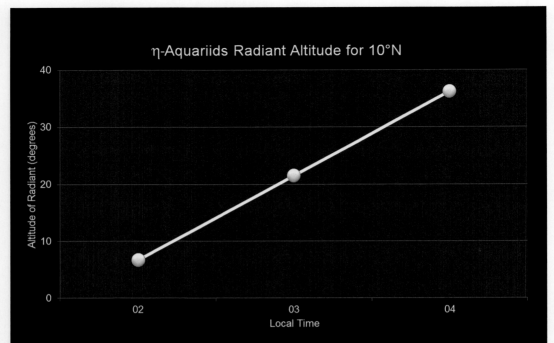

Conakry, Guinea; Kaduna, Nigeria; Addis Ababa, Ethiopia; Ho Chi Minh City, Vietnam; Manila, Philippines; San Jose, Costa Rica; Caracas, Venezuela.

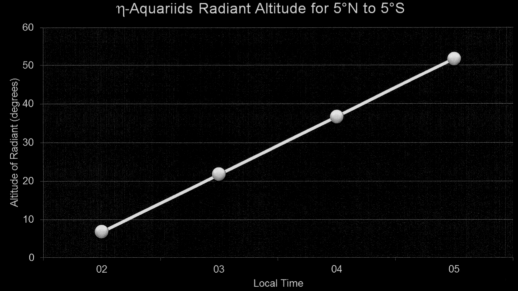

North: Accra, Ghana; Colombo, Sri Lanka; George Town, Malaysia; Bogota, Colombia; Cayenne, French Guiana.

South: Kinshasa, D.R.Congo; Jakarta, Indonesia; Piura, Peru; Teresina, Brazil.

The η-Aquariids

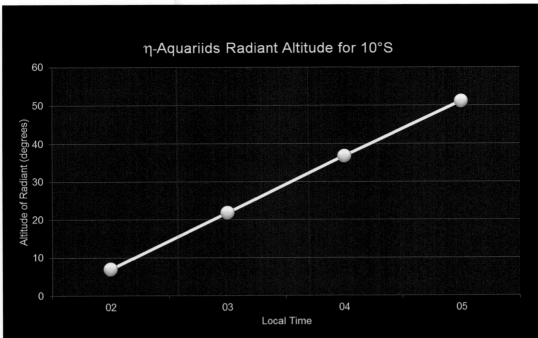

Luanda, Angola; Kolwetzi, D.R.Congo; Kupang, Indonesia, Maceio, Brazil.

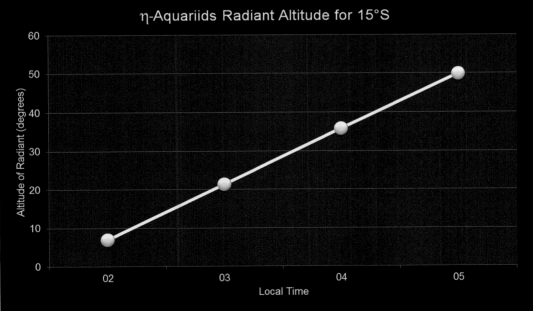

Lusaka, Zambia; Blantyre, Malawi; Coronation Island, Western Australia.

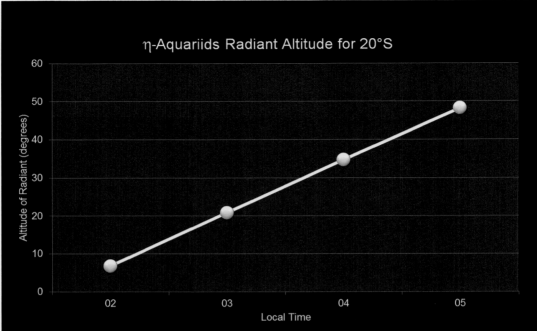

Bulawayo, Zimbabwe; Beira, Mozambique; Port Hedland, Western Australia; Bowen, Queensland, Australia; Iquique, Chile; Cariacica, Brazil.

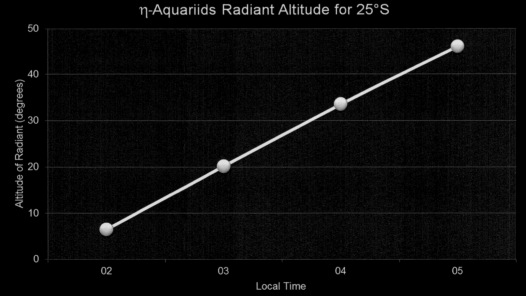

Gibeon, Namibia; Pretoria, South Africa; Carnarvon, Western Australia; Bundaberg Central, Queensland, Australia; Taltal, Chile; Asuncion, Paraguay.

The η-Aquariids

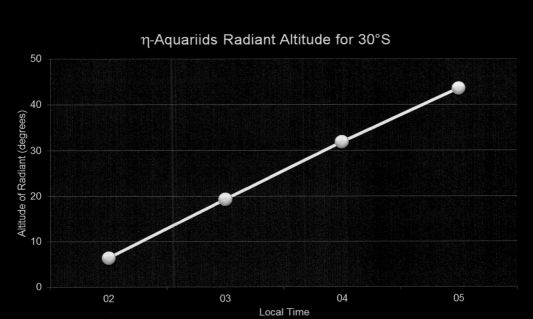

Durban, South Africa; Perth, Western Australia; Coffs Harbour, NSW, Australia; Coquimbo, Chile; Poto Alegre, Brazil.

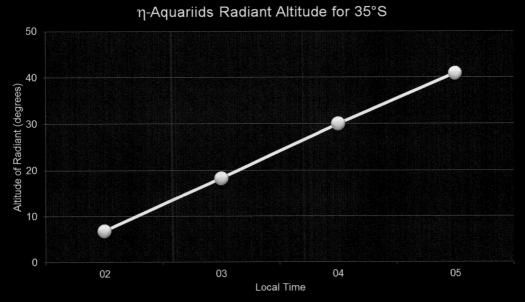

Curico, Chile; Buenos Aires, Argentina; Montevideo, Uruguay; Mangonui, New Zealand.

118 5 The Major Meteor Showers: January to August

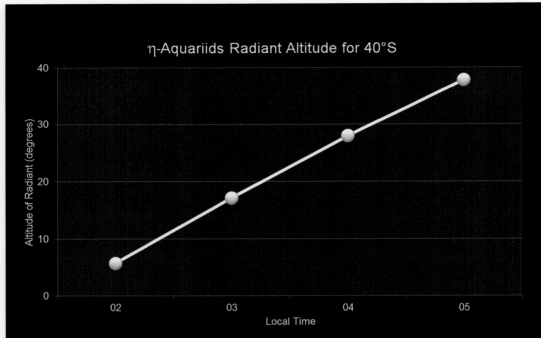

Valdivia, Chile; El Cuy, Argentina; Wanganui, New Zealand.

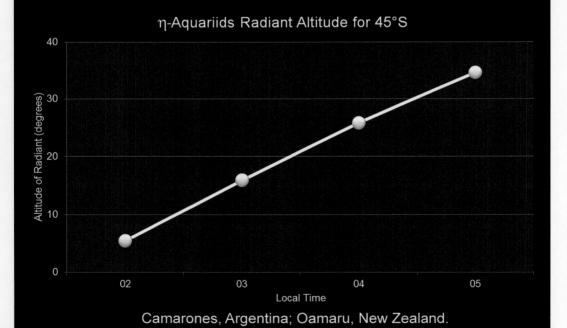

Camarones, Argentina; Oamaru, New Zealand.

The η-Aquariids

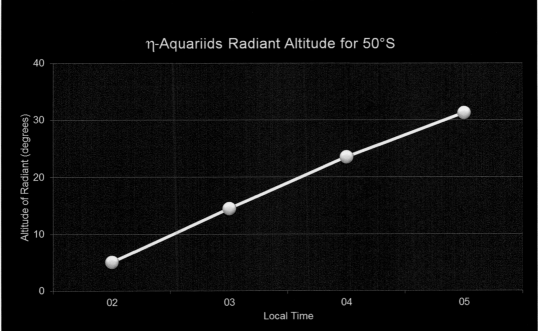

Puerto Santa Cruz, Argentina; Stanley, Falkland Islands.

The Southern δ-Aquariids

The Southern delta-Aquariids

The Southern δ-Aquariids seem to have become more active in the last decade, or is it just contamination?

Observing Notes

Like the η-Aquariids in May, the Southern δ-Aquariids favour observers in the Southern Hemisphere. Active from about July 12 through to August 18, the shower reaches maximum around July 27–28 at 126°. The radiant first makes an appearance in Capricornus, almost centred on the third magnitude star γ Capricorni, then drifts eastwards by about 0.95°/day towards the star δ Aquarii[4] but not quite making it before reaching maximum. For the first few days, the ZHR varies between 2 and 5 meteors per hour but can often be higher until, a little less than a week before maximum, rates pick up and steadily climb to an ill-defined peak that can occur any time between $\lambda_\odot$ 125° to 128°; a four-day spread. After maximum, activity tails off more slowly and rates of 5 meteors per hour can still be detected almost two weeks later.

Anyone living farther north than 45°N—Bordeaux, Minneapolis, Toronto, Montreal –will struggle to observe the Southern δ-Aquariids. Even at this latitude the radiant never quite reaches 30° altitude, at around 2 am LT, and dawn occurs just a couple of hours later. A little farther south at 30°N, the radiant reaches a respectable 43° altitude from Madeira, Basra, Tokyo, Austin and Jacksonville, and dawn is about 6 am, by which time the radiant is setting. From there on, south is definitely the way to go. For observers 15°S to 20°S, the radiant is about 30° above the horizon at 10 pm LT, almost directly overhead at 2 am and is still about 40° altitude at 6 am, an hour or so before dawn, giving sky watchers a good 8 hours of observation. So anyone in Zimbabwe to Mozambique, Australia, or southern Chile to Brazil will get the best view.

The meteoroids that produce the Southern δ-Aquariids enter the Earth's atmosphere at about 41.3 km/s, or 148,690 km/h, so appear moderately swift. Around 10% produce persistent trains, and the brightest are often said to be yellow or bluish-yellow [37]. However, the meteors generally tend to be on the faint side with an average magnitude of $\overline{m}_v + 2.8$.

Observers in the Southern Hemisphere see a radiant that is at best 5° in diameter, though some report a much smaller radiant of about 1° across [38]. By contrast, in the Northern Hemisphere the radiant is often reported to be large and diffuse at 15° to 20°. This is an artefact of a phenomenon called *zenith attraction*. When the Earth encounters a meteoroid stream, it gravitationally deflects the individual particles onto curved paths and increases their velocity. From the Earth's surface, it looks as though the radiant is being displaced towards the zenith and, in some situations, smeared across a wide area.

In 2004, Audrius Dubietis and Rainer Arlt undertook a thorough study of the shower using data compiled by the International Meteor Organization for the years 1997–2002 [39]. One of their findings was that the shower's maximum ZHR was 18 meteors/h. Over a much longer period, 1990–2020, the ZHR averaged 22 and in the past decade has shot up to 26, which seems unusual. So has the shower become more active in recent years or is there another explanation?

The Southern δ-Aquariids suffer from two main issues: poor distribution of observers and poor publicity.

---

[4]An interesting fact about δ-Aquarii is that its Arabic name, *Skat*, means 'a place whence something falls'

The Southern δ-Aquariids

Table 5.17a: Southern η-Aquariid Activity Details				
		Start	Maximum	End
Dates (approx.)		Jul 12	July 29/30	Aug 19
Solar longitude:	$\lambda_\odot$	109°	126.5°	147°
Right Ascension at max:	α		$22^h\,44^m$ (341°)	
Declination at max:	δ		-16°	
Time of transit :			7.7^h L.T.	
ZHR (2010-2020 average):			28	
Radiant diameter:	$\varnothing$	1° to 5°	1° to 5°	1° to 5°

Table 5.17b: Southern η-Aquariid Other Data		
IAU Abbreviation: SDA	**IAU Code:** 005	**AKA:** η-Aquariids
Mean daily motion of radiant:	Δ	$\Delta\alpha = +0.95°$ $\Delta\delta = +0.38°$
Geocentric velocity:	V_g	41.3 km/s (148,680 km/h)
Population index:	r	2.3
Mean magnitude:	$\overline{m}_v$	+2.8
Parent body: Possibly 96P/Maccholz 1 or C/1490 Y1		
Best visibility: Post-midnight in the Southern Hemisphere.		
The radiant is circumpolar from 74°S.		

Table 5.18: Southern δ-Aquariid Maxima 2020-2030				
	Maximum ($\lambda_\odot$ 126.5°)			Moon
year	month	day	hour (UT)	Age (d)
2020	07	29	10:50	10
2021	07	29	17:06	21
2022	07	29	23:03	2
2023	07	30	05:13	14
2024	07	29	11:31	24
2025	07	29	17:33	6
2026	07	29	23:46	16
2027	07	30	05:59	27
2028	07	29	12:08	8
2029	07	29	18:24	19
2030	07	30	00:22	1

Despite its low altitude, most observations of the Southern δ-Aquariids are made by observers in the Northern Hemisphere. As discussed above, the radiant appears more diffuse and its location can lead to an activity overestimation as non-shower meteors are erroneously assigned to the radiant. But it is perhaps the second issue—poor publicity—that causes the most problems.

During July and August, this region of the sky is highly active. Apart from the Southern δ-Aquariids, there is also the Northern δ-Aquariids, the Piscis Austrinids and the α-Capricornids. The Northern δ-Aquariids lie just a few degrees to the north of the Southern δ-Aquariids, while the Piscis Austrinids lie about 7° to the south. All three showers are active during the same period. As the Southern δ-Aquariids reach maximum, the α-Capricornids are just 30° away—easy enough to spot from the Southern Hemisphere but easily confused by observers in the north. If an observer is unaware of the existence of the other showers, then misidentifying meteors is an easy mistake to make. Unfortunately, many websites, as well as more traditional media, often highlight the Southern δ-Aquariids but totally ignore the other showers because of their low ZHRs. The trouble is that the existence of the minor showers contaminates the Southern δ-Aquariids' figures. Anyone wishing to observe the Southern δ-Aquariids also needs to be aware of activity from the other radiants in the area and to double-check that they are correctly assigning each meteor (see Table 5.19).

Results published by the IMO show a gradual increase in the shower's popularity over the past 35 years (Fig. 5.44).

Discovery

If the Southern δ-Aquariids are associated with the Quadrantid Complex, as suspected (see The Quadrantids) then it is unlikely that the shower is more than about 500 years old, so the lack of Aquariid meteors in ancient annals is only to be expected.

In all probability, it was Eduard Heis and his colleague Georg Balthazar von Neumayer who discovered the shower and who published details in their work *On Meteors in the Southern Hemisphere* in 1867 [40]. More than 10 years later, W.F. Denning discovered what he called the β-Piscids but which are almost certainly the Northern δ-Aquariids [41]. The low radiant, faint meteors and the lack of observers in the Southern Hemisphere made the shower unpopular, so it was a further 57 years before Alphonse King of Ashby, England managed to pinpoint the radiant, which was given as α 340° (22^h 40^m), δ -16°, almost identical to today's accepted values [42]. Considering that only four meteors were used in the analysis, the quality of King's observations cannot be faulted.

The appearance of Ronald A. McIntosh on the scene transformed meteor astronomy in the Southern Hemisphere, and the Southern δ-Aquariids was one of the showers to benefit [43]. Using observations by members of the New Zealand Astronomical Society between 1926 and 1933, McIntosh published the first detailed report of the shower in 1934 and was the first person to attempt to calculate the stream's orbit. Nearly 20 years later, Mary Almond at Jodrell Bank used radio observations to link the shower to the Daytime Arietids, which occur in June.

In 1963, Alexandra Terentjeva noted that the meteoroid stream has a small perihelion distance of just 0.06 AU (9 million km), which means the temperatures of the meteoroids could be up to 827 °C, which is a couple of hundred degrees above the melting point of silicate minerals [44]. Terentjeva commented that this may account for "…the peculiar general appearance of the shower meteors which are sharp, show no wakes, and give off no sparks".

The Southern δ-Aquariids

Origins

In 1963, Hamid and Whipple pointed out that the stream's orbital elements were also very similar to the Quadrantids, and the concept of a Quadrantid Complex began to take shape. The parent comet, however, remained elusive. Bruce McIntosh suggested in 1990 that Comet 96P/Machholz was a likely candidate, but 10 other comets and two apparent asteroids have since been associated with the complex (see The Quadrantids).

Table 5.19: Other Active Meteor Showers During the Southern δ-Aquariids Calendar dates are approximate. Use solar longitude $\lambda_\odot$								
Shower	IAU Code	Start $\lambda_\odot$	Max $\lambda_\odot$	End $\lambda_\odot$	At Max. α	At Max. δ	V_g (km/s)	ZHR
φ-Piscids	PPS	Jun 8 77°	Jul 4 102°	Aug 2 130°	$02^h 06^m$ 32°	+31°	67	<1
c-Andromedids	CAN	Jun 24 93°	Jul 9 106°	Jul 21 119°	$02^h 10^m$ 32.4°	+48.4°	57.8	<1
ε-Pegasids	EPG	Jun 28 97°	Jul 11 109°	Jul 13 111°	$22^h 28^m$ 337°	+16°	28.4	<1
ς-Cassiopeiids[1]	ZCS	Jul 2 100°	Jul 15 113°	Jul 27 124°	$00^h 56^m$ 14°	+53°	57	1
July ξ-Arietids	JXA	Jul 2 100°	Jul 13 111°	Aug 1 129°	$02^h 44^m$ 41°	+10.2°	69	1
α-Capricornids	CAP	Jul 3 101°	Jul 25-30 122°- 127°	Aug 15 142°	$20^h 28^m$ 307°	-10°	22.7	5
July Pegasids	JPE	Jul 3 101°	Jul 17 115°	Aug 5 133°	$23^h 40^m$ 355°	+13°	63.3	1
ψ-Cassiopeiids	PCA	Jul 4 102°	Jul 21 119°	Aug 7 135°	$02^h 08^m$ 32°	+73°	42.1	1
49 Andromedids	FAN	Jul 5 103°	Jul 20 118°	Aug 13 140°	$01^h 36^m$ 24°	+48°	60.2	1
Piscis Austrinids	PAU	Jul 15 113°	Aug 1 129°	Aug 10 138°	$22^h 44^m$ 341°	-22.8°	43.1	5
Northern δ-Aquariids	NDA	July 16 114°	Aug 11 139°	Sep 9 167°	$23^h 06^m$ 346.4°	+1.4°	39.1	3
Perseids	PER	Jul 18 115°	Aug 12 140°	Aug 25 153°	$03^h 11^m$ 48°	+58°	59.1	130
July γ-Draconids	GDR	Jul 24 121°	Jul 27 125°	Aug 1 129°	$18^h 41^m$ 280.2°	+50.9°	27.4	Var.
κ-Cygnids	KCG	Jul 26 123°	Aug 17 145°	Sep 1 159°	$18^h 59^m$ 284.7°	+59°	23	3
η-Eridanids	ERI	Aug 2 130°	Aug 9 137°	Sep 16 176°	$02^h 58^m$ 44.4°	-12.2°	64	<1
Northern ι-Aquariids	NIA	Aug 11 139°	Aug 18 146°	Sep 10 168°	$22^h 17^m$ 334.3°	-5°	29.9	<5
August Draconids	AUD	Aug 13 141°	Aug 15 143°	Aug 19 147°	$18^h 08^m$ 272.1°	+59°	19.2	1
β-Hydrusids	BHY	?	Aug 16 144°	?	$02^h 25^m$ 36.3°	-74.5°	22.8	<1

[1]May be early Perseids

Table 5.20: Southern δ-Aquariid Timeline	
1858 Aug	Probable discovery of the shower by E. Heis and G. Neumayer, published in *On Meteors in the Southern Hemisphere*
1879	W.F. Denning probably discovers the Northern δ-Aquariids, which he calls the β-Piscids.
1922	A. King, Ashby, England, plots four meteors and determines the radiant to be at α 340° ($23^h 40^m$), δ -16°
1934	R.A. McIntosh publishes the first detailed analysis of the shower by members of the New Zealand Astronomical Society between 1926 and 1933. He determines the stream's orbit.
1949	First radio observations of the shower, by D.W.R. McKinley, Ottawa, Canada, who detects both the Northern and Southern branches but does not recognize the Northern component.
1952	Mary Almond, Jodrell Bank, determines the shower's orbit from radio observations and notes its similarity to the Daytime Arietids in June.
1952-54	F.W. Wright, L.G. Jacchia and F.L. Whipple rediscover the Northern component and link it to the Southern branch.
1963	A.K. Terentjeva notes that the stream has a small perihelion distance of 0.06 AU (9 million km).
1963	S.E. Hamid and F.L. Whipple note that the stream's orbital elements are similar to that of the Quadrantids 1,300 to 1,400 years ago.
1990	B. McIntosh suggests Comet 96P/Machholz as the parent body.
2003	Peter Jenniskens announces object 2003 EH_1 to be the most likely parent of the Quadrantids, and by association, the Southern δ-Aquariids, suggesting the shower is very young at about 500 years old.

The Southern δ-Aquariids 125

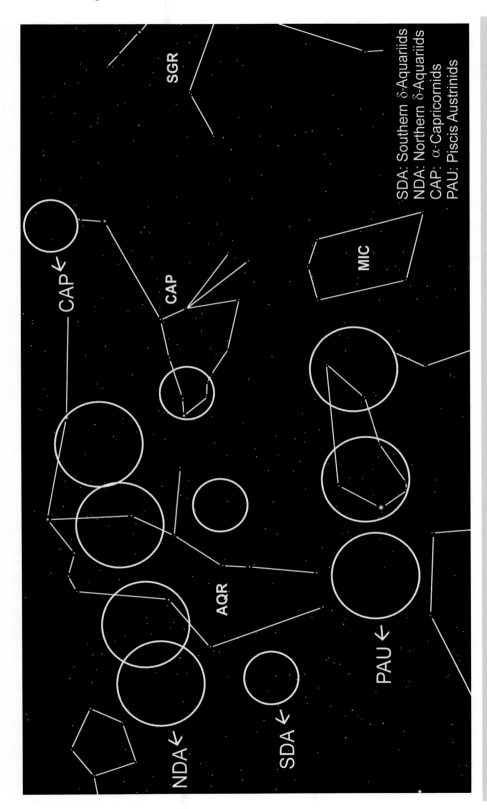

Fig. 5.36 Southern δ-Aquariids sky view

126 5 The Major Meteor Showers: January to August

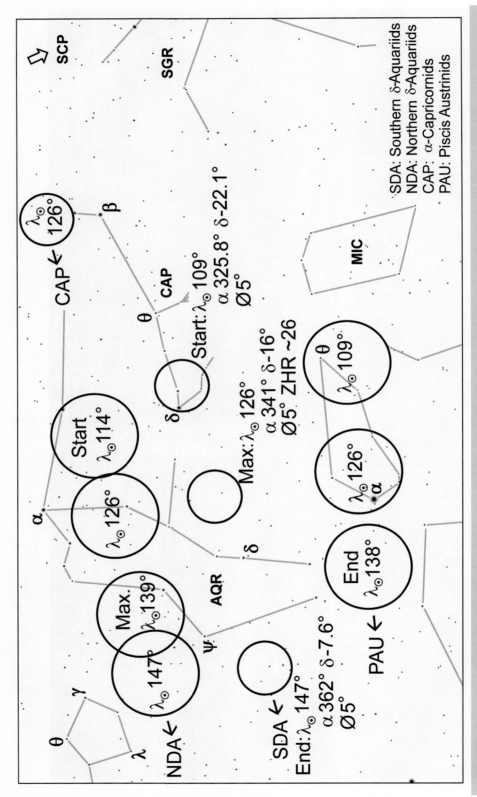

Fig. 5.37 Southern δ-Aquariids radiant

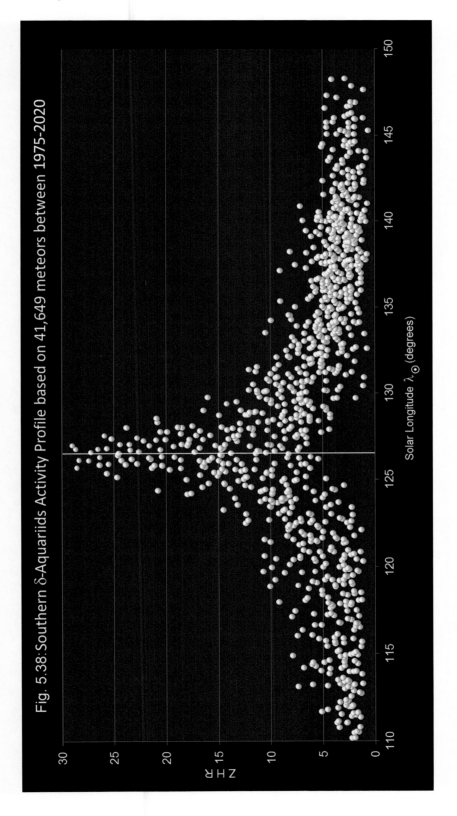

Fig. 5.38: Southern δ-Aquariids Activity Profile based on 41,649 meteors between 1975-2020

128 5 The Major Meteor Showers: January to August

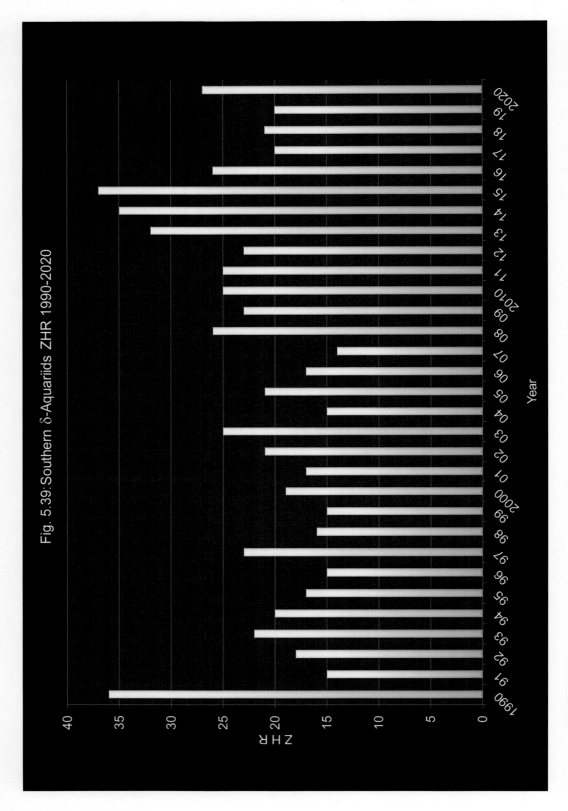

Fig. 5.39: Southern δ-Aquariids ZHR 1990-2020

The Southern δ-Aquariids

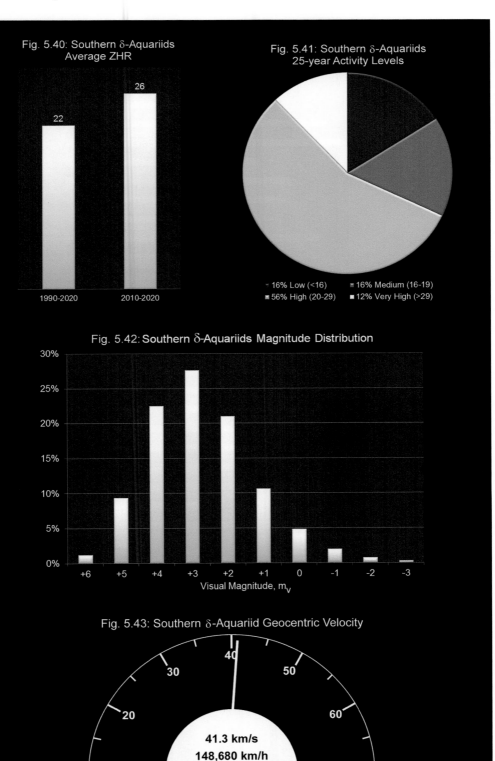

Fig. 5.40: Southern δ-Aquariids Average ZHR

Fig. 5.41: Southern δ-Aquariids 25-year Activity Levels

Fig. 5.42: Southern δ-Aquariids Magnitude Distribution

Fig. 5.43: Southern δ-Aquariid Geocentric Velocity

5 The Major Meteor Showers: January to August

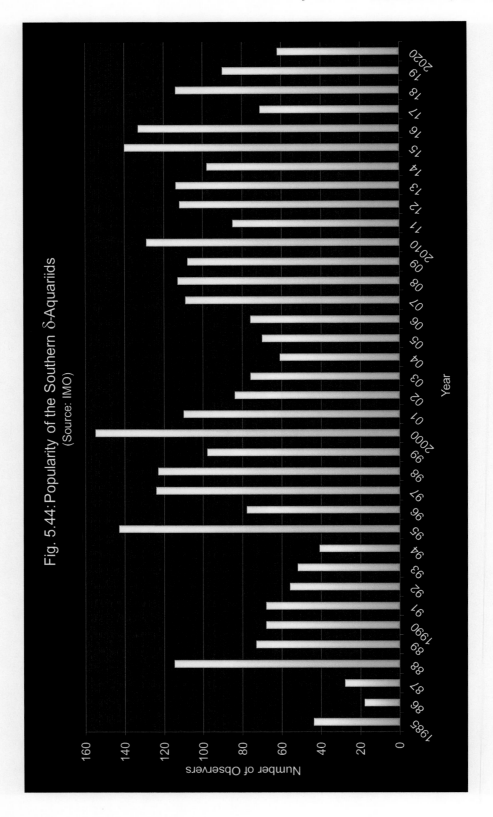

Fig. 5.44: Popularity of the Southern δ-Aquariids
(Source: IMO)

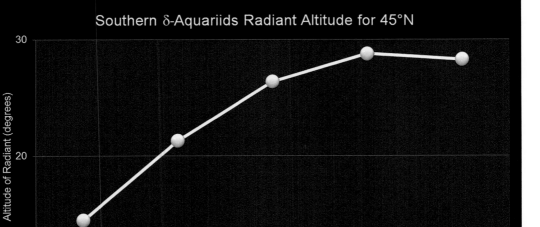

Bordeaux, France; Zagreb, Croatia; Ulaanbaatar, Mongolia; Harbin, China; Portland and Minneapolis, USA; Toronto and Montreal, Canada.

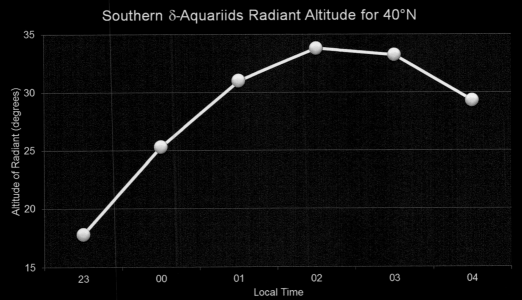

Porto, Portugal; Rome, Italy; Istanbul, Turkey; Toshkent, Uzbekistan; Beijing, China; Sapporo, Japan; Sacramento, Salt Lake City, Denver, Kansas City, Indianapolis, Pittsburgh and Philadelphia, USA.

Adjust Local Time for Daylight Saving if applicable

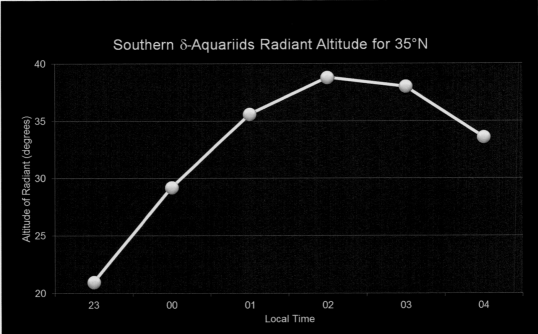

Casablanca, Morocco; Malaga, Spain; Athens, Greece; Tehran, Iran; Jinan, China; Soul, South Korea; Los Angeles, Memphis and Charlotte, USA.

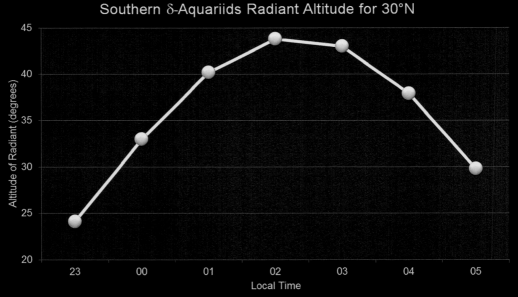

Madeira, Portugal; Tripoli, Libya; Cairo, Egypt; Basra, Iraq; Multan, Pakistan; Lahore, India; Shanghai, China; Tokyo, Japan; San Diego, Austin and Jacksonville, USA.

The Southern δ-Aquariids

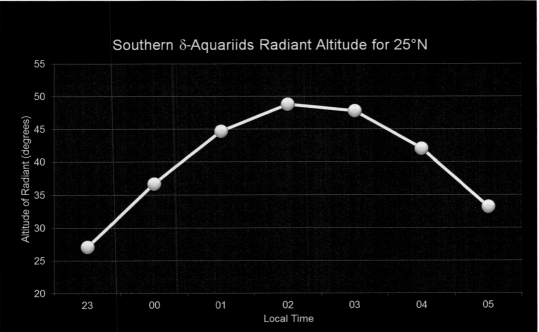

Las Palmas, Canary Islands; Aswan, Egypt; Karachi, Pakistan; Varanasi, India; Kunming, China; T'aipei, Taiwan; Torreon, Mexico; Miami, USA.

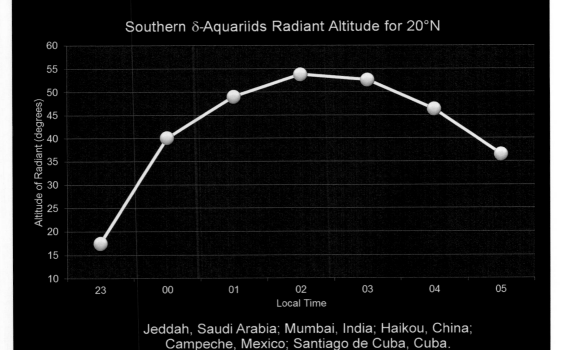

Jeddah, Saudi Arabia; Mumbai, India; Haikou, China; Campeche, Mexico; Santiago de Cuba, Cuba.

134   5   The Major Meteor Showers: January to August

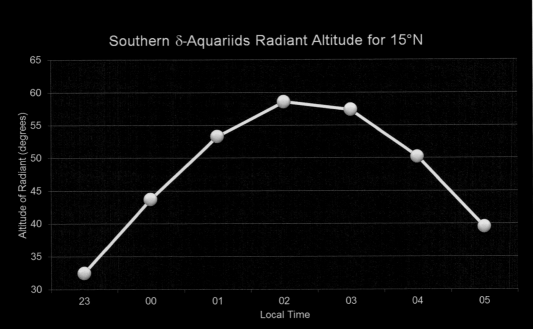

Dakar, Senegal; Khartoum, Sudan; Hyderabad, India; Yangon, Myanmar; Manila, Philippines; Guatemala City, Guatemala

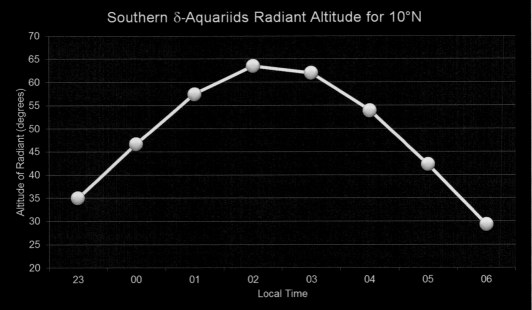

Conakry, Guinea; Kaduna, Nigeria; Addis Ababa, Ethiopia; Ho Chi Minh City, Vietnam; Manila, Philippines; San Jose, Costa Rica; Caracas, Venezuela.

The Southern δ-Aquariids

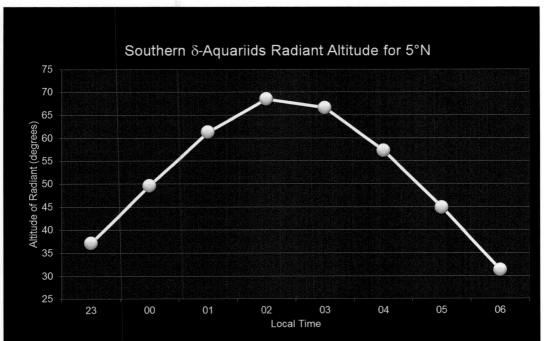

Accra, Ghana; Colombo, Sri Lanka; George Town, Malaysia; Bogota, Colombia; Cayenne, French Guiana.

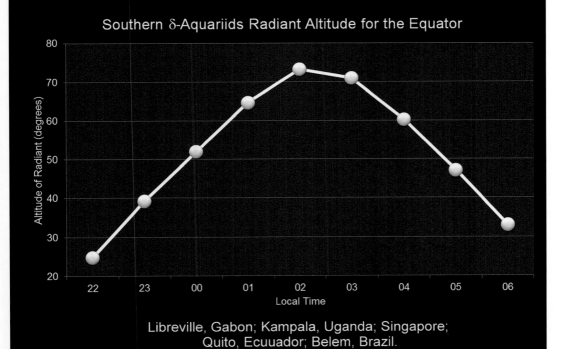

Libreville, Gabon; Kampala, Uganda; Singapore; Quito, Ecuuador; Belem, Brazil.

136　　　　　　　　　　　　　　5　The Major Meteor Showers: January to August

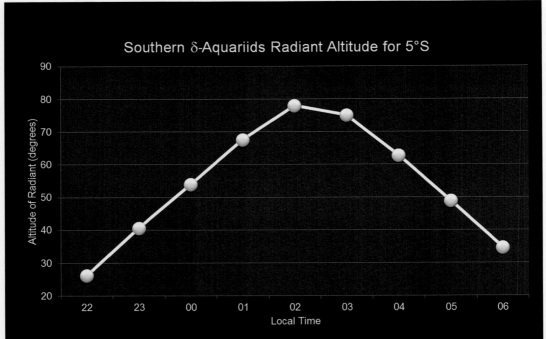

Kinshasa, D.R.Congo; Jakarta, Indonesia; Piura, Peru; Teresina, Brazil.

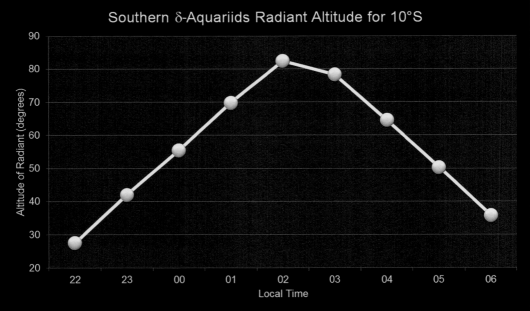

Luanda, Angola; Kolwetzi, D.R.Congo; Kupang, Indonesia, Maceio, Brazil.

The Southern δ-Aquariids

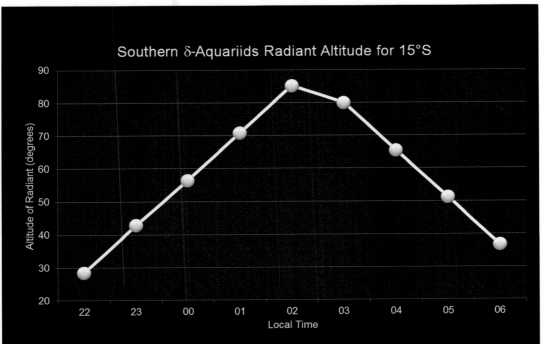

Lusaka, Zambia; Blantyre, Malawi; Coronation Island, Western Australia.

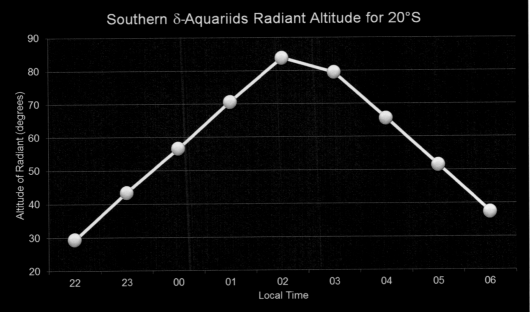

Bulawayo, Zimbabwe; Beira, Mozambique; Port Hedland, Western Australia; Bowen, Queensland, Australia; Iquique, Chile; Cariacica, Brazil.

138 5 The Major Meteor Showers: January to August

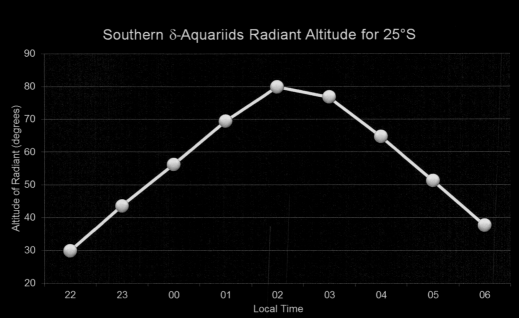

Gibeon, Namibia; Pretoria, South Africa; Carnarvon, Western Australia; Bundaberg Central, Queensland, Australia; Taltal, Chile; Asuncion, Paraguay.

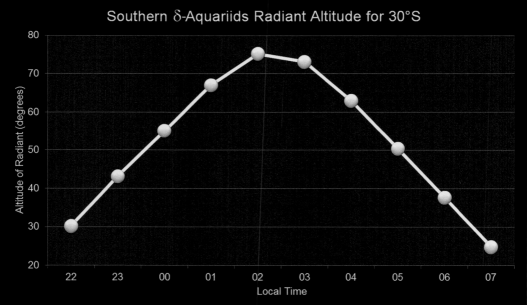

Durban, South Africa; Perth, Western Australia; Coffs Harbour, NSW, Australia; Coquimbo, Chile; Poto Alegre, Brazil.

The Southern δ-Aquariids

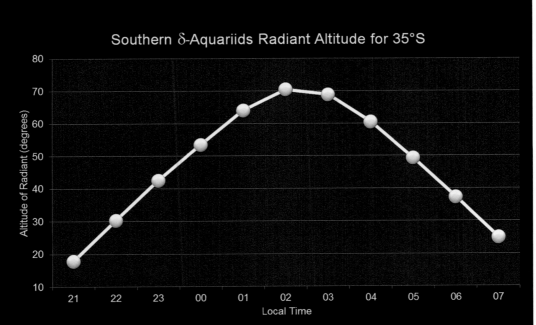

Curico, Chile; Buenos Aires, Argentina; Montevideo, Uruguay; Mangonui, New Zealand.

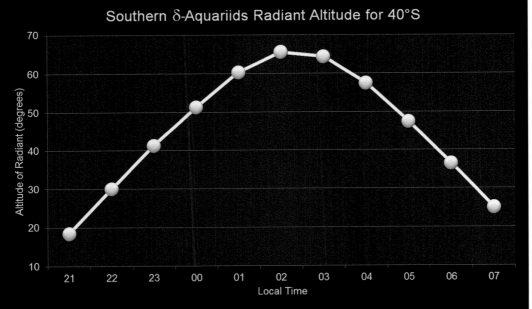

Valdivia, Chile; El Cuy, Argentina; Wanganui, New Zealand.

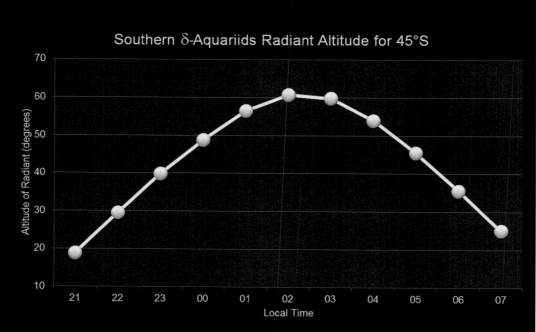

Camarones, Argentina; Oamaru, New Zealand.

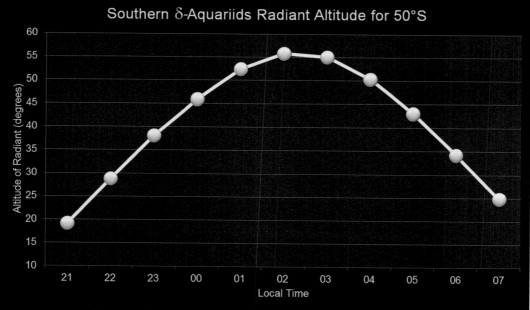

Puerto Santa Cruz, Argentina; Stanley, Falkland Islands.

The Perseids

We are living in the Golden Age of the Perseids with rates consistently higher than at any time in the last seven decades. Little wonder that they are the most popular of all meteor showers.

Observing Notes

More people observe the Perseids than all the other meteor showers put together. A combination of high rates, warm nights and good weather, sprinkled with a fair amount of media cover and public awareness, encourages whole families to seek out these celestial fireworks.

Activity from the Perseids begins about a month before maximum at around $\lambda_\odot$ 115° (July 18–19). Expect rates of just a couple of meteors per hour at first, but two weeks later the rate typically climbs to about 8–10 meteors/h, and a week before maximum the numbers will have often doubled. Then they start to take off: 5 days before max, 20–23; 4 days, 25–28, 3 days, about 30; 2 days, 35; 1 day, 50-ish. Peak activity finally arrives on August 12–13 at a very well-defined $\lambda_\odot$ 140°. If you miss maximum, don't fret. A day later, rates can still be around 40–60 meteors/h, but they then fall off quite rapidly. At $\lambda_\odot$ 142°—two days after maximum—the rate is typically 20 meteors/h, and ten days later they have all but disappeared.

During the past decade, the ZHR has averaged 130, but that is unusually high. Looking back over the past 25 years, the average ZHR was only 106, and in the 1950s it was around 50. Some of this increase is down to changes in the way we calculate the ZHR, but even taking this into account, the shower is considerably more active today than it was 70 years ago. Nowadays, there is a 36% chance you will see a ZHR in the range of 100–159 and a 12% chance in excess of 159, if the recent past is anything to go by.

The Perseids are very much a northern shower. Anyone living in the Southern Hemisphere can be forgiven for wondering what all the fuss is about. From 10°S—Luanda in Angola, Kupang in Indonesia and Maceio in Brazil—the radiant just about reaches an altitude of 22° above the horizon at 5 am LT and observers will see only about one-third of the activity, and considerably less if there is haze, light pollution from towns and cities and obstructions like forests and mountains. And then dawn breaks.

From the Northern Hemisphere, the Perseids don't really get going until after midnight, when the radiant climbs to 40° altitude at mid-northern latitudes. Observers then have only 3 or 4 hours of observation before sunrise, but for many, they are often the busiest and most rewarding few hours of the year. By comparison, members of the general public are often disappointed by the Perseids. The popular press encourages people to go outside and watch the spectacular Perseids but often fail to mention that the best displays are post-midnight, so they end up seeing very little.

It is easy to get overawed by a good Perseid display and forget about the science. In addition to the main radiant, there have been reports of at least four active sub-radiants. It was probably William F. Denning who first drew attention to sub-radiant activity in 1879 when he noted meteors coming from the areas around χ and γ Persei [45]. Subsequent observations revealed another two sub-radiants close to α and β Persei. A study by Jodrell Bank in the 1950s, however, could only identify the α Persei sub-radiant with any degree of certainty.

The Perseids

Table 5.21a: Perseid Activity Details

		Start	Maximum	End
Dates (approx.)		Jul 18	Aug 12/13	Aug 25
Solar longitude:	$\lambda_\odot$	115°	140°	153°
Right Ascension at max:	α		$03^h\ 11^m$ (48°)	
Declination at max:	δ		+58°	
Time of transit:			5.9^h L.T.	
ZHR (2010-2020 average):			130	
Radiant diameter:	$\varnothing$	5°	5°	5°

Table 5.21b: Perseid Additional Data

IAU Abbreviation: PER **IAU Code:** 007 **AKA:** Tears of St. Lawrence	
Mean daily motion of radiant: Δ	$\Delta\alpha = +1.34°$ $\Delta\delta = +0.25°$
Geocentric velocity: V_g	59.1 km/s (212,760 km/h)
Population index: r	2.3
Mean magnitude: $\bar{m}_v$	+1.7
Parent body: Comet 109P/Swift-Tuttle	
Best visibility: Post-midnight in the Northern Hemisphere	
The radiant is circumpolar from 32°N, i.e. north of Madeira, Portugal; Casablanca, Morocco; Tel Aviv, Israel; An Najaf, Iraq; Amritsar, India; Nanjing, China; Phoenix, Arizona and Atlanta, Georgia, USA.	

Table 5.22: Perseid Maxima 2020-2030

	Maximum ($\lambda_\odot$ 140°)			Moon
year	month	day	hour (UT)	Age (d)
2020	08	12	13:11	23
2021	08	12	19:17	4
2022	08	13	01:20	15
2023	08	13	07:33	26
2024	08	12	13:42	7
2025	08	12	19:51	17
2026	08	13	02:00	1
2027	08	13	08:11	11
2028	08	12	14:28	21
2029	08	12	20:36	2
2030	08	13	02:38	14

Table 5.23: Perseid Sub-Radiants

Component	Active Period $\lambda_\odot$	R.A. α	Dec. δ	Diameter $\varnothing$	Notes
χ-Persei	Aug 7 – 16 135° - 144°	$2^h\ 20^m$ 35°	+56°	2°	Maximum between $\lambda_\odot$ 137° - 139° (Aug 9 to 11)
α-Persei	Aug 7 – 24 135° - 152°	$3^h\ 24^m$ 51°	+50°	1.5°	Maximum anytime between $\lambda_\odot$ 140° - 145° (Aug 12 to 17)
γ-Persei	Aug 11 – 16 139° - 144°	$2^h\ 44^m$ 41°	+55°	2°	Activity mimics main radiant.
β-Persei	Aug 12 - 18 140° - 146°	$3^h\ 8^m$ 47°	+40°	1°	Weakest component. Rates irregular.

In the late 1960s and early 70s, observers in the Crimea, led by Vasily Vasil'evich Martynenko, undertook a study of the sub-radiants [46]. Martynenko was a highly experienced observer and great meteor enthusiast. He led several meteor expeditions and wrote countless articles and papers on meteor astronomy. The data he and his team assembled are summarized in Table 5.23. Figures 5.48b–5.48e are based on Martynenko's work, but more extensive and careful observations are certainly required to confirm the details. Various other sub-radiants have appeared from time to time, all of which hint at the complexity of the Perseid meteoroid stream.

In addition to sub-radiants, filaments of material have evolved within the stream that can produce a sudden jump in the ZHR, often outnumbering activity from the main radiant [47]. Figure 5.50d shows activity from the main stream and two separate filaments between 1988 and 2007. In 1991, the main stream put on a fairly decent shower of 100 meteors/h, but the Earth also encountered a filament and the ZHR suddenly shot up to more than 300 for a brief period [48]. Filaments tend to be narrow, so enhanced activity only lasts for a short while—typically tens of minutes. Filament bursts can appear either before or after the main Perseid peak.

A sobering thought about the Perseids is that no one ever sees the shower at its best. Even for observers at 60°N, the radiant only attains an altitude on 70°. So for sky watchers in Lerwick, Scotland, Oslo, Stockholm, Helsinki, St. Petersburg in Russia and Nanontalik, Greenland, the best they will see is a little over 90% of activity before sunrise snatches the radiant from the sky.

The Perseids have a geocentric velocity of 59.1 km/s (212,760 km/h), which is pretty close to the maximum attainable velocity of 72 km/s, so they appear very swift. Overall, the Perseids tend to be brighter than most showers, which no doubt helps with their popularity. About 12% of Perseids are brighter than zero magnitude with the average being $\overline{m}_v + 1.8$. Almost half of Perseid meteors have trains. The colour of meteors is a moot point, but the author's impression is that many of the brighter meteors tend to be bluish.

Discovery

The earliest reports of Perseid meteors were made by Chinese Court Astrologers on CE 36 Jul 15 [49], when they noted,"At dawn, more than 100 small meteors flew in all directions."[5] Strong activity was also noted between the eighth and 11th centuries by Chinese, Japanese and Korean observers. The first western account was not until the eighteenth century when, in 1762, Petrus van Musschenbroeck and J. Lulofs noted, "Falling stars are more numerous in August than at any other time of the year." [50] It took another 74 years before Adolphe Quetelet, Director of the Brussels Observatory, finally recognized the periodicity of the shower in 1836 [51]. In one of those strange twists of fate, the following year saw a very strong display that captured the attention of several European astronomers. It was G.C. Schaeffer who made the first attempt to pinpoint the radiant [52]. Unfortunately, he was observing part of the sky that was 90° from the radiant, which made it difficult to trace back the meteors with any degree of accuracy. He put the right ascension at α 55° ($3^h\ 40^m$)— some 7° out—and declination at δ +30°, an error of 28°.

---

[5] If the date is confusing it is because the stream's node progresses by 0.00038°/year or 1 day every 70.59 years. So, 2000 years / 70.59 years = 28 days, added to July 15 equates to August 12.

In 1862 July 16, American astronomer Lewis Swift discovered a new comet.[6] Three days later, on July 19, another prolific American comet hunter, Horace P. Tuttle, spotted the same comet. Together they shared the credit and Comet Swift-Tuttle 1862 III entered the history books. The story would have ended there, had it not been for Giovanni Schiaparelli who, as we'll see in *Origins*, found a link between the comet and the Perseid meteoroid stream.

Between 1869 and 1898, Denning managed to record a total of 2409 Perseids, an average of 80 per year, and had enough data to publish the first ephemeris in 1901 running from July 27 to August 16. Then rates started to decline, rapidly reaching an all-time low of just 1 or 2 meteors per hour in 1911 and 1912. Denning thought for a while that he had seen his last great Perseid shower. Eventually however, rates started to climb again, putting on a strong display in 1920; 1931, when Charles P. Olivier saw more than 1 meteor per minute; and 1945, when Olivier reported a rate in excess of 120 meteors/hour. Denning died in 1931.

By the 1950s, rates had fallen again to typically 50 meteors per hour, but they eventually started to increase again to the levels we see today. In 1988 Paul Roggemans detected a pre-maximum peak due to a filament within the stream [53]. It had first been observed three years earlier but was not recognized as a real feature. In 1991, the filament's ZHR reached 350 at $\lambda_\odot$ 139.58° while the main stream was just below 100. The filament was detected several times in subsequent years between 1988 and 2007, often outperforming the main stream.

The next big event for the Perseids is predicted to be 2028, when the Earth will pass within 60,000 km (0.0004 AU) of the dust trail that was ejected by the comet in 1479. It should favour observers in North America.

Origins

In 1866, Giovanni Schiaparelli became the first person to associate a meteoroid stream with a comet: the Perseids with Comet Swift-Tuttle (1862 III) [28]. Using observations from 1864–1866, Schiaparelli showed that the Perseids' orbit was similar to Comet Swift-Tuttle, which had an estimated orbital period of 120 years. However, it was Daniel Kirkwood—of Kirkwood Gaps fame—who first suggested a link between comets and meteor showers [54]. Kirkwood was fascinated by comets that had split into two or more pieces. He theorized that fragmentation was a continuous process and concluded, "May not our periodic meteors be the debris of ancient but now disintegrated comets, whose matter has become distributed around their orbits?" Schiaparelli, having learnt of Kirkwood's theory, set about proving the American professor right. In the process, Schiaparelli made a second major contribution to meteor science. Before Schiaparelli, meteor showers did not have names. They were simply referred to the time when they appeared, e.g. *the August meteors*. It was Schiaparelli who named them the Perseids and inadvertently launched a new trend.

The orbital period of Swift-Tuttle proved to be problematic. Smithsonian astronomer Brian G. Marsden issued a prediction in 1973 that the comet would return to perihelion between 1980 and 1982 [55]. However, he also pointed out that a comet observed in the eighteenth century by Ignatius Kegler[7] might actually be Swift-Tuttle. If that was the case,

---

[6] Swift discovered a total of 13 comets, the last when he was 79 years of age.
[7] Kegler was a German astronomer and missionary who worked as the Director of the Imperial Office of Astronomy at the court of the Chinese Emperors Xuanye and Yinzhen during the Qing Dynasty.

then the orbital period was somewhat longer by about a dozen years, which would mean that the comet was more likely to reach perihelion on 1992 November 25. 1980 came and went without any sign of the comet, although the Perseids' ZHR did reach 180 after several years of gradual increases and remained high until 1983. With still no sign of the comet, it looked as though Marsden was correct in his assumption that Comet 1737 II was indeed Comet Swift-Tuttle. It turned out that Marsden's second prediction was just 17 days adrift: the comet reached perihelion on 1992 December 12.

Swift-Tuttle's orbit, like Halley's, lies mostly below the plane of the ecliptic, where it spends 97% of its time (Fig. 5.46). The comet eventually breaks through the plane at the ascending node 13.2 AU from the Sun—somewhere between Saturn and Uranus—heading north. It then takes four years to arrive at perihelion, swings around the Sun heading south again and crosses the Earth's orbit just 6 months later. The debris stream it leaves behind is vast, but Earth passes only through its outermost edge. Even so, it takes 40 days to journey through the 102-million-kilometre segment.

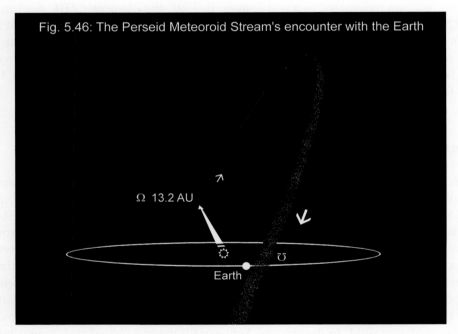

Comet Swift-Tuttle's ascending node will be about 13.2 AU from the Sun when it next approaches perihelion.

The age of the stream is not known with any certainty. Estimates range from 2600 to 27,000 years, with the wise money being on 6000 years. Individual particles are unlikely to exist for more than about 3000 years. David Hughes and Neil McBride calculated the mass of the stream to be 3.1×10^{17}g (about 240 times the mass of the Quadrantids) [15].

Table 5.24: Other Active Meteor Showers During the Perseids
Calendar dates are approximate. Use solar longitude $\lambda_\odot$

Shower	IAU Code	Start $\lambda_\odot$	Max $\lambda_\odot$	End $\lambda_\odot$	At Max. α	At Max. δ	V_g (km/s)	ZHR
φ-Piscids	PPS	Jun 8 77°	Jul 4 102°	Aug 2 130°	$02^h\ 06^m$ 32°	+31°	67	<1
c-Andromedids	CAN	Jun 24 93°	Jul 9 106°	Jul 21 119°	$02^h\ 10^m$ 32.4°	+48.4°	57.8	<1
ε-Pegasids	EPG	Jun 28 97°	Jul 11 109°	Jul 13 111°	$22^h\ 28^m$ 337°	+16°	28.4	<1
ς-Cassiopeiids[1]	ZCS	Jul 2 100°	Jul 15 113°	Jul 27 124°	$00^h\ 56^m$ 14°	+53°	57	1
July ξ-Arietids	JXA	Jul 2 100°	Jul 13 111°	Aug 1 129°	$02^h\ 44^m$ 41°	+10.2°	69	1
α-Capricornids	CAP	Jul 3 101°	Jul 25-30 122°-127°	Aug 15 142°	$20^h\ 28^m$ 307°	-10°	22.7	5
July Pegasids	JPE	Jul 3 101°	Jul 17 115°	Aug 5 133°	$23^h\ 40^m$ 355°	+13°	63.3	1
ψ-Cassiopeiids	PCA	Jul 4 102°	Jul 21 119°	Aug 7 135°	$02^h\ 08^m$ 32°	+73°	42.1	1
49 Andromedids	FAN	Jul 5 103°	Jul 20 118°	Aug 13 140°	$01^h\ 36^m$ 24°	+48°	60.2	1
Southern δ-Aquariids	SDA	Jul 12 109°	Jul 28 126°	Aug 19 147°	$22^h\ 44^m$ 341°	-16°	41.3	28
Piscis Austrinids	PAU	Jul 15 113°	Aug 1 129°	Aug 10 138°	$22^h\ 44^m$ 341°	-22.8°	43.1	5
Northern δ-Aquariids	NDA	July 16 114°	Aug 11 139°	Sep 9 167°	$23^h\ 06^m$ 346.4°	+1.4°	39.1	3
July γ-Draconids	GDR	Jul 24 121°	Jul 27 125°	Aug 1 129°	$18^h\ 41^m$ 280.2°	+50.9°	27.4	Var.
κ-Cygnids	KCG	Aug 3 131°	Aug 17 145°	Aug 24 152°	$18^h\ 59^m$ 284.7°	+59°	23	3
η-Eridanids	ERI	Aug 2 130°	Aug 9 137°	Sep 16 176°	$02^h\ 58^m$ 44.4°	-12.2°	64	<1
Northern ι-Aquariids	NIA	Aug 11 139°	Aug 18 146°	Sep 10 168°	$22^h\ 17^m$ 334.3°	-5°	29.9	<5
August Draconids	AUD	Aug 13 141°	Aug 15 143°	Aug 19 147°	$18^h\ 08^m$ 272.1°	+59°	19.2	1
β-Hydrusids	BHY	?	Aug 16 144°	?	$02^h\ 25^m$ 36.3°	-74.5°	22.8	<1

[1] May be early Perseids

Table 5.25: Perseid Timeline

CE 36 Jul 15	First reports of the Perseids by Chinese Court Astrologers: 'At dawn more than 100 small meteors flew in all directions.'
8th – 11th Centuries	Several accounts found in Chinese, Japanese and Korean annals.
1762	Petrus van Musschenbroeck notes in *Introductio ad Philosophiam Naturalem* that 'Falling stars are more numerous in August than at any other time of the year.'
1836	Adolphe Quetelet is the first person to recognize the Perseids as an annual shower.
1837	Strong display captures the attention of several European astronomers. G.C. Schaeffer attempts to pinpoint the radiant, but observing at 90° from the radiant causes significant errors.
1839	Another strong display with a ZHR in excess of 160.
1861-63	Strong displays.
1862 Jul 16	Lewis Swift and, on July19, Horace P. Tuttle, discover a new comet later shown to be the parent body of the Perseids.
1866	Giovanni Schiaparelli confirms Kirkwood's 1861 hypothesis by demonstrating that the orbit of Comet Swift-Tuttle is very similar to the meteors that appeared in mid-August, which he calls the 'Perseids', thereby inadvertently introducing a method of naming annual meteor showers.
1869-1898	W.F. Denning records 2,409 Perseids.
1879	Denning detects two sub-radiants near χ and γ Persei.
1900	Rates start to rapidly decline.
1901	W.F. Denning publishes the first ephemeris of the radiant position.
1911-1912	After years of declining numbers, the Perseids reach an all-time low of just 1 or 2 meteors per hour.
1920	Denning reports high rates after years of little activity.
1931	Olivier reports rates of more than one meteor per minute.
1945	Olivier reports rates in excess of 120 meteors/hour.
1950s-1960s	Rates in the order of 40-60 per hour, reaching a ZHR of 80 in 1956.
1969-71	Vasily Vasil'evich Martynenko and his team make detailed measurements of the Perseid sub-radiants.
1973	Brian G. Marsden suggests that Comet Swift-Tuttle will reach perihelion between 1980-82, but he also points out that if it is the same comet as seen in the 18th century, 1737 II, then perihelion would be 1992 November 25.
1980	Comet Swift-Tuttle fails to reach perihelion, but ZHR attains 180 after several years of gradual increases.
1983	ZHR 187
1985	A pre-max peak goes unrecognized as a real feature.
1988	Pre-max peak detected by Paul Roggemans.
1989	Pre-max peak put down to the predicted perihelion passage of Swift-Tuttle in 1980-81.
1989	David Hughes and Neil McBride estimate the mass of the stream to be 3.1×10^{17} g
1991	Pre-max peak produces a short-lived ZHR of 350 for observers in Japan and the Far East.
1992 Dec 12	Comet Swift-Tuttle reaches perihelion, confirming it is the same comet as 1737 II.
2028	Earth is predicted to pass within 0.0004 AU (60,000 km) of a dust trail released in 1479, favoring North American observers.

The Perseids 149

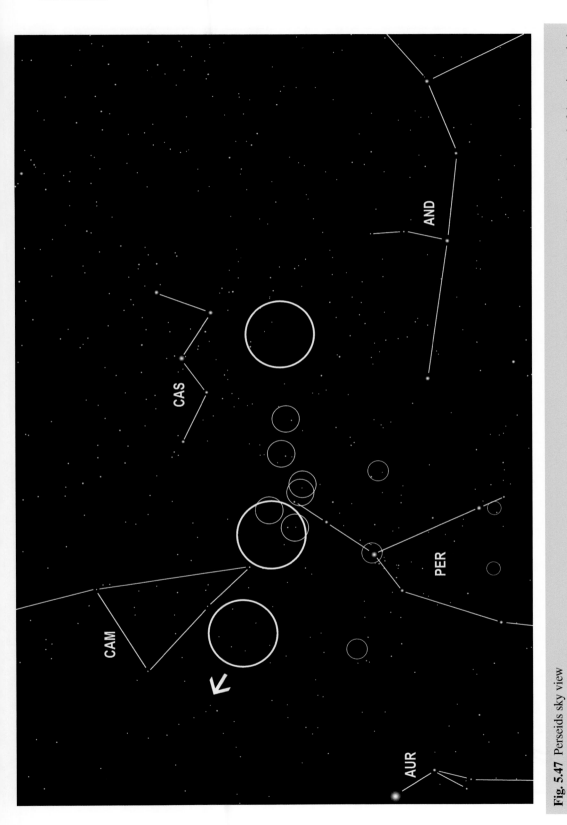

Fig. 5.47 Perseids sky view
The Perseids' route from Cassiopeia through Perseus to Camelopardalis, showing the radiant at the start at maximum activity, and at the end of the active period. The smaller circles are sub-radiants (see Figs. 5.48b–5.48e)

150 5 The Major Meteor Showers: January to August

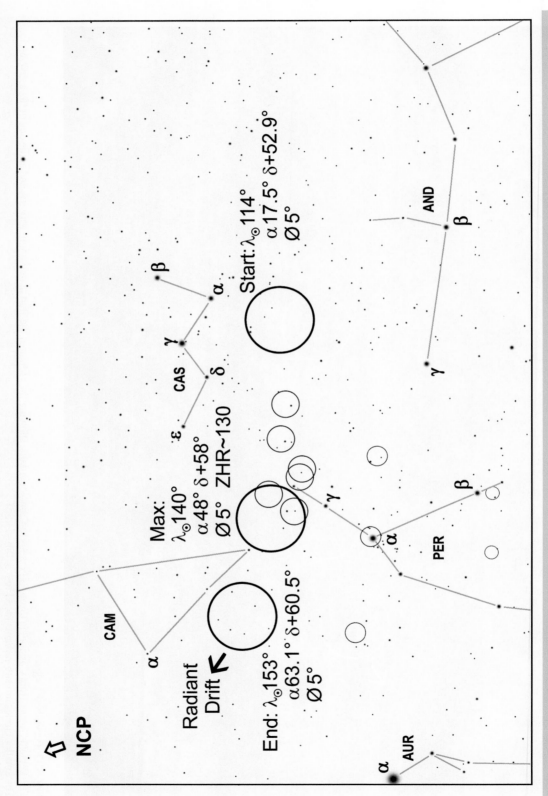

Fig. 5.48a Perseids main radiant

The Perseids

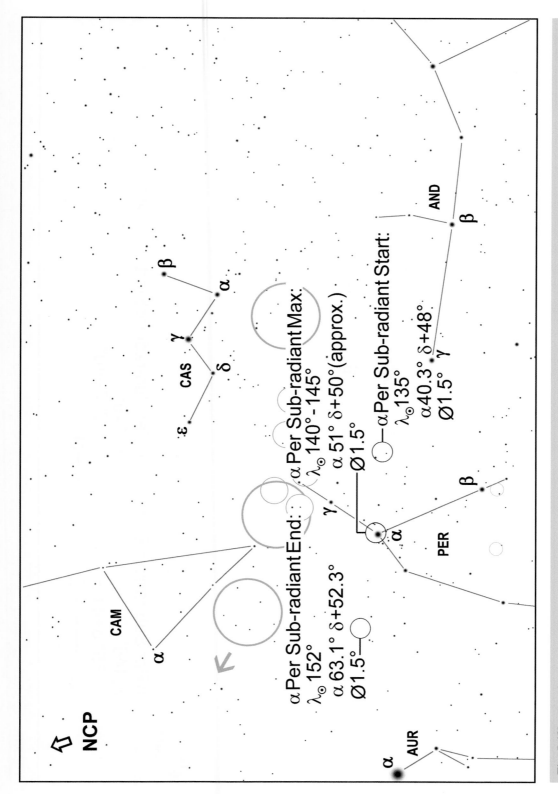

Fig. 5.48b Perseids α sub-radiant

152   5 The Major Meteor Showers: January to August

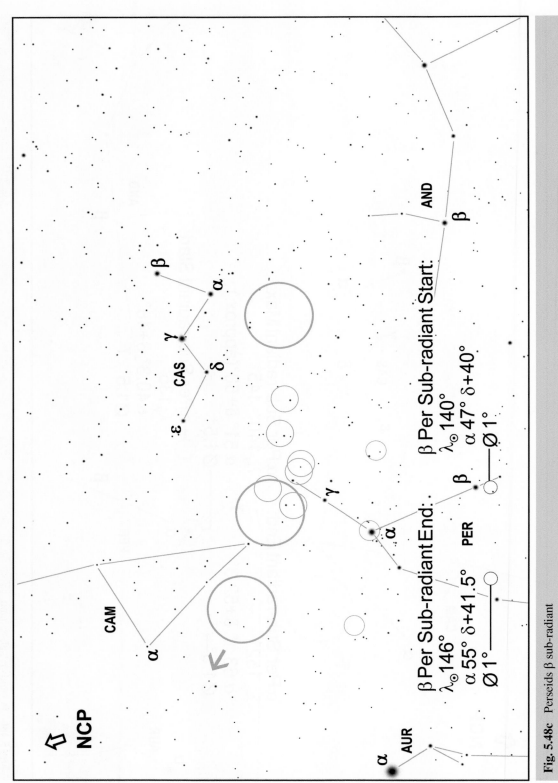

Fig. 5.48c Perseids β sub-radiant

The Perseids 153

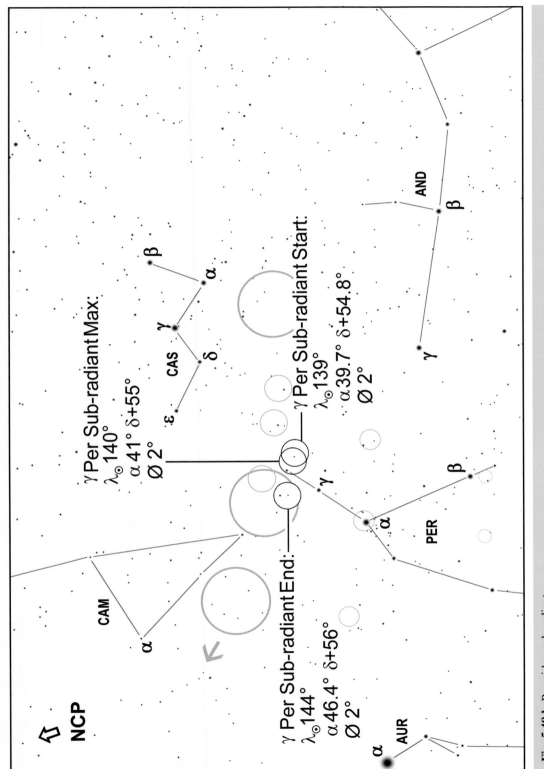

Fig. 5.48d Perseids γ sub-radiant

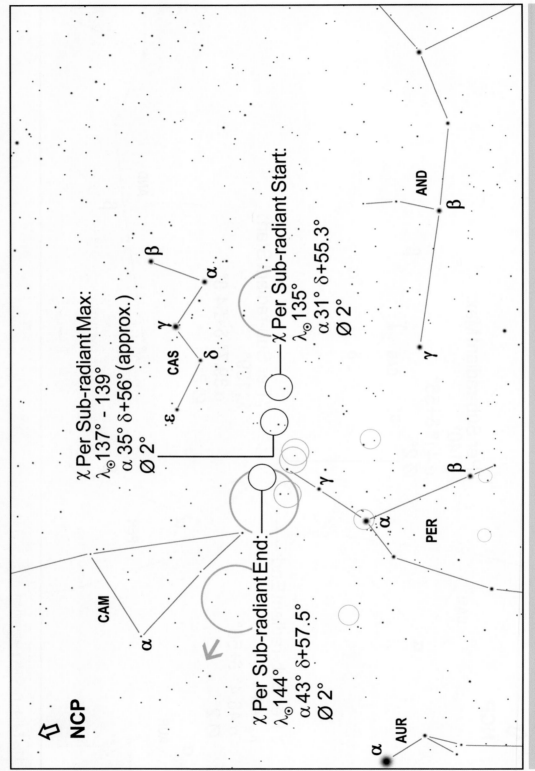

Fig. 5.48e Perseids χ sub-radiant

The Perseids

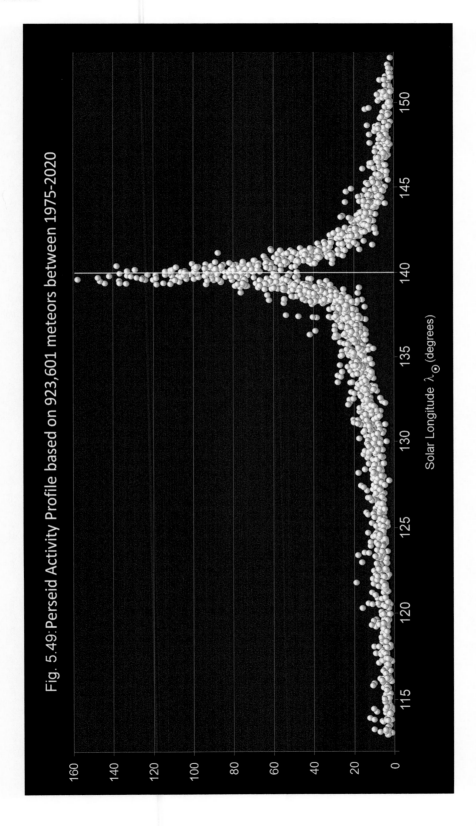

Fig. 5.49: Perseid Activity Profile based on 923,601 meteors between 1975-2020

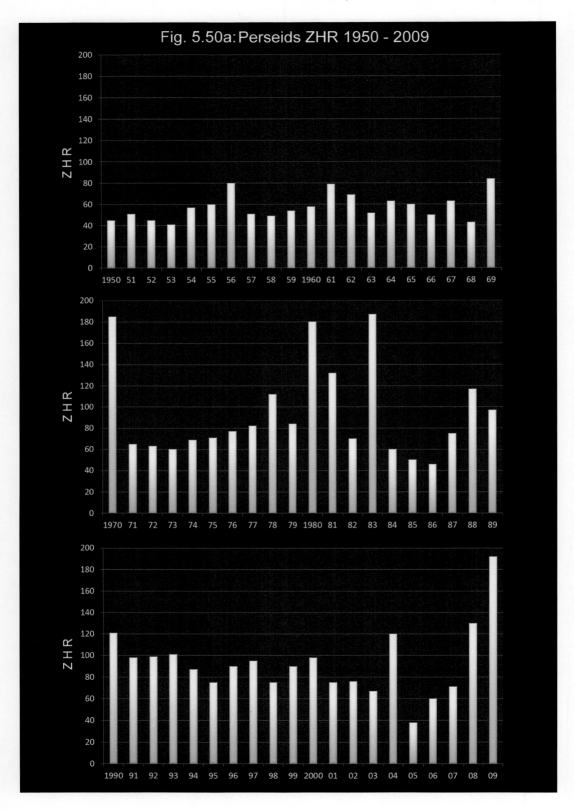

The Perseids

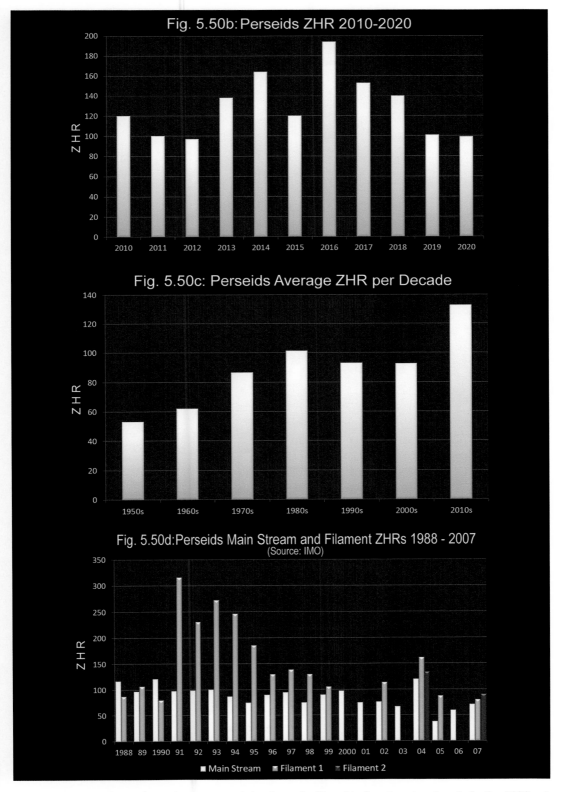

A filament outshone the main stream activity from the Perseids for almost a decade in the 1990s. A secondary filament was also detected on a couple of occasions.

158   5   The Major Meteor Showers: January to August

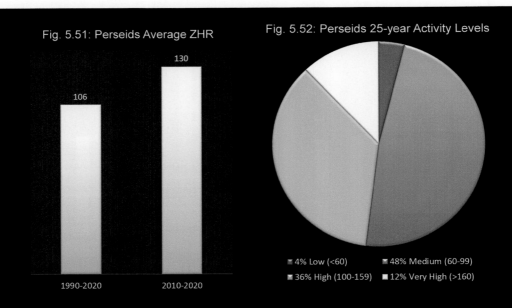

Fig. 5.51: Perseids Average ZHR

Fig. 5.52: Perseids 25-year Activity Levels

- 4% Low (<60)
- 48% Medium (60-99)
- 36% High (100-159)
- 12% Very High (>160)

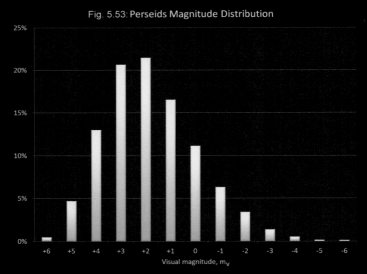

Fig. 5.53: Perseids Magnitude Distribution

Visual magnitude, m_v

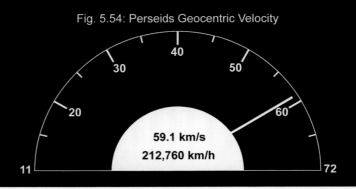

Fig. 5.54: Perseids Geocentric Velocity

59.1 km/s
212,760 km/h

The Perseids

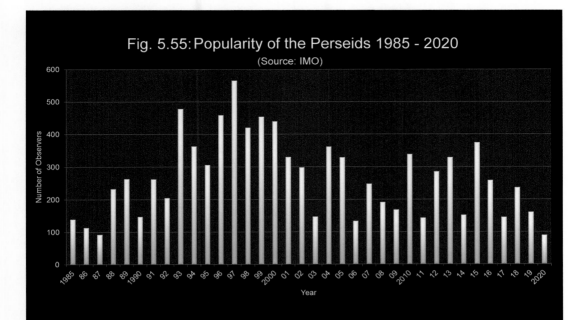

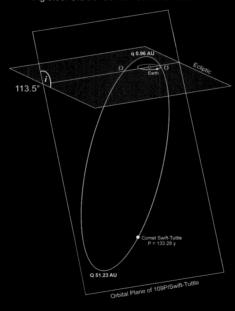

Fig 5.56: Orbit of Comet 109/Swift-Tuttle

Table 5.26: Orbital Data for the Perseid Parent Body*								
Comet 109P/Swift-Tuttle								
q	e	a	ω	Ω	i	P	Q	b
AU		AU	°	°	°	y	AU	AU
0.9595	0.9632	26.0921	152.9822	139.3812	113.4538	133.28	51.2246	7.0132

*For Epoch 1995-10-10.0

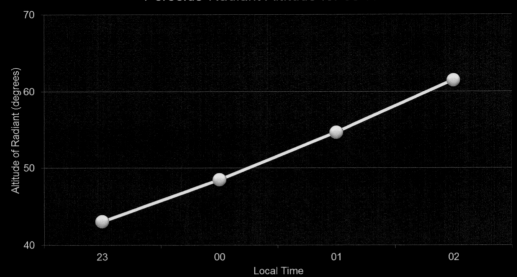

Lerwick, Scotland; Oslo, Norway; Stockholm, Sweden; Helsinki, Finland; St. Petersburg, Russia; Nanontalik, Greenland.

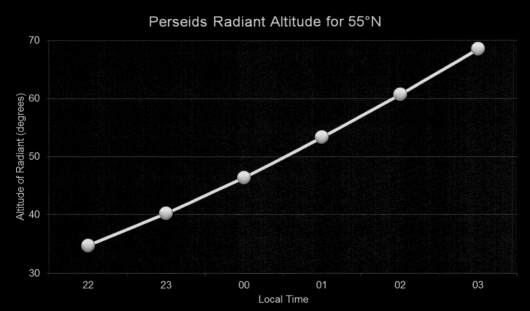

Newcastle upon Tyne, England; Copenhagen Denmark; Malmö, Sweden; Vilnius, Lithuania; Moscow, Russia; Chelyabinsk, Russia; Edmonton, Canada.

Adjust Local Time for Daylight Saving if applicable.

The Perseids

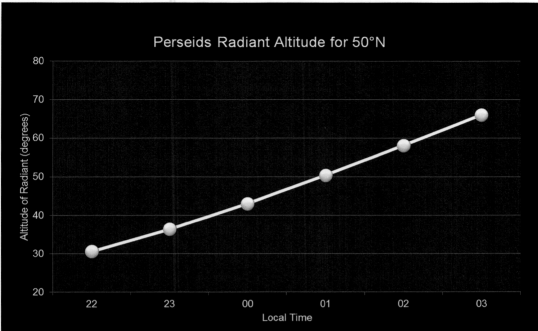

London, England; Frankfurt, Germany; Bern, Switzerland; Kiev, Ukraine; Qaraghandy, Kazakhstan, Qiqihar, China, Vancouver and Winnipeg, Canada.

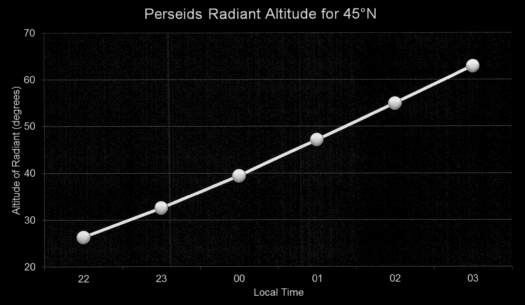

Bordeaux, France; Zagreb, Croatia; Ulaanbaatar, Mongolia; Harbin, China; Portland and Minneapolis, USA; Toronto and Montreal, Canada.

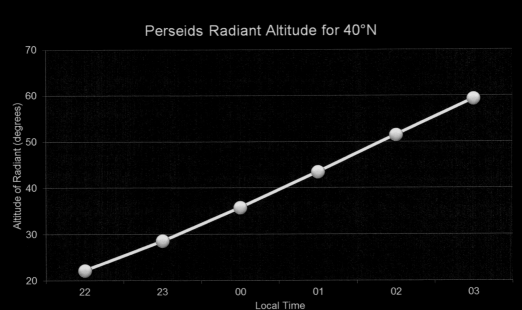

Porto, Portugal; Rome, Italy; Istanbul, Turkey; Toshkent, Uzbekistan; Beijing, China; Sapporo, Japan; Sacramento, Salt Lake City, Denver, Kansas City, Indianapolis, Pittsburgh and Philadelphia, USA.

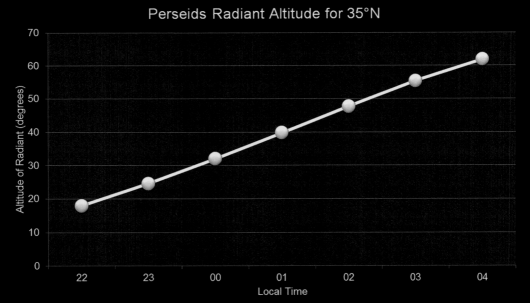

Casablanca, Morocco; Malaga, Spain; Athens, Greece; Tehran, Iran; Jinan, China; Soul, South Korea; Los Angeles, Memphis and Charlotte, USA.

The Perseids

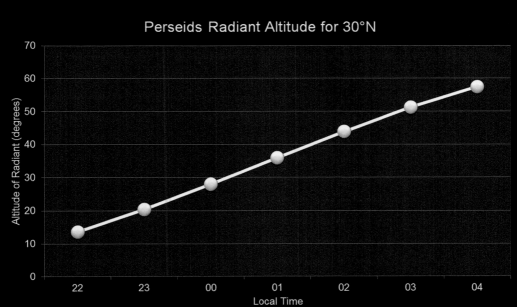

Perseids Radiant Altitude for 30°N

Madeira, Portugal; Tripoli, Libya; Cairo, Egypt; Basra, Iraq; Multan, Pakistan; Lahore, India; Shanghai, China; Tokyo, Japan; San Diego, Austin and Jacksonville, USA.

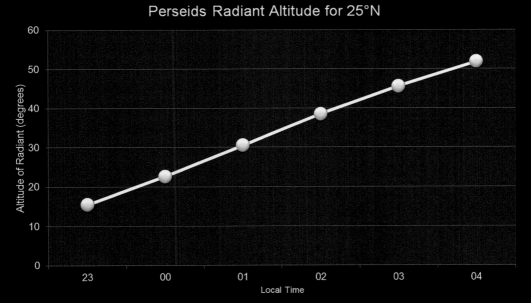

Perseids Radiant Altitude for 25°N

Las Palmas, Canary Islands; Aswan, Egypt; Karachi, Pakistan; Varanasi, India; Kunming, China; T'aipei, Taiwan; Torreon, Mexico; Miami, USA.

164   5 The Major Meteor Showers: January to August

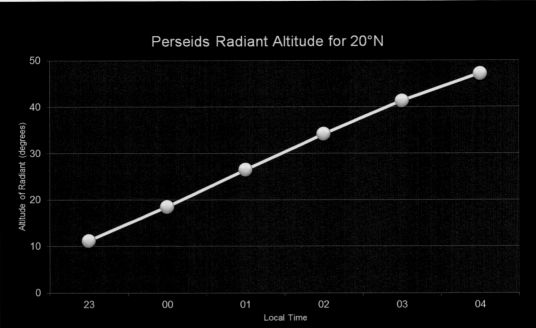

Jeddah, Saudi Arabia; Mumbai, India; Haikou, China; Campeche, Mexico; Santiago de Cuba, Cuba.

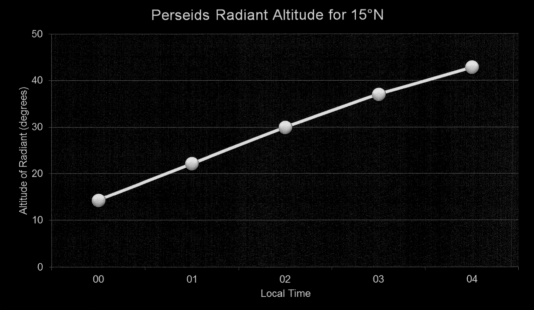

Dakar, Senegal; Khartoum, Sudan; Hyderabad, India; Yangon, Myanmar; Manila, Philippines; Guatemala City, Guatemala

The Perseids

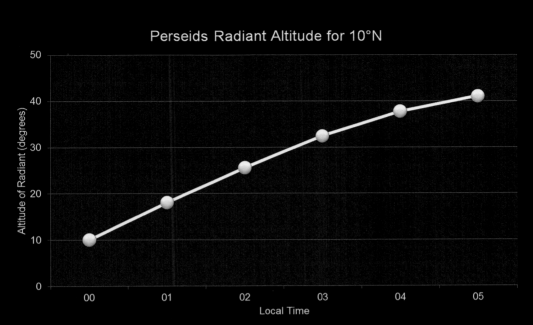

Conakry, Guinea; Kaduna, Nigeria; Addis Ababa, Ethiopia; Ho Chi Minh City, Vietnam; Manila, Philippines; San Jose, Costa Rica; Caracas, Venezuela.

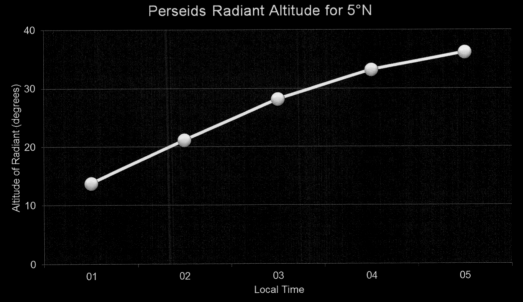

Accra, Ghana; Colombo, Sri Lanka; George Town, Malaysia; Bogota, Colombia; Cayenne, French Guiana.

166   5   The Major Meteor Showers: January to August

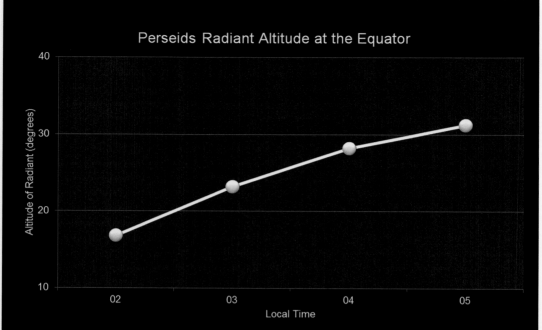

Libreville, Gabon; Kampala, Uganda; Singapore; Quito, Ecuuador; Belem, Brazil.

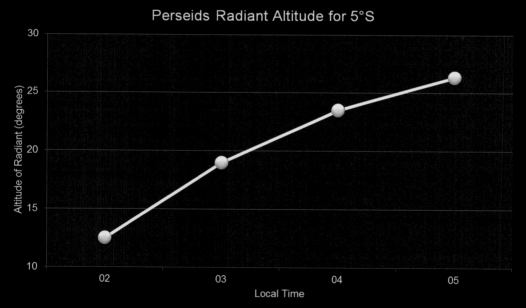

Kinshasa, D.R.Congo; Jakarta, Indonesia; Piura, Peru; Teresina, Brazil.

The Perseids

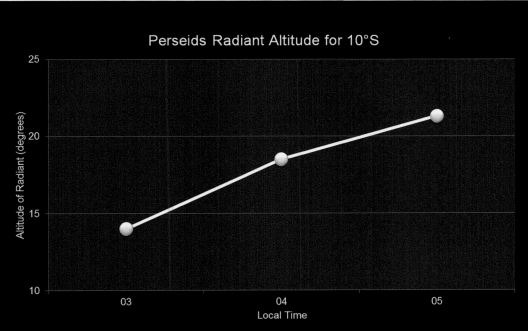

Luanda, Angola; Kolwetzi, D.R.Congo; Kupang, Indonesia, Maceio, Brazil.

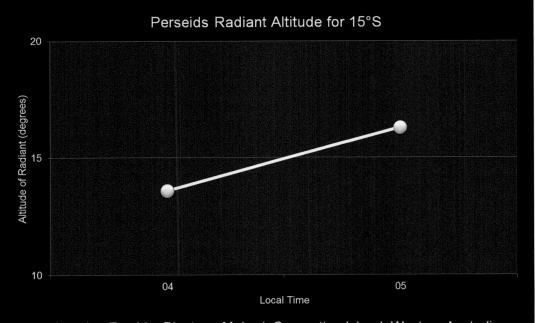

Lusaka, Zambia; Blantyre, Malawi; Coronation Island, Western Australia.

References

1. Jenniskens, P., Betlem, H., de Lignie, M., Langbroek, M., & van Vliet, M. Meteor stream activity. V. The quardantids. *Astronomy & Astrophysics, 327*, 1242.
2. Retrieved from https://www.skyatnightmagazine.com/advice/skills/quadrantid-meteor-shower-when-how-to-see/. Accessed 20 Dec 2020.
3. Retrieved from https://skyandtelescope.org/astronomy-news/catch-the-quadrantid-meteors-if-you-can/. Accessed 20 Dec 2020.
4. Brucalassi. (1825). A. Antologia. 135.
5. Quetelet, L. A. J. (1845). Catalogue of the main appearances of shooting stars. *Nouveaux Mémoires de l'Académie Royale des Sciences et Belles-Lettres de Bruxelles, 12*, 26.
6. Herrick, E. C. (1839). *American Journal of Science and the Arts., 35*, 366.
7. Masterman, S. (1863). *American Journal of Science and the Arts (2nd Series), 35*, 149.
8. Denning, W. F., & Wilson, F. (1918). The meteor shower of January. *Monthly Notices of the Royal Astronomical Society, 78*, 198.
9. Hamid, S. E., & Youssef, M. N. (1963). A short note on the origin and age of the Quadrantids. *Smithsonian Contributions to Astrophysics., 7*, 309.
10. Hamid, S. E., & Whipple, F. L. (1963). Common origin between the quadrantids and the δ-aquarids streams. *The Astronomical Journal, 68*, 537.
11. Kronk, G. (2008). *Cometography*. Cambridge University Press.
12. McIntosh, B. A. (1990). Comet P/machholz and the quadrantid meteor stream. *Icarus, 86*, 299.
13. Hasegawa, I. (1979). *Orbits of ancient and medieval comet* (Vol. 31, p. 263). Publications of the Astronomical Society of Japan.
14. Marsden, B. G. (1983). *IAU Circular 3881: 1983 TB and the Geminid Meteors; 1983 SA; KR Aur*. Central Bureau for Astronomical Telegrams.
15. Hughes, D. W., & McBride, N. (1989). The mass of meteoroid streams. *Monthly Notices of the Royal Astronomical Society, 240*, 7.
16. Bird, K. (1979). The lyrids 1974-78. *Northern Meteor Network Bulletin, 6*, 17.
17. Jenniskens, P. (2006). *Meteor showers and their parent comets* (p. 617). Cambridge University Press.
18. Verniani, F. (1967). Meteor masses and luminosity. *Smithsonian Contributions to Astrophysics., 10*, 181.
19. Newton, H. A. (1863). *American Journal of Science and the Arts (2nd Series)., 36*, 146.
20. Olivier, C. P. (1925). *Meteors* (p. 61). Williams and Wilkins.
21. Herrick, E. C. (1839). *American Journal of Science and the Arts., 36*, 358.
22. Herschel, A. S. (1865). *Report of the Annual Meeting of the British Association for the Advancement of Science., 34*, 40.
23. Corder, H. (1880). Meteor showers 1870-79. *Monthly Notices of the Royal Astronomical Society, 40*, 131.
24. Denning, W. F. (1923). Radiant points of shooting stars observed at Bristol chiefly from 1912 to 1922 inclusive. *Monthly Notices of the Royal Astronomical Society, 84*, 46.
25. Hindley, K. B. (1969). The April Lyrid meteor stream in 1969. *Journal of the British Astronomical Association, 79*, 477.
26. Lindblad, B. A., & Porubčan, V. (1992). *Asteroids, comets and meteors, 1991* (p. 367). Lunar and Planetary Institute.
27. Porubčan, V., & Kornoš, L. (2008). The lyrid meteor stream: Orbit and structure. *Earth, Moon and Planets, 102*, 91.
28. Schiaparelli, G. V. (1866). *Bullettino Meteorologico dell'Osservatorio del Collegio Romano, 5*, 127.
29. Weiss, E. (1867). *Astronomische Nachrichten, 69*, 33.
30. Newton, H. A. (1863). *American Journal of Science and the Arts (Series 2), 36*, 148.
31. Tupman, G. L. (1873). *Report of the Annual Meeting of the British Association for the Advancement of Science., 33*, 301.
32. Denning, W. F. (1883). The Aquariids of April 29 to May 3 (Tupman, No. 33). *Monthly Notices of the Royal Astronomical Society, 43*, 111.
33. McIntosh, R. A. (1929). The meteor swarm of Halley's comet. *Monthly Notices of the Royal Astronomical Society, 90*, 158.

References

34. McIntosh, B. A., & Hajduk, A. (1983). Comet Halley meteor stream: A new model. *Monthly Notices of the Royal Astronomical Society, 205*, 931.
35. Molau, S., et al. (2013). Results of the IMO Video Meteor Network — May 2013. *Journal of the International Meteor Organization, 41*, 133.
36. Falb, R. (1868). The Comet Halley and its Meteors. *Astronomische Nachrichten, 72*, 361.
37. Frossard, F. (1979). Les pluies de météores de l'été. *Le Ciel Déchu., 16*, 64.
38. Norman, V. (1982). The eta Aquariids: Prospects and Planning. *Meteor!, 22*, 12.
39. Dubietis, A., & Arlt, R. (2004). Observational characteristics of meteor showers associated with the Aquarid-Capricornid complex. *Journal of the International Meteor Organization, 32*, 69.
40. Heis, E. and von Neumayer, G.B. (1867) On meteors in the southern hemisphere. Mannheim.
41. Denning, W. F. (1899). General catalogue of the radiant points of meteoric showers and of fireballs and shooting stars observed at more than one station. *Monthly Notices of the Royal Astronomical Society, 53*, 203.
42. Kronk, G. W. (2014). *Meteor showers: An annotated catalogue*. Springer.
43. McIntosh, R. A. (1934). Ephemeris of the radiant-point of the delta aquarid meteor stream. *Monthly Notices of the Royal Astronomical Society, 94*, 583.
44. Terentjeva, A. (1963). On the structure of the δ-aquarid meteor stream. *Smithsonian Contributions to Astrophysics., 7*, 293.
45. Denning, W. F. (1879). The august perseids. *Nature, 20*, 457.
46. Martynenko, V. V., & Smirnov, N. V. (1971). The structure of the perseids radiants in 1971. *Solar System Research, 7*, 104.
47. Arlt, R., & Rendtel, J. (1997). First Analysis of the 1997 perseids. *Journal of the International Meteor Organization, 25*, 207.
48. Roggemans, P., Gyssens, M., & Rendtel. (1991). One-hour outburst of the 1991 perseids surprises Japanese observers! *Journal of the International Meteor Organization, 19*, 181.
49. Pankenier, D. W., Xu, Z., & Jiang, Y. (2008). *Archaeoastronomy in East Asia*. Cambria Press.
50. van Musschenbroeck, P., & Lulofs, J. (1762). Introductio ad philosophiam naturalem. *Vol., 2*.
51. Quetelet, L. A. J., Brandes, H. W., Chladni, E. F. F., & Sauveur, D. (1836). *Bulletins de l'Académie Royale des Sciences et Belles-Lettres de Bruxelles, 3*, 410.
52. Schaeffer, G. C. (1838). *American Journal of Science and the Arts., 33*, 133.
53. Roggemans, P. (1988). The Perseid meteor stream in 1988: A double maximum! *Journal of the International Meteor Organization, 17*, 127.
54. Kirkwood, D. (1861). *Danville Quarterly Review, 1*, 636.
55. Marsden, B. G. (1973). The next return of the comet of the Perseid meteors. *Astronomical Journal, 78*, 654.

Chapter 6

The Major Meteor Showers: October to December

The Orionids

Half of the major meteor showers occur in the last 3 months of the year. October sees the Earth pass through Comet Halley's debris for the second time in a year, with meteors radiating from Orion. Unlike its sister stream, the η-Aquariids, the Orionids favour observers in the Northern Hemisphere, but there is still plenty for Southern observers to see.

Observing Notes

For mid-northern observers, October can be a good time to spend a night meteor hunting. The short summer nights have long gone, but temperatures are often still reasonable and the weather has not yet turned. Two months have passed since the Perseids and the appearance of the Orionids heralds a busy period in the meteor shower calendar.

The Orionids last a little over a month. First appearance is around October 3 ($\lambda_\odot$ 190°) when rates are just one or two meteors per hour. Over the next 10 days, they slowly but surely start to climb. This is a great time to get back into the swing of things. You'll have plenty of time to hone your skills: backtracking individual meteors to the radiant, perhaps using plotting charts, and estimating magnitudes by comparison with nearby stars—see Appendix 7. Two weeks after they first make their appearance, the ZHR is anywhere between 8 and 20, but still rising to a broad, flat maximum that lasts perhaps 4 or 5 days centred on $\lambda_\odot$ 209°, which is around Oct 22/23 (see Fig. 6.5).

Over the past 25 years, the ZHR has averaged 29 and has only fallen slightly to 26 during the past decade (Figs. 6.7 and 6.8). Rates do vary, however, and between 2005 and 2010 were regularly around 40 with a peak of 82 in 2007. It's not the first time that there has been a sudden spike in activity; 1924, 1936 and 1948 were all good years. Although few and far between, there's always the possibility that rates will suddenly jump again.

Table 6.1a: Orionid Activity Details				
		Start	Maximum	End
Dates (approx.)		Oct 3	Oct 22/23	Nov 7
Solar longitude:	λ☉	190°	209°	220°
Right Ascension at max:	α		06^h 24^m (96°)	
Declination at max:	δ		+15.7°	
Time of transit :			4.5^h L.T.	
ZHR (2010-2020 average):			26	
Radiant diameter:	∅	10°	10°	10°

Table 6.1b: Orionid Additional Data	
IAU Abbreviation: QUA IAU Code: 008 AKA: October Halleyids	
Mean daily motion of radiant: Δ	Δα = +0.71° Δδ = +0.07°
Geocentric velocity: V$_g$	65.7 km/s (236,520 km/h)
Population index: r	2.5
Mean magnitude: $\bar{m}_v$	+2.4
Parent body: Comet 1P/Halley	
Best visibility: Post-midnight in the Northern Hemisphere	
The radiant is circumpolar from 74°N, i.e. Spitsbergen, Norway, Novosibirskiye Ostrova, Russia, Grise Fiord, Ellesmere Island, Canada and Qaanaaq (Thule), Greenland.	

Table 6.2: Orionid Maxima 2020-2030				
Maximum (λ☉ 209°)				Moon
year	month	day	hour (UT)	Age (d)
2020	10	22	05:37	6
2021	10	22	11:49	16
2022	10	22	17:56	26
2023	10	23	00:03	9
2024	10	22	06:19	20
2025	10	22	12:18	1
2026	10	22	18:27	11
2027	10	23	00:44	23
2028	10	22	06:46	4
2029	10	22	13:01	15
2030	10	22	19:14	25

The Orionids

The average visual magnitude is $\bar{m}_v$ +2.4, with around 13% being m_v 0 or brighter. Like the meteors from its sister stream, the η-Aquariids, the Orionids are very swift at 65.7 km/s (236,520 km/h). If the past 25 years are anything to go by, then there is a 60% chance that in any one year, the ZHR will be in the range of 20–35, and a 24% chance that it will be higher (Fig. 6.8).

The Earth encounters the meteoroid stream at the ascending node as the stream particles travel from below to above the plane of the ecliptic (Fig. 6.1). The Orionid radiant is large and diffuse at about 10° in diameter. Various observers in the past have thought they had detected two radiants (see Fig. 6.4b). Denning named his two radiants the *Orionids* and the *Geminids* and thought they were both about 5° in diameter (1912) [1]. Some twenty-seven years later, J.P.M. Prentice also claimed to have detected two radiants but considered them to be branches of the same stream, and called them *Primary* and *Secondary* [2]. A telescopic

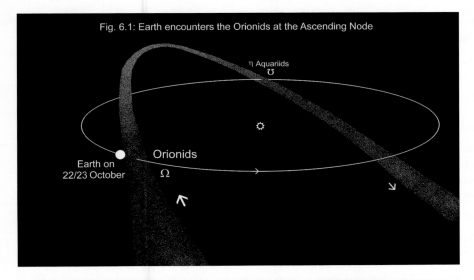

Fig. 6.1: Earth encounters the Orionids at the Ascending Node

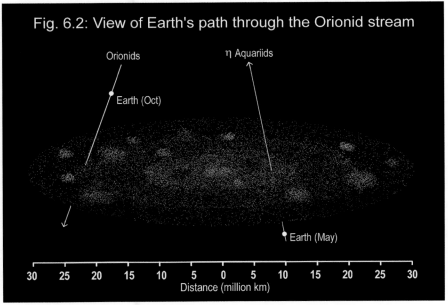

Fig. 6.2: View of Earth's path through the Orionid stream

study in 1968 by V. Znojil also identified two radiants, which were named the *Northern* and the *Southern components*, all of which hints at the complexity of the stream's structure [3].

The Orionids are best after midnight, when the radiant climbs to a decent height above the horizon. Mid-northern observers will see the radiant reach 60°–70° in the pre-dawn hours, but by far the best-placed observers are at 10°N to 25°N, where the radiant culminates at 80°+ and about 98% of activity is visible.

The farther south you go, the more limited your observing window. At 35°S—Buenos Aires in Argentina, Montevideo in Uruguay and Mangonui in New Zealand—the radiant still manages to climb to a height of about 40°, but just before dawn. From these southern locations observation is only really possible from about 1 am LT, with the radiant at 20° altitude, to about 4 am at most.

Table 6.3: Other Active Meteor Showers During the Orionids
Calendar dates are approximate. Use solar longitude $\lambda_\odot$

Shower	IAU Code	Start $\lambda_\odot$	Max $\lambda_\odot$	End $\lambda_\odot$	At Max. α	At Max. δ	V_g (km/s)	ZHR
October Capricornids[(1)]	OCC	Sep 21 178°	Oct 2 + 9 189° +196°	Oct 12 199°	$20^h\ 12^m$ 303°	$-09°$	11	Var.
Southern Taurids	STA	Sep 10 168°	Nov 5 223°	Nov 20 238°	$03^h\ 33^m$ 53°	$+12.9°$	26.1	5
ε-Geminids	EGE	Sep 30 187°	Oct 11-18 198°-205°	Oct 27 214°	$05^h\ 57^m$ 89°	$+28°$	70	3
October Camelopardalids	OCT	Oct 4 191°	Oct 5 192.6°	Oct 9 196°	$10^h\ 56^m$ 164°	$+79$	46.6	5
October Draconids[(2)]	DRA	Oct 6 193°	Oct 8 195.4°	Oct 10 197°	$17^h\ 28^m$ 262°	$+54$	21	Var.
Leonis Minorids	LMI	Oct 12 199°	Oct 24 211°	Nov 5 223°	$10^h\ 48^m$ 170°	$+37°$	62	2
October Ursae Majorids	OCU	Oct 14 201°	Oct 15 202°	Oct 19 206°	$09^h\ 43^m$ 145.8°	$+63.8°$	38.8	<1
o-Eridanids	OER	Oct 16 203°	Nov 5 223°	Nov 24 242°	$03^h\ 00^m$ 45°	$-04°$	29	1
Northern Taurids	NTA	Oct 20 207°	Nov 12 230°	Dec 10 258°	$03^h\ 57^m$ 59°	$+22.3°$	28.7	5
χ-Taurids	CTA	Oct 20 207°	Nov 4 222°	Nov 17 235°	$03^h\ 32^m$ 53°	$+25°$	41	1
ξ-Draconids	XDR	Oct 21 208°	Oct 23 210°	Oct 24 211°	$11^h\ 21^m$ 170.3°	$+73°$	35.8	<1
Andromedids	AND	Oct 26 213°	Nov 6 224°	Nov 17 235°	$00^h\ 38^m$ 10°	$+24°$	19	<1
λ-Ursae Majorids	LUM	Oct 27 214°	Oct 28 215°	Oct 29 216°	$10^h\ 24^m$ 156°	$+49°$	61	<1
Southern λ-Draconids	SLD	Nov 1 219°	Nov 3 221°	Nov 4 222°	$10^h\ 48^m$ 162°	$+68°$	49	<1
κ-Ursae Majorids	KUM	Nov 3 221°	Nov 7 225°	Nov 10 228°	$09^h\ 49^m$ 147°	$+45°$	66	<1
ρ-Puppids	RPU	Nov 10 228°	Nov 14 232°	Nov 20 238°	$08^h\ 23^m$ 126°	$-25°$	58	<1

[(1)]Note double maximum.
[(2)]Previously called the Draconids (hence the IAU Code: DRA) and also known as the Giacobinids after the stream's parent body, Comet 21 P/Giacobini-Zinner.

Table 6.4: Orionid Timeline	
CE 902	Possibly the first record of the Orionids according to Adolphe Quetelet. He also believes the shower was recorded in 1202.
1798	Heinrich Wilhelm Brandes and Johann Friedrich Benzenberg, two students at the University of Göttingen, use a 16-km baseline to triangulate the heights of meteors. They found the heights to start at between 50 and 171 km. They also accidentally discover the Orionids.
1839-40	E.C. Herrick reports activity from the Orionids in *Silliman's Journal*.
1864 Oct 18	A.S. Herschel makes first precise measurement of the radiant position.
1868	R. Falb first suggests link between Halley's Comet, the η-Aquariids and the Orionids.
1892	Hourly rate determined to be 15 meteors/h.
1900	Rate drops to just 7 meteors/h.
Early 1900s	British observers unable to detect radiant drift. W.F. Denning takes the view that the radiant is stationary, with which C.P. Olivier disagrees.
1912	Denning believes that two radiants are active, which he calls the *Orionids* and the *Geminids*.
1922	Rate reaches a new high of 35 meteors/h.
1922	Olivier and others demonstrate that the Orionid radiant does drift.
1924	High levels of activity observed.
1936	Rates of about 50 meteors/h observed.
1938	High activity level.
1939	J.P.M. Prentice locates two radiants, which he calls the *Primary* and the *Secondary*.
1939-1959	Average ZHR 18.
1944-1950	Czech astronomers detect a secondary maximum, but their results are not published until 1982.
1960-1974	Average ZHR 24 with highs of between 30 and 40 on four occasions.
1968	V. Znojil's telescopic studies suggest two radiants, which he identifies as a *Northern* and a *Southern* component.
1975-1985	Average ZHR 18.
1985-86	The stream's parent comet, 1P/Halley, returns to perihelion but there is no noticeable increase in meteor rates.
1989	David W. Hughes and Neil McBride estimate the mass of the Orionid and η-Aquariid meteoroid stream to be 3.3×10^{16} g.
2010-2020	Average ZHR 26.

There are at least 16 other showers active during the Orionid period, and observers should particularly note the risk of contamination by the ε-Geminids and the Taurids, both the Northern and Southern branches (Table 6.3).

Despite the fact that the Orionids consistently put on a good display and at a good time of year, the shower has fallen out of favour with IMO visual observers. Just before the turn of the century, about 80–150 observers would lodge their reports with the IMO. Nowadays, it is just around 35 (Fig. 6.11).

Discovery

The Orionids were discovered quite by accident. In 1798, two German students at the University of Göttingen, Heinrich Wilhelm Brandes and Johann Friedrich Benzenberg, set themselves the task of estimating the heights of meteors [4].[1] Using a 16-km baseline, they estimated the altitudes of 17 simultaneously observed meteors to be between 50 and 171 km, but they also noticed that on the night in question, October 14/15, there were far more meteors than on previous nights. And so the Orionids were discovered. Or perhaps not. The Director of the Brussels Observatory, Adolphe Quetelet, found records of activity in Oriental annals dating back to CE 902 and 1202, which he believed were the first sightings of the Orionids [5].

The first attempt to accurately locate the shower's radiant had to wait more than half a century until 1864, when Alexander Herschel used 11 meteors to pinpoint the radiant at α 90° ($06^h\ 00^m$), δ +15° (today's estimates are α 96° ($06^h\ 24^m$), δ +15.7°) [6].

It was the German astronomer R. Falb who first suggested a link between Halley's Comet, the η-Aquariids and the Orionids [7]. Apart from the orientation of the orbit and the resulting radiant positions, additional evidence came in the similarity between the two showers, such as the average magnitude and velocity. The only really big difference is in the levels of activity: the η Aquariids are about three times more active than the Orionids. But how can this be if they are the same stream? The answer lies in where the Earth passes through the meteoroid stream.

Meteoroid streams are not a tube of evenly spaced particles. Quite the opposite: they are really messy. They are full of clumps of material, gaps, holes, kinks, twists and lanes or filaments where material has become concentrated because of mainly gravitational effects. Generally, the highest concentration of material lies towards the middle, but this somewhat depends on the age of the meteoroid stream and how often it has been replenished by the parent comet returning to perihelion.

In the case of the Orionids and η-Aquariids, the stream is probably somewhere between 50 and 60 million kilometres across. Seen from an external perspective (Fig. 6.2), the Earth passes through the stream at about 24 million kilometres from the centre in October, where dust levels are relatively low, to produce the Orionids. During the η-Aquariids encounter, however, the Earth passes within just nine million kilometres of the centre where dust levels are much higher. Overall, the stream is rich in small particles with masses of less than 10^{-5} g.

In the early 1900s, British observers failed to detect any movement in the Orionid radiant. William Denning seized upon this as proof that his hypothesis of stationary radiants was true, even though the geometry dictated that radiants must move eastwards across the sky. Thereafter followed a war of words, with Denning & Co on the side of stationary radiants and Charles Olivier and several other American astronomers in favour of radiant drift [9]. Olivier won the day, with clear evidence of the Orionids moving across the celestial sphere at somewhat under 1° per day [10].

During the twentieth century, rates fluctuated between a low of 7 (1900) and a high of 50+ meteors per hour (1936). In recent years, the rate has been fairly stable with a ZHR of 26 (Figure 6.7).

Origins

Please refer to the section on the η-Aquariids.

---

[1] This was not the first attempt at determining the altitude of a meteor. More than a century earlier, in 1676, Geminiano Montanari estimated the height and speed of the Great Meteor observed over Italy on March 31, which he calculated to be 34–40 Italian miles (55–65 km) and travelling at least 160 miles per minute, or 4.3 km/s [8].

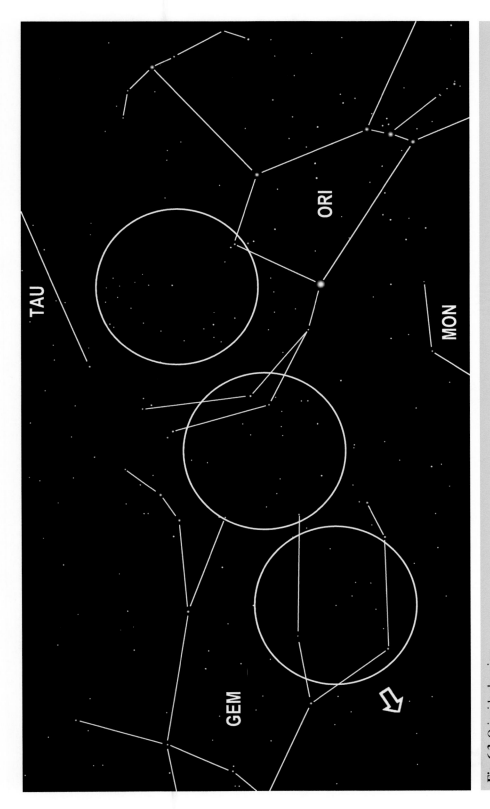

Fig. 6.3 Orionids sky view

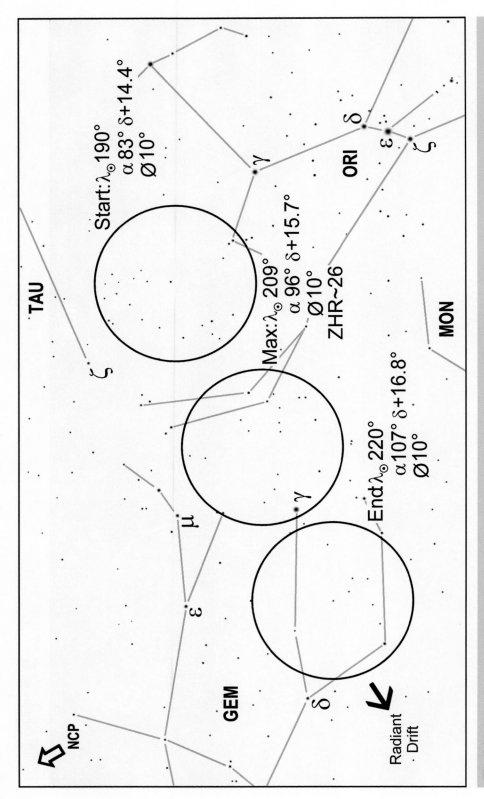

Fig. 6.4a Orionids radiant chart

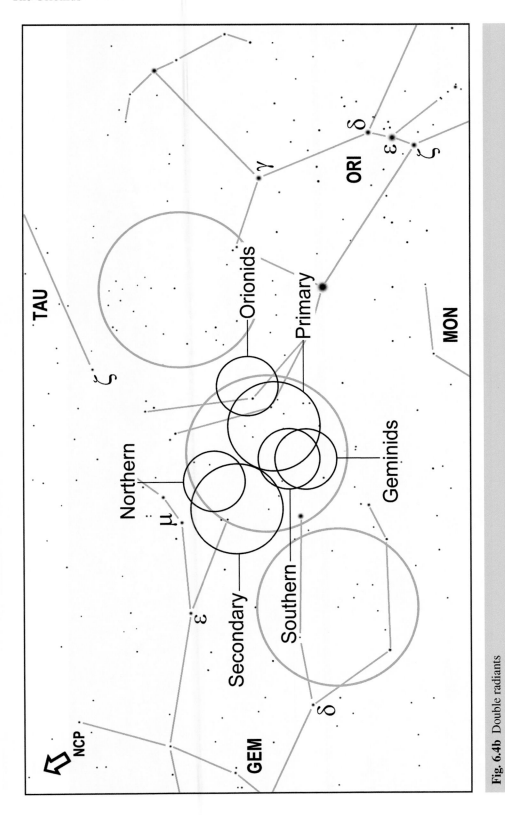

Fig. 6.4b Double radiants
William F. Denning claimed to have detected two radiants, which he called the Orionids and the Geminids. J.P.M. Prentice thought he found a Northern and a Southern component, while V. Znojil's telescopic studies revealed a Primary and a Secondary radiant

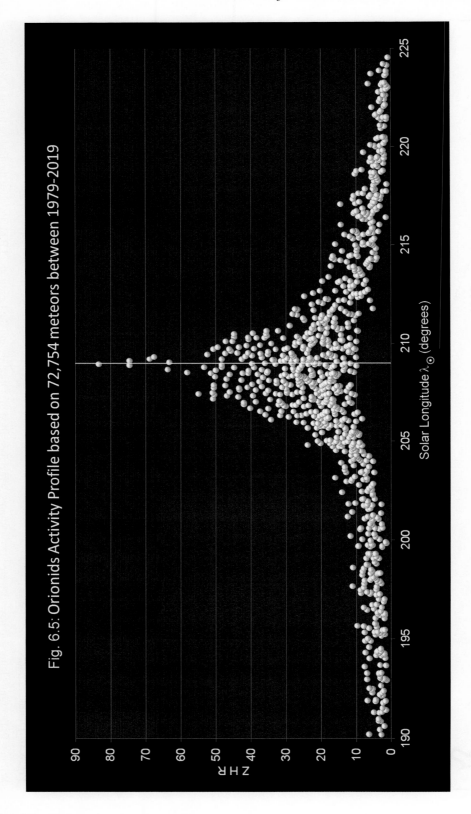
Fig. 6.5: Orionids Activity Profile based on 72,754 meteors between 1979-2019

The Orionids

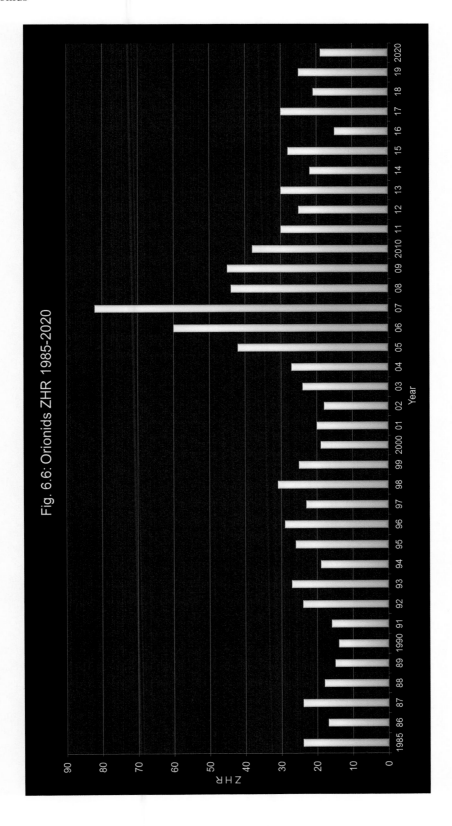

Fig. 6.6: Orionids ZHR 1985-2020

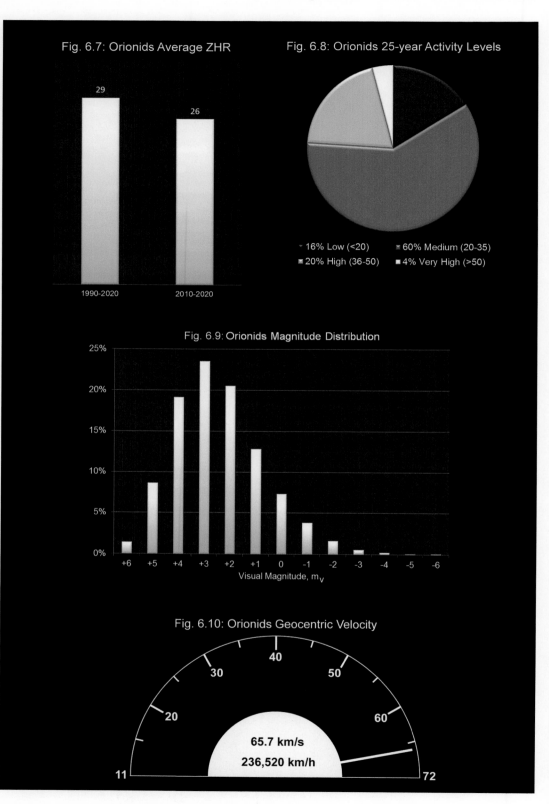

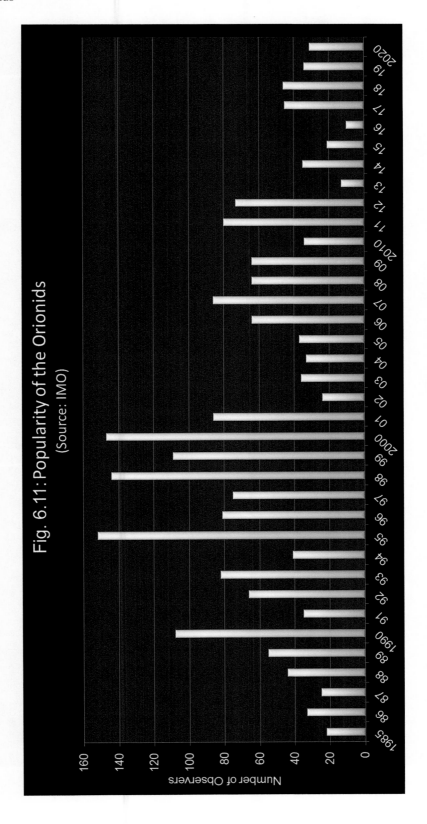

Fig. 6.11: Popularity of the Orionids
(Source: IMO)

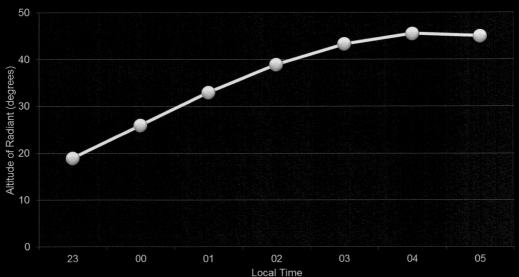

Lerwick, Scotland; Oslo, Norway; Stockholm, Sweden; Helsinki, Finland; St. Petersburg, Russia; Nanontalik, Greenland.

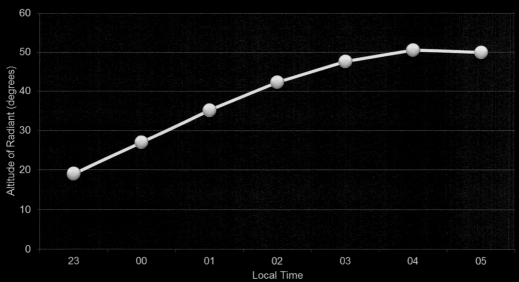

Newcastle upon Tyne, England; Copenhagen Denmark; Malmö, Sweden; Vilnius, Lithuania; Moscow, Russia; Chelyabinsk, Russia; Edmonton, Canada.

Adjust local time for daylight saving if applicable

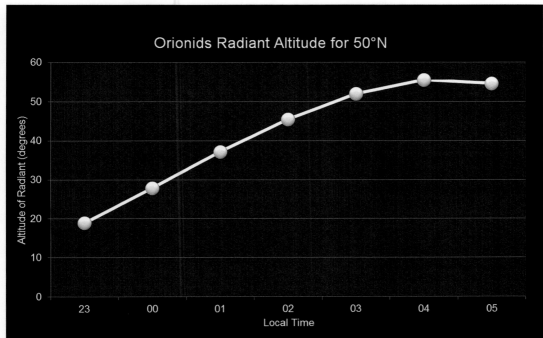

London, England; Frankfurt, Germany; Bern, Switzerland; Kiev, Ukraine; Qaraghandy, Kazakhstan, Qiqihar, China, Vancouver and Winnipeg, Canada.

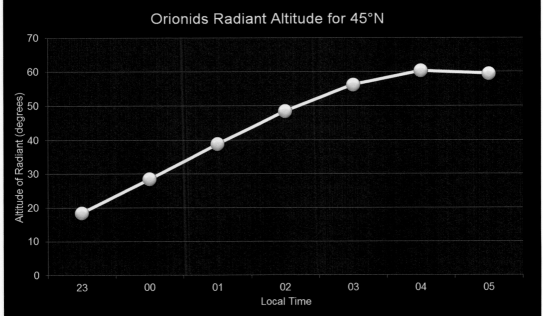

Bordeaux, France; Zagreb, Croatia; Ulaanbaatar, Mongolia; Harbin, China; Portland and Minneapolis, USA; Toronto and Montreal, Canada.

186   6   The Major Meteor Showers: October to December

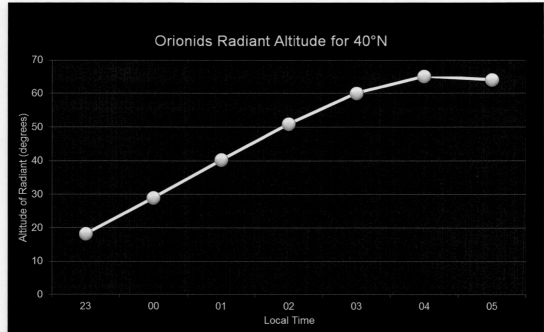

Porto, Portugal; Rome, Italy; Istanbul, Turkey; Toshkent, Uzbekistan; Beijing, China; Sapporo, Japan; Sacramento, Salt Lake City, Denver, Kansas City, Indianapolis, Pittsburgh and Philadelphia, USA.

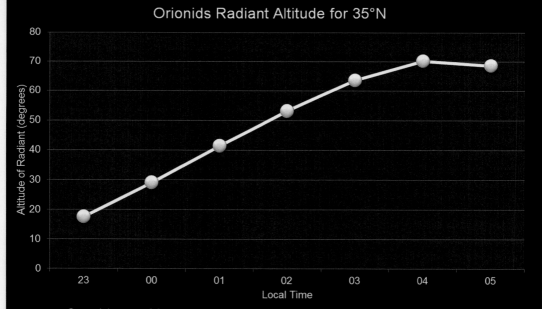

Casablanca, Morocco; Malaga, Spain; Athens, Greece; Tehran, Iran; Jinan, China; Soul, South Korea; Los Angeles, Memphis and Charlotte, USA.

The Orionids

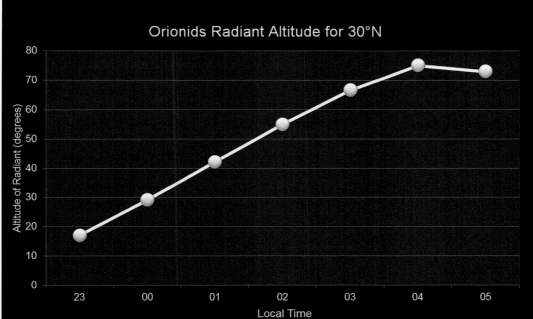

Madeira, Portugal; Tripoli, Libya; Cairo, Egypt; Basra, Iraq; Multan, Pakistan; Lahore, India; Shanghai, China; Tokyo, Japan; San Diego, Austin and Jacksonville, USA.

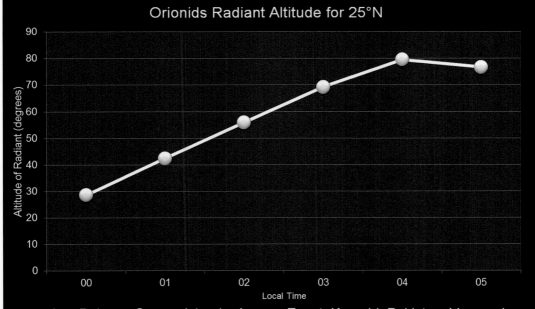

Las Palmas, Canary Islands; Aswan, Egypt; Karachi, Pakistan; Varanasi, India; Kunming, China; T'aipei, Taiwan; Torreon, Mexico; Miami, USA.

188   6   The Major Meteor Showers: October to December

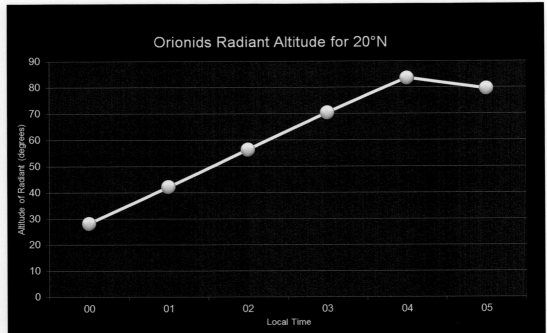

Jeddah, Saudi Arabia; Mumbai, India; Haikou, China; Campeche, Mexico; Santiago de Cuba, Cuba.

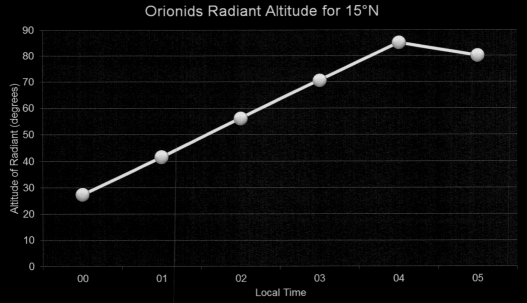

Dakar, Senegal; Khartoum, Sudan; Hyderabad, India; Yangon, Myanmar; Manila, Philippines; Guatemala City, Guatemala

The Orionids

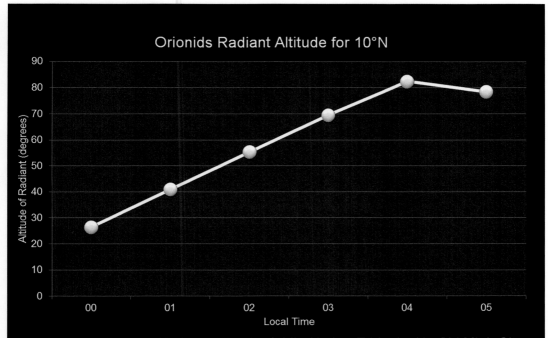

Conakry, Guinea; Kaduna, Nigeria; Addis Ababa, Ethiopia; Ho Chi Minh City, Vietnam; Manila, Philippines; San Jose, Costa Rica; Caracas, Venezuela.

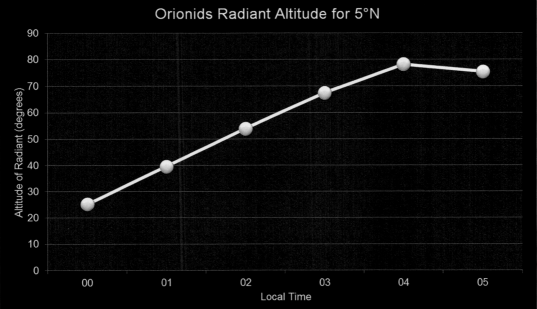

Accra, Ghana; Colombo, Sri Lanka; George Town, Malaysia; Bogota, Colombia; Cayenne, French Guiana.

190   6 The Major Meteor Showers: October to December

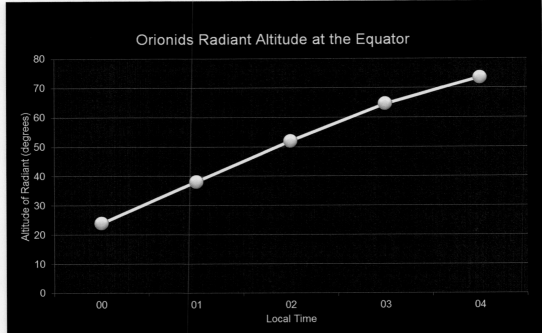

Libreville, Gabon; Kampala, Uganda; Singapore; Quito, Ecuuador; Belem, Brazil.

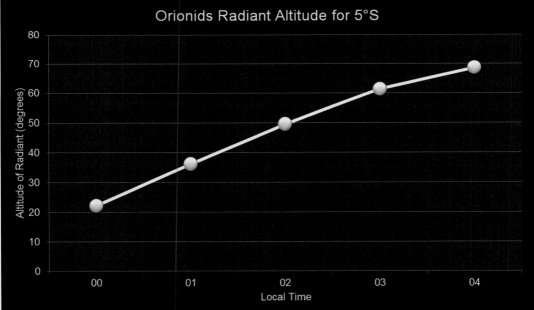

Kinshasa, D.R.Congo; Jakarta, Indonesia; Piura, Peru; Teresina, Brazil.

The Orionids

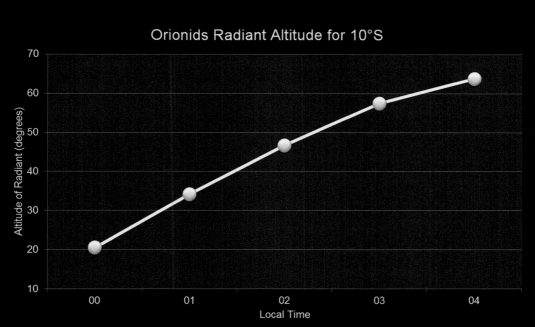

Luanda, Angola; Kolwetzi, D.R.Congo; Kupang, Indonesia, Maceio, Brazil.

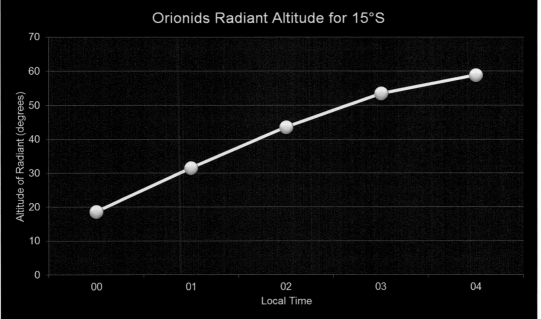

Lusaka, Zambia; Blantyre, Malawi; Coronation Island, Western Australia.

192   6   The Major Meteor Showers: October to December

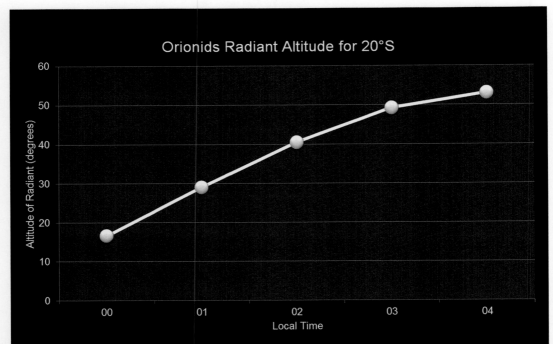

Bulawayo, Zimbabwe; Beira, Mozambique; Port Hedland, Western Australia; Bowen, Queensland, Australia; Iquique, Chile; Cariacica, Brazil.

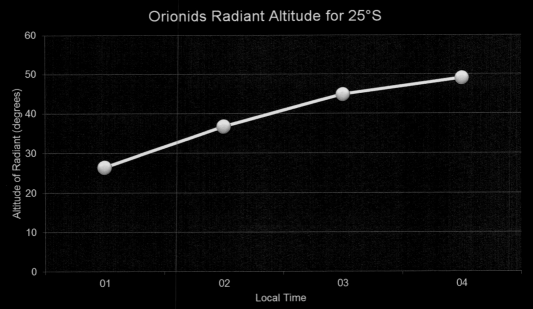

Gibeon, Namibia; Pretoria, South Africa; Carnarvon, Western Australia; Bundaberg Central, Queensland, Australia; Taltal, Chile; Asuncion, Paraguay.

The Orionids

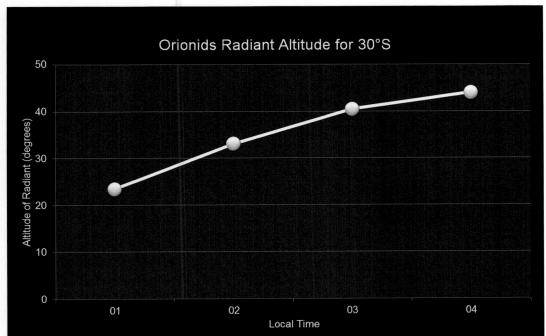

Durban, South Africa; Perth, Western Australia; Coffs Harbour, NSW, Australia; Coquimbo, Chile; Poto Alegre, Brazil.

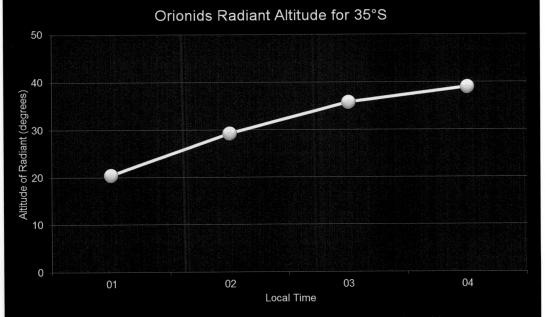

Curico, Chile; Buenos Aires, Argentina; Montevideo, Uruguay; Mangonui, New Zealand.

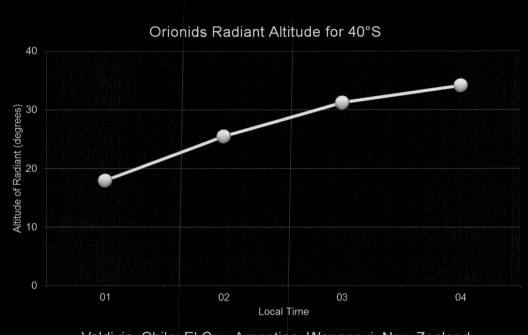

Valdivia, Chile; El Cuy, Argentina; Wanganui, New Zealand.

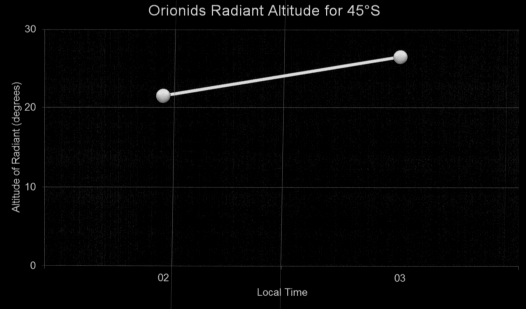

Camarones, Argentina; Oamaru, New Zealand.

The Orionids

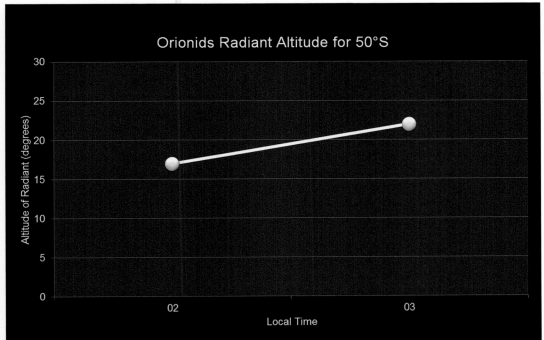

Puerto Santa Cruz, Argentina; Stanley, Falkland Islands.

The Taurids

A shower that can enthral and disappoint in equal measure, the Taurids are part of a huge complex that covers more than half the year.

Observing Notes

If you want to test your skill as an observer, then the Taurids are for you. This shower has two branches, the Northern branch (NTA) and the Southern branch (STA), which are so close together at maxima that sky watchers struggle to separate the two. And when rates are low, the two peaks, at $\lambda_\odot$ 223° (STA) and $\lambda_\odot$ 230° (NTA), seem to merge, giving the appearance of a flat maximum lasting up to 3 weeks. But there's often a prize for your efforts: the shower can suddenly start hurling brilliant fireballs across the sky. Although rates are generally low with a combined ZHR <12, the prospect of possibly witnessing several fireballs on the same night should be enough to tempt you out into the cold!

The Taurids are about as close as you can get to a global shower. Unless you happen to be doing a bit of midnight sunbathing in Antarctica, you can watch the Taurids from every country on Earth, and for most of the night. For people in mid-northern latitudes, 40°N–50°N, the radiants climb to a pretty decent 60° to 70° altitude by midnight, where about 85–95% of activity can be seen in dark sky areas. If you are lucky enough to live at about 15°N, you will see the radiants culminate at about 80°, almost directly overhead, and see 98% of activity. In the Southern Hemisphere at 25°S—Pretoria, South Africa; Carnarvon, Western Australia and Asuncion, Paraguay—the radiants are about 40° altitude around midnight, and you'll catch about two-thirds of the display. The radiants are quite diffuse and large. Some analysts suggest that observers should regard the radiants as being oval—20° in right ascension by 10° in declination—in order to capture the maximum number of Taurids whilst minimizing contamination from other sources, but it is questionable whether the results justify this strategy.

The Southern Taurids are said to begin their activity cycle around September 10, when the Sun is at solar longitude $\lambda_\odot$168° and the radiant is in Cetus. They peak around November 5 near the star ε Tauri—adding a few celestial fireworks to *Bonfire Night* in the UK—and fizzle out around November 20, a 10-week display.

Table 6.5: Taurid Swarm Years

1900	1910	1920	1930	1940	1950	1960	1970	1980	1990	2000	2010	2020	2030
1901	1911	1921	1931	1941	1951*	1961	1971	1981	1991	2001	2011	2021	2031
1902	1912	1922	1932	1942	1952	1962	1972	1982	1992	2002	2012*	2022	2032
1903	1913	1923	1933	1943	1953	1963	1973	1983	1993	2003	2013	2023	2033
1904	1914	1924	1934*	1944	1954	1964	1974	1984	1994	2004	2014	2024	2034
1905	1915	1925	1935	1945	1955	1965	1975	1985	1995*	2005	2015	2025	2035
1906	1916	1926	1936	1946	1956	1966	1976	1986	1996	2006	2016	2026	2036
1907	1917*	1927	1937	1947	1957	1967	1977	1987	1997	2007	2017	2027	2037
1908	1918	1928	1938	1948	1958	1968	1978	1988	1998	2008	2018	2028	2038
1909	1919	1929	1939	1949	1959	1969	1979	1989	1999	2009	2019	2029	2039

Legend: **1954** Taurids   **1982** ζ Perseids / β Taurids   **1934*** Taurids /ζ Perseids / β Taurids

The Taurids

Table 6.6a: Activity Details – Southern Taurids

		Start	Maximum	End
Dates (approx.)		Sep 10	Nov 5/6	Nov 20
Solar longitude:	$\lambda_\odot$	168°	223°	238°
Right Ascension at max:	α		$03^h\ 33^m\ (53°)$	
Declination at max:	δ		+12.9°	
Time of transit :	L.T.		0.6^h	
ZHR (10 year average):			5	
Radiant diameter:	$\varnothing$	10°	10°	10°

Table 6.6b: Additional Data – Southern Taurids

IAU Abbreviation: STA		IAU Code: 002	AKA: Taurids
Mean daily motion of radiant:	Δ		$\Delta\alpha = +0.74°\quad \Delta\delta = +0.13°$
Geocentric velocity:	V_g		26.1 km/s (93,960 km/h)
Population index:	r		2.4
Mean magnitude:	m_v		+2.7
Parent body: Comet 2P/Encke			
Best visibility: All night.			
The radiant is circumpolar from 77°N, i.e. Spitsbergen, Norway, Novosibirskiye Ostrova, Russia, Grise Fiord, Ellesmere Island, Canada and Qaanaaq (Thule), Greenland.			

Table 6.6c: Activity Details – Northern Taurids

		Start	Maximum	End
Dates (approx.)		Oct 20	Nov 12/13	Dec 10
Solar Longitude:	$\lambda_\odot$	207°	230°	258°
Right Ascension at max:	α		$03^h\ 57^m\ (59°)$	
Declination at max:	δ		+22.3°	
Time of Transit :	L.T.		0.6^h	
ZHR (10 year average):			5	
Radiant diameter:	$\varnothing$	10°	10°	10°

Table 6.6d: Additional Data – Northern Taurids

IAU Abbreviation: NTA		IAU Code: 017	AKA: Taurids
Mean daily motion of radiant:	Δ		$\Delta\alpha = +0.79°\quad \Delta\delta = +0.15°$
Geocentric velocity:	V_g		28.7 km/s (103,320 km/h)
Population index:	r		2.4
Mean magnitude:	$\bar{m}_v$		+2.7
Parent body: Comet 2P/Encke			
Best visibility: All night.			
The radiant is circumpolar from 77°N, i.e. Spitsbergen, Norway, Novosibirskiye Ostrova, Russia, Grise Fiord, Ellesmere Island, Canada and Qaanaaq (Thule), Greenland.			

The Northern Taurids lag behind their southern counterpart by 6 weeks, not really showing themselves until around October 20, when the radiant is midway between Aries and Cetus. Maximum activity falls on November 12 or 13 at $\lambda_\odot$ 230°, after which the rates fall off fairly rapidly and disappear altogether a month later, with the radiant near the bluish star β Tauri (Fig. 6.14). With atmospheric entry velocities of between 26.1 km/s and 28.7 km/s, the Taurids are the slowest of the major showers. The magnitude distribution of both branches is similar, with an average visual magnitude of $\overline{m}_v$ +2.7.

One of the things you must bear in mind about the Taurids is that the dates are very fluid. The Taurids are part of a vast complex of meteoroid streams that spread more than halfway around the Earth's orbit. There is no real start or end date for either of the two branches, and even maxima can be tricky to find.

Keith Hindley, the former Director of the BAA's Meteor Section, pointed out in the early 1970s that the activity levels of the two components switch around the time of maximum [11]. In the run-up to maximum, it is the Northern branch that is the most active. Then, around $\lambda_\odot$ 215° (Oct 28), the Southern branch suddenly becomes very active. This lasts for about 2 weeks until $\lambda_\odot$ 230° when they switch back again just as the Northern branch reaches its peak. In years when activity levels are low, the two maxima seem to merge into one, and it is often just about impossible to determine when maximum activity starts and finishes (Fig. 6.15).

The Taurid radiants overlap and are in fact part of the diffuse Antihelion radiant, a sprawling conglomerate of minor meteor showers that once had their own identities. Activity from the Antihelion source is low—a ZHR of just a few meteors per hour though rising towards dawn—but it can still mess up the Taurid ZHR, with observers becoming confused as to which meteors are Taurids and which are Antihelion. Trying to separate the various meteors can be challenging. Then there are the Swarm Years to look forward to.

A typical meteoroid stream consists mainly of dust no bigger than a grain of sand at most. Not the Taurids. Embedded within the stream are swarms of larger particles, up to the size of small pebbles, which can produce some brilliant, beautiful fireballs—but not every year. *Swarm Years* occur at intervals of 3, 4 or 7 years [12]. Some Swarm Years seem to be more active than others, but this could be because maximum is short and occurs over either the Atlantic or the Pacific.[2] Normally, fireballs make up about 1% of all Taurids but, in Swarm Years, it can be anywhere between 2.4% and, exceptionally, 11%. What makes the Taurid fireballs doubly impressive is their relatively slow passage, travelling across the sky at less than half the speed of the Perseids. Table 6.5 shows the Swarm Years for 1900–2039.

The Taurid Complex stretches from early September to the end of December and possibly into January. Table 6.8 details the various components, while Table 6.9 lists a number of other showers active during the Taurid period. Note that the Taurid Complex also includes two daytime showers, the ζ-Perseids and the β-Taurids, that occur in June, post-perihelion.

Discovery

The German astronomer Eduard Heis [13] is credited with the discovery of the Taurids in 1867 after studying observational records for the period 1839–49, although Taurid fireballs seem to have been prolific in the eleventh century. Two years after his announcement,

---

[2] These are often referred to as the Atlantic Gap and the Pacific Gap because of the lack of observers.

T.W. Backhouse, G.L. Tupman and G. Zezioli independently discovered the two branches of the shower. The first detailed analysis had to wait until 1920, when William F. Denning published his findings and, in 1924, noted the variation in activity from the Southern Taurids (which he referred to as the λ-Taurids) [14, 15]. Denning would have been aware of the Swarm Years of 1910, 1917 and 1920.

The first radio studies of the Taurids were undertaken by Jodrell Bank between 1946 and 1950. Even though Jodrell had the best equipment in the world at that time, it could not resolve the two branches; that had to wait until 1952. A year earlier, Mary Almond used the Jodrell Bank data to compute the stream's daylight components and discovered that the β-Taurids in June have orbits very similar to the Northern Taurids [16]. In 1978, the Slovak astronomer Ľubor Kresák suggested that the massive explosion that had occurred at Tunguska, Russia, on 1908 June 30 may have been a fragment of Comet Encke from the β-Taurids [17]. That particular fragment exploded at an altitude of about 10 km, but the resulting pressure wave flattened about 2000 km^2 of dense Siberian forest. A study by Peter Brown, Valerie Marchenko, Danielle Moser, Robert Weryk and William Cooke in 2013 confirmed that Taurid meteoroids may be sufficiently large and robust to land as meteorites, provided the retardation point is below 35 kilometres and the velocity is less than 10 km/s [18].

In 1986, J. Štohl suggested that Comet Encke was once part of a much larger body, perhaps up to 30 km in diameter, and its fragmentation not only resulted in Encke but also gave rise to both the Helion and Antihelion meteor sources.

The Taurids were most popular among IMO contributors towards the end of the last century but have declined in popularity since then, perhaps as a result of more observers opting for automated camera monitoring (Fig. 6.21).

Origins

By the early 1930s, the long duration of the shower and slow radiant drift was interpreted by O.H.J. Knopf, and later by Cuno Hoffmeister, as being indicative of an interstellar stream [20, 21]. This view was dismissed by Fred Whipple in 1940, who showed not only that the stream was not interstellar, but that it in fact had a remarkably short orbital period and was associated with Encke's Comet. Whipple also suggested that a daylight shower would occur in June or July, post-perihelion, and that the meteoroid stream should also intersect Mercury, Venus and Mars, producing meteor showers on the two latter planets.

In 1984, Clube and Napier developed the *Taurid Complex Giant Comet Hypothesis* [22]. In this hypothesis, a giant comet of about 100 km in diameter fragmented 10,000 to 20,000 years ago. The result was a series of meteoroid streams, dust and some fairly large bodies, including Encke's Comet. The two researchers also confirmed Ľubor Kresák's suggestion that the object that caused the Tunguska event of 1908 June 30 was part of the Taurid Complex.

Comet 2P/Encke was actually discovered by the French astronomer Pierre Méchain in 1786 January 17 and independently by his compatriot Charles Messier (of *Messier Catalogue* fame). However, it is named after the German astronomer Johann Franz Encke who, in 1819, recognized it as a periodic comet. Only one other periodic comet was known at that time: 1P/Halley.

Table 6.7a: Southern Taurid Maxima 2020 - 2030

year	Maximum ($\lambda_\odot$ 223°)			Moon Age (d)
	month	day	hour (UT)	
2020	11	05	06:04	20
2021	11	05	12:11	1
2022	11	05	18:17	12
2023	11	06	00:31	22
2024	11	05	06:38	4
2025	11	05	12:42	15
2026	11	05	18:54	25
2027	11	06	01:02	8
2028	11	05	07:11	18
2029	11	05	13:22	28
2030	11	05	19:32	10

Table 6.7b: Northern Taurid Maxima 2020 - 2030

year	Maximum ($\lambda_\odot$ 230°)			Moon Age (d)
	month	day	hour (UT)	
2020	11	12	05:21	27
2021	11	12	11:26	8
2022	11	12	17:39	19
2023	11	12	23:45	28
2024	11	12	05:55	11
2025	11	12	12:03	22
2026	11	12	18:08	3
2027	11	13	00:21	15
2028	11	12	06:28	25
2029	11	12	12:36	6
2030	11	12	18:53	17

Table 6.8: Components of the Taurid Complex
Calendar dates are approximate. Use solar longitude $\lambda_\odot$

Shower	IAU Code	Start $\lambda_\odot$	Max $\lambda_\odot$	End $\lambda_\odot$	At Max. α	At Max. δ	V_g (km/s)	ZHR
Northern δ-Piscids	NPI	Sep 3 161°	Sep 17 175°	Sep 21 178°	$00^h\ 29^m$ 7.3°	+6.9°	65.7	<1
Southern δ-Piscids	SPI	Sep 8 166°	Sep 24 182°	Sep 21 178°	$01^h\ 17^m$ 19.2°	+1°	46.6	<1
Southern October δ-Arietids	SOA	Sep 22 179°	Oct 10 197°	Oct 27 214°	$02^h\ 10^m$ 32.6°	+9.6°	21	<1
Northern October δ-Arietids	NOA	Sep 22 179°	Oct 16 203°	Oct 22 209°	$02^h\ 27^m$ 36.7°	+19°	62	<1
Southern χ-Orionids	ORS	Nov 13 231°	Dec 2 250°	Dec 21 269°	$05^h\ 29^m$ 82°	+18°	26	1
Northern χ-Orionids	ORN	Nov 23 241°	Dec 12 260°	Dec 18 266°	$05^h\ 52^m$ 88.1°	+25.7°	29	<1
Daytime ζ-Perseids	ZPE	May 20 59°	Jun 9/14 78°/83°	Jul 5 103°	$04^h\ 18^m$ 64.5°	+27.5°	11	
Daytime β-Taurids	BTA	Jun 5 75°	Jun 28 96.7°	Jul 17 115°	$05^h\ 42^m$ 85.5°	+20.5°	70	

The Taurids

Table 6.9: Other Active Meteor Showers During the Taurids
Calendar dates are approximate. Use solar longitude $\lambda_\odot$

Shower	IAU Code	Start $\lambda_\odot$	Max $\lambda_\odot$	End $\lambda_\odot$	At Max. α	At Max. δ	V_g (km/s)	ZHR
ν-Eridanids	NUE	Aug 24 151°	Sep 24 181°	Nov 16	$08^h 28^m$ 127°	+16	67	<1
October Capricornids[1]	OCC	Sep 20 177°	Oct 2 189°	Oct 19 206°	$20^h 28^m$ 307°	−09°	11	Var.
ε-Geminids	EGE	Sep 30 187°	Oct 11 198°	Oct 25 212°	$05^h 57^m$ 89°	+28°	70	1
Orionids	ORI	Oct 3 190°	Oct 22 209°	Nov 7 220°	$06^h 24^m$ 96°	+15.7	65.7	26
October Camelopardalids	OCT	Oct 4 191°	Oct 5 192.6°	Oct 9 196°	$10^h 56^m$ 164°	+79	46.6	5
October Draconids[2]	DRA	Oct 6 193°	Oct 8 195.4°	Oct 10 197°	$17^h 28^m$ 262°	+54	21	Var.
Leonis Minorids	LMI	Oct 12 199°	Oct 24 211°	Nov 5 223°	$10^h 48^m$ 170°	+37°	62	2
October Ursae Majorids	OCU	Oct 14 201°	Oct 15 202°	Oct 19 206°	$09^h 43^m$ 145.8°	+63.8°	38.8	<1
o-Eridanids	OER	Oct 16 203°	Nov 5 223°	Nov 24 242°	$03^h 00^m$ 45°	−04°	29	1
χ-Taurids	CTA	Oct 20 207°	Nov 4 222°	Nov 17 235°	$03^h 32^m$ 53°	+25°	41	1
ξ-Draconids	XDR	Oct 21 208°	Oct 23 210°	Oct 24 211°	$11^h 21^m$ 170.3°	+73°	35.8	<1
Andromedids[3]	AND	Oct 26 213°	Nov 6 224°	Nov 17 235°	$00^h 38^m$ 10°	+24°	19	<1
λ-Ursae Majorids	LUM	Oct 27 214°	Oct 28 215°	Oct 29 216°	$10^h 24^m$ 156°	+49°	61	<1
Southern λ-Draconids	SLD	Nov 1 219°	Nov 3 221°	Nov 4 222°	$10^h 48^m$ 162°	+68°	49	<1
κ-Ursae Majorids	KUM	Nov 3 221°	Nov 7 225°	Nov 10 228°	$09^h 49^m$ 147°	+45°	66	<1
Leonids	LEO	Nov 6 224°	Nov 18 236°	Nov 30 248°	$10^h 17^m$ 154°	+21.4	69.7	15
November Orionids	NOO	Nov 7 225°	Nov 28 246°	Dec 17 265°	$06^h 04^m$ 91°	+16	43	3
ρ-Puppids	RPU	Nov 10 228°	Nov 14 232°	Nov 20 238°	$08^h 23^m$ 126°	−25°	58	<1
σ-Hydrids	HYD	Dec 3 251°	Dec 9 257°	Dec 21 269°	$08^h 20^m$ 125°	+2°	62	7
α-Monocerotids[4]	AMO	Nov 15 233°	Nov 21 239.32°	Nov 25 243°	$07^h 48^m$ 117°	+0.9°	63	Var.
November θ-Aurigids	THA	Nov 17 235°	Nov 26 244°	Dec 1 249°	$06^h 13^m$ 93°	+35	33	<1
December Monocerotids	MON	Nov 27 245°	Dec 9 257°	Dec 20 268°	$06^h 40^m$ 100°	+8°	41	3
December α-Draconids	DAD	Nov 30 248°	Dec 8 256°	Dec 15 263°	$13^h 34^m$ 204°	+58	44	1

Shower	Code	Begins	Maximum	Ends	RA	Dec	V∞	ZHR
Puppids-Velids[5]	PUP	Dec 1 249°	Dec 7 255°	Dec 26 274°	08^h 12^m 123°	-45	40	10
December ϕ-Cassiopeiids[6]	DPC	Dec 1 249°	Dec 6 254°	Dec 8 256°	01^h 36^m 24°	+50	16	Var.
December κ-Draconids	DKD	Dec 2 250°	Dec 3 251°	Dec 7 255°	12^h 29^m 187°	+70	41	2
ψ-Ursae Majorids	PSU	Dec 2 250°	Dec 5 253°	Dec 10 258°	11^h 16^m 169°	+42	62	1
Phoenicids[7]	PHO	Dec 6 254°	Dec 7 255°	Dec 8 256°	01^h 02^m 16°	-45	12	<1

[1] May be the remnants of the lost comet D/1978 R1 Haneda-Campos.
[2] Previously called the Draconids (hence the IAU Code: DRA) and also known as the Giacobinids after the stream's parent body, Comet 21 P/Giacobini-Zinner.
[3] Referred to in older texts as the Bielids after the parent body, Comet 3D/Biela.
[4] Outbursts in 1925, 1935, 1985, 1995, possibly 2016 and 2019.
[5] Several radiants active around this central point.
[6] This is the original 'Andromedids' radiant prior to the breakup of the parent body, Comet 3D/Biela, in the 1840s.
[7] ZHR ~100 in 1956.

	Table 6.10: Taurid Timeline
11th Century	First records of Taurid fireballs uncovered by I.S. Astapovich and A.K. Terentjeva in 1968.
1867	Eduard Heis discovers the Taurids based on observations made between 1839-49.
1869	T.W. Backhouse, G.L. Tupman and G. Zezioli independently discover two branches.
1918	A. King begins to recognize the long duration of the Earth's encounter with the stream starting in August and ending in November.
1920	In the first detailed analysis of the Taurids, W.F. Denning comments on the large number of fireballs.
1924	W.F. Denning notes how the Southern Taurids vary from one year to the next.
1931	O.H.J. Knopf suggests the long duration of the stream indicates an interstellar origin.
1940	F.L. Whipple discovers that the stream has a short period and suggests an association with Comet 2P/Encke.
1946-50	First radio-echo studies of the Taurids by Jodrell Bank Experimental Station. The equipment does not resolve the two branches.
1950	F.W. Wright and F.L. Whipple associate the Northern and Southern October δ-Arietids with the Taurids.
1950	F.L. Whipple and S.E. Hamid suggest that Encke's Comet was once part of a much larger body that had fragmented.
1951	M. Almond computes orbits for daylight showers discovered by Jodrell Bank and finds that the β-Taurids in June have orbits similar to the Northern Taurids.
1952	Jodrell Bank resolves the two branches.
1972	K.B. Hindley discovers variability within the Taurid Complex, with the Northern branch the most active until $\lambda_\odot 215°$, when the Southern branch suddenly becomes active, so the bulk of the activity comes from the Southern component. After $\lambda_\odot 230°$, the Northern branch becomes more active.
1978	L. Kresák speculates that the Tunguska Event of 1908 June 30, which destroyed 2,000 km² of Siberian forest, was a fragment of 2P/Encke from the β-Taurids.
1986	J. Štohl suggests the breakup of the Encke progenitor – which may have been up to 30 km in diameter – produced the sporadic background that gives rise to the Helion and Antihelion sources.
1993	D.J. Asher and S.V.M. Clube suggest that the stream is in a 7:2 resonance with Jupiter, which produces concentrations of material that encounter the Earth at 3-, 4- and 7- year intervals.
2013	P.G. Brown, V. Marchenko, D.E. Moser, R. Weryk and W. Cooke suggest that Taurid meteoroids may be sufficiently large and robust to land as meteorites, provided that the retardation point is below 35 kilometres and the velocity is less than 10 km/s.

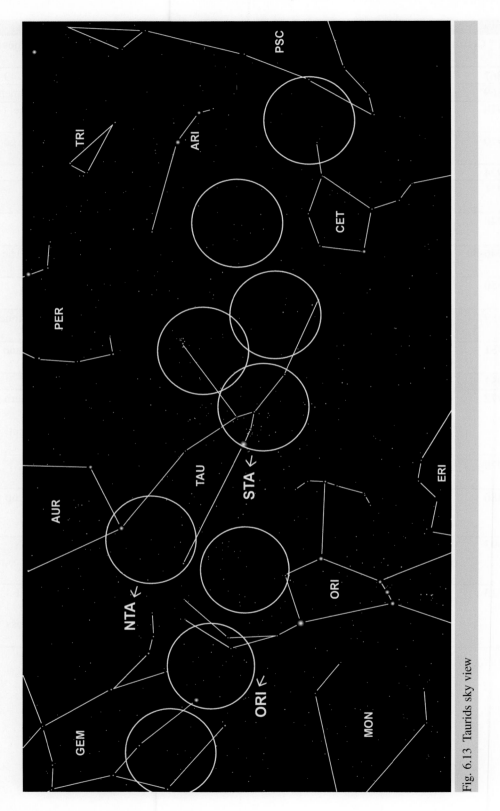

Fig. 6.13 Taurids sky view

The Taurids

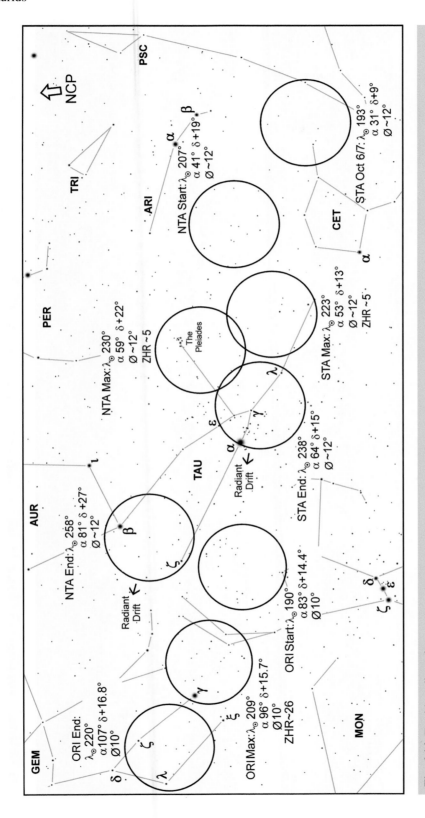

Fig. 6.14 Taurids radiant chart

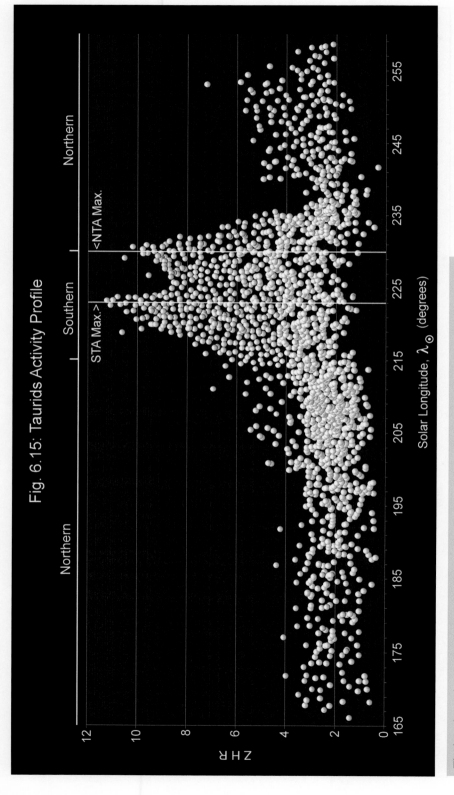

Fig. 6.15: Taurids Activity Profile

The bar above the graph, marked Northern and Southern, indicates the most active component.

The Taurids

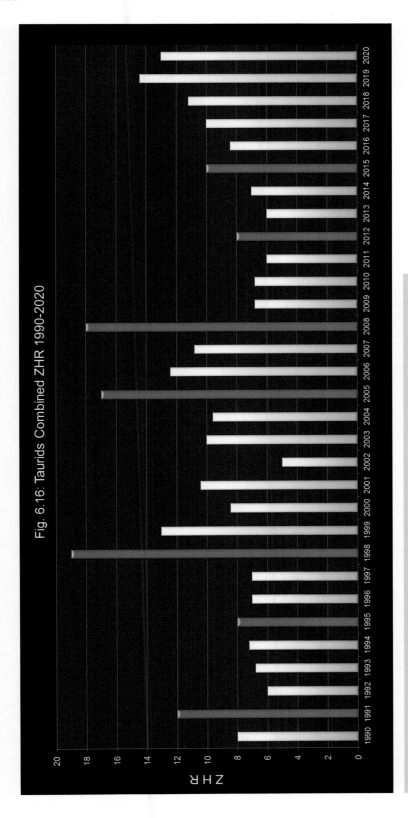

Fig. 6.16: Taurids Combined ZHR 1990-2020

The darker bars represent Swarm Years

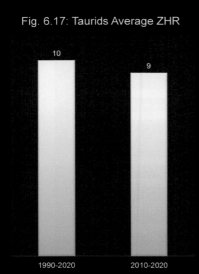

Fig. 6.17: Taurids Average ZHR

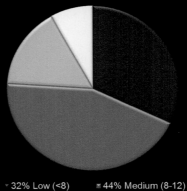

Fig. 6.18: Taurids 25-year Activity Levels

- 32% Low (<8)
- 44% Medium (8-12)
- 16% High (13-17)
- 8% Very High (>17)

The Taurids ZHR has averaged 9 during the period 2010–2020, slightly down on the period 1990–2020

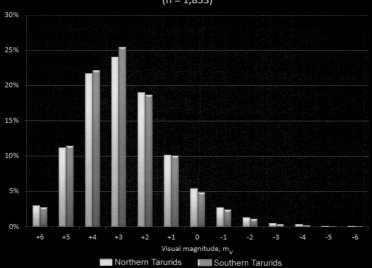

Fig. 6.19: Taurids Magnitude Distribution in non-Swarm Years (n = 1,853)

The Taurids

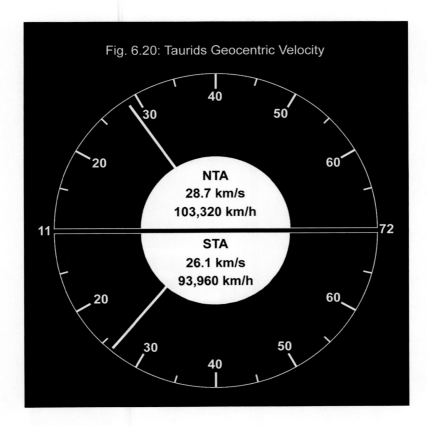

210 6 The Major Meteor Showers: October to December

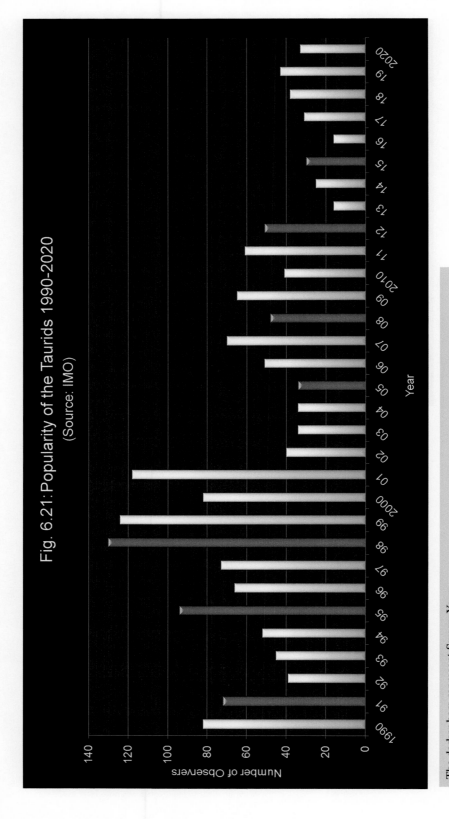

Fig. 6.21: Popularity of the Taurids 1990-2020 (Source: IMO)

The darker bars represent Swarm Years.

The Taurids

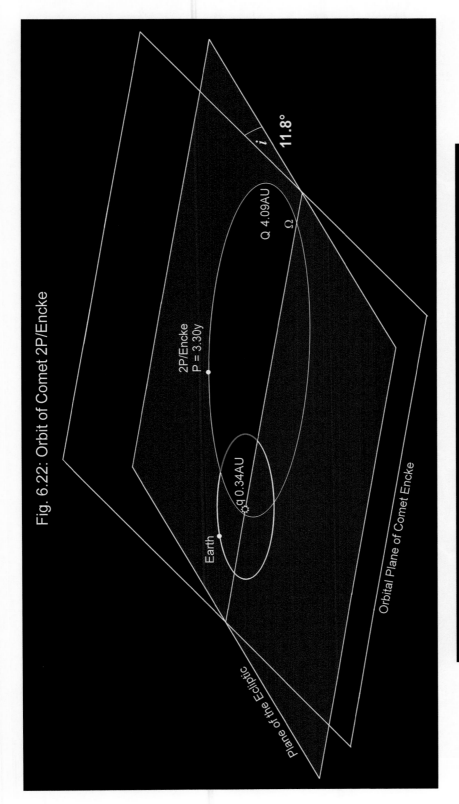

Fig. 6.22: Orbit of Comet 2P/Encke

Table 6.11: Orbital Data for the Taurid Parent Body*

		Comet 2P/Encke				
q	e	a	ω	Ω	i	P	Q	b
AU		AU	°	°	°	y	AU	AU
0.3359	0.8483	2.2152	186.5459	334.5679	11.7816	3.30	4.0944	1.1730

*For Epoch 2015-07-07.0

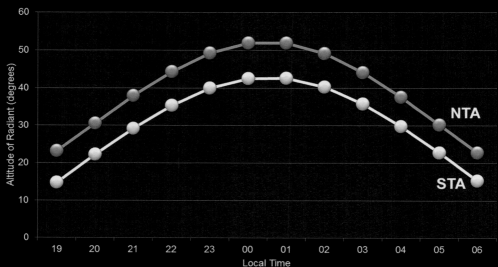

Lerwick, Scotland; Oslo, Norway; Stockholm, Sweden; Helsinki, Finland; St. Petersburg, Russia; Nanontalik, Greenland.

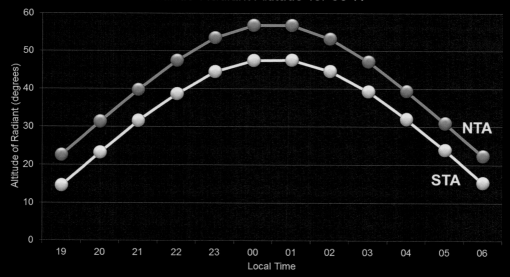

Newcastle upon Tyne, England; Copenhagen Denmark; Malmö, Sweden; Vilnius, Lithuania; Moscow, Russia; Chelyabinsk, Russia; Edmonton, Canada.

Adjust local time for daylight saving if applicable

The Taurids

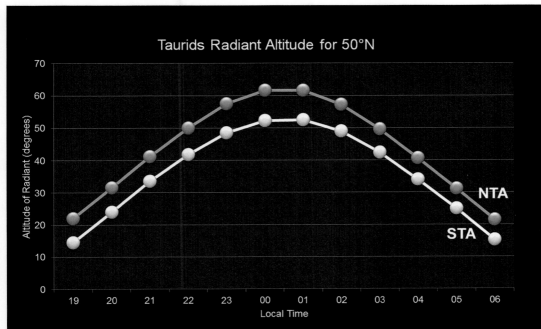

London, England; Frankfurt, Germany; Bern, Switzerland; Kiev, Ukraine; Qaraghandy, Kazakhstan, Qiqihar, China, Vancouver and Winnipeg, Canada.

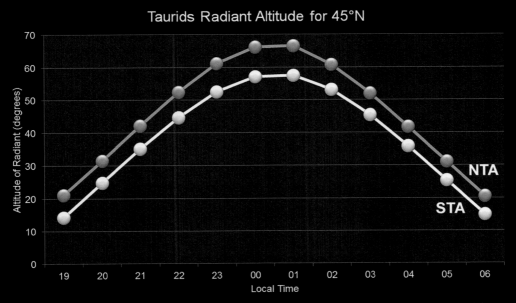

Bordeaux, France; Zagreb, Croatia; Ulaanbaatar, Mongolia; Harbin, China; Portland and Minneapolis, USA; Toronto and Montreal, Canada.

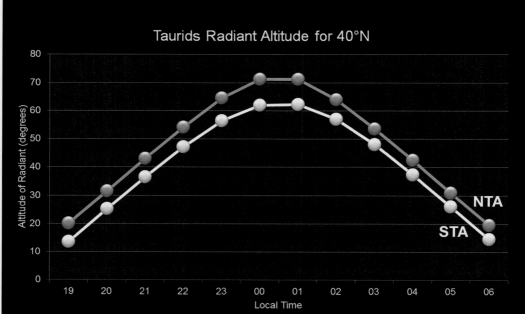

Porto, Portugal; Rome, Italy; Istanbul, Turkey; Toshkent, Uzbekistan; Beijing, China; Sapporo, Japan; Sacramento, Salt Lake City, Denver, Kansas City, Indianapolis, Pittsburgh and Philadelphia, USA.

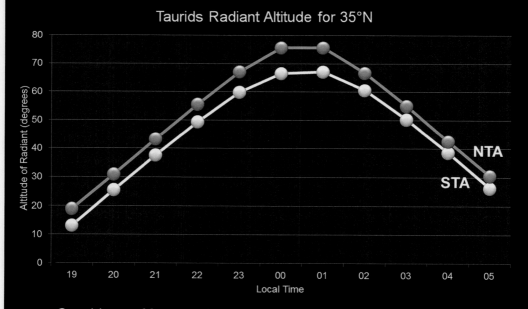

Casablanca, Morocco; Malaga, Spain; Athens, Greece; Tehran, Iran; Jinan, China; Soul, South Korea; Los Angeles, Memphis and Charlotte, USA.

The Taurids

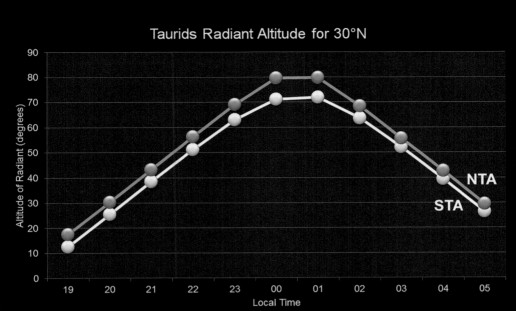

Madeira, Portugal; Tripoli, Libya; Cairo, Egypt; Basra, Iraq; Multan, Pakistan; Lahore, India; Shanghai, China; Tokyo, Japan; San Diego, Austin and Jacksonville, USA.

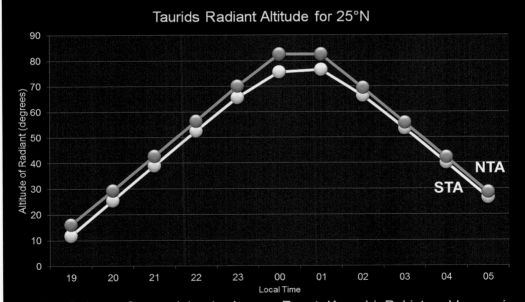

Las Palmas, Canary Islands; Aswan, Egypt; Karachi, Pakistan; Varanasi, India; Kunming, China; T'aipei, Taiwan; Torreon, Mexico; Miami, USA.

216   6 The Major Meteor Showers: October to December

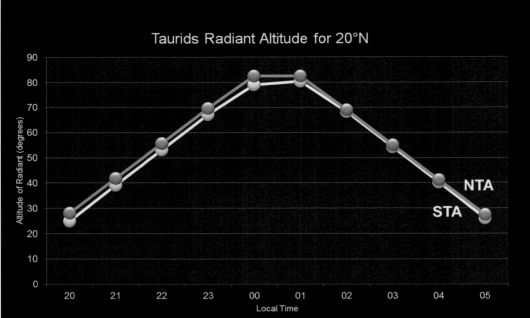

Jeddah, Saudi Arabia; Mumbai, India; Haikou, China;
Campeche, Mexico; Santiago de Cuba, Cuba.

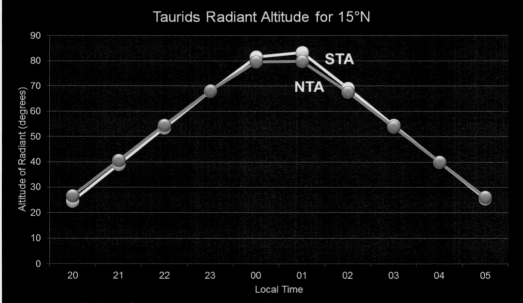

Dakar, Senegal; Khartoum, Sudan; Hyderabad, India; Yangon,
Myanmar; Manila, Philippines; Guatemala City, Guatemala

The Taurids

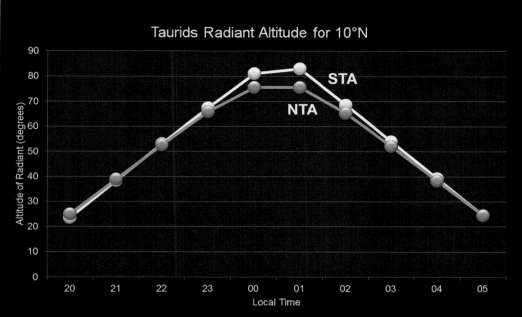

Conakry, Guinea; Kaduna, Nigeria; Addis Ababa, Ethiopia; Ho Chi Minh City, Vietnam; Manila, Philippines; San Jose, Costa Rica; Caracas, Venezuela.

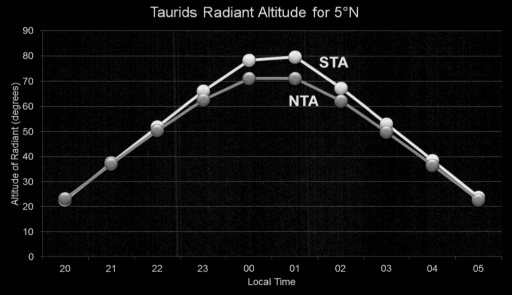

Accra, Ghana; Colombo, Sri Lanka; George Town, Malaysia; Bogota, Colombia; Cayenne, French Guiana.

218   6   The Major Meteor Showers: October to December

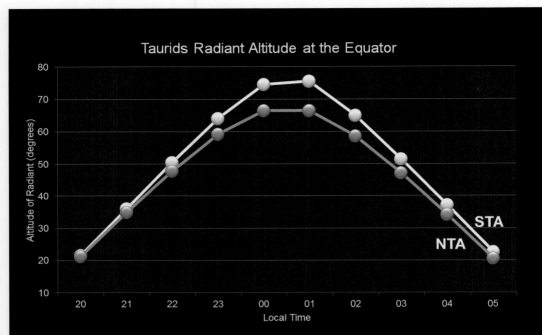

Libreville, Gabon; Kampala, Uganda; Singapore; Quito, Ecuuador; Belem, Brazil.

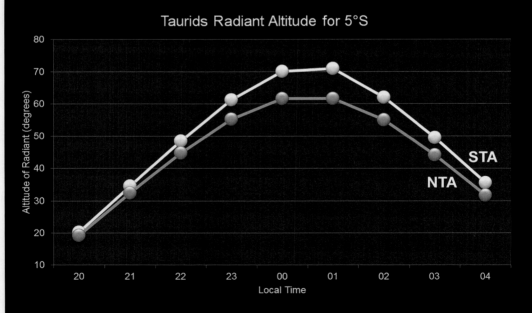

Kinshasa, D.R.Congo; Jakarta, Indonesia; Piura, Peru; Teresina, Brazil.

The Taurids

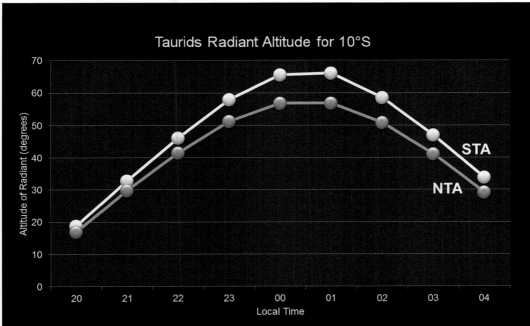

Luanda, Angola; Kolwetzi, D.R.Congo; Kupang, Indonesia, Maceio, Brazil.

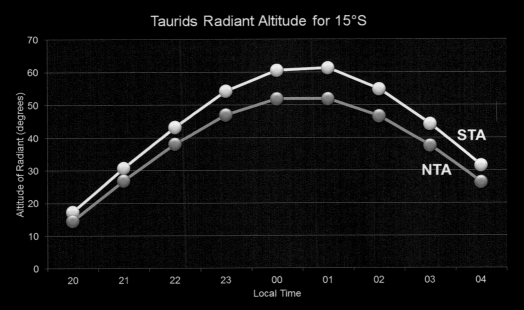

Lusaka, Zambia; Blantyre, Malawi; Coronation Island, Western Australia.

220 6 The Major Meteor Showers: October to December

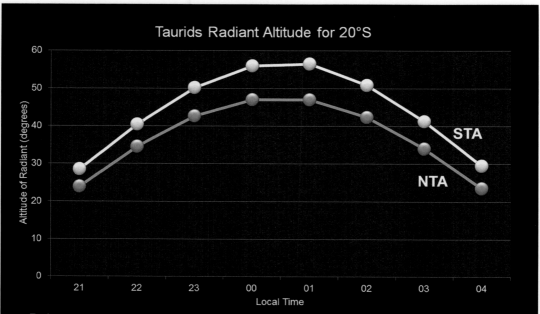

Bulawayo, Zimbabwe; Beira, Mozambique; Port Hedland, Western Australia; Bowen, Queensland, Australia; Iquique, Chile; Cariacica, Brazil.

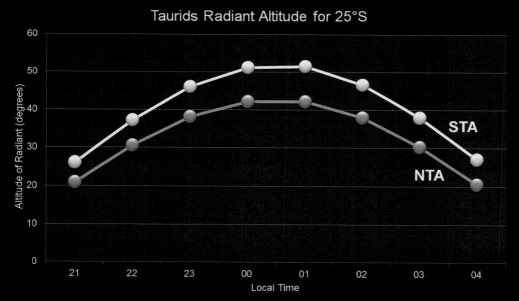

Gibeon, Namibia; Pretoria, South Africa; Carnarvon, Western Australia; Bundaberg Central, Queensland, Australia; Taltal, Chile; Asuncion, Paraguay.

The Taurids

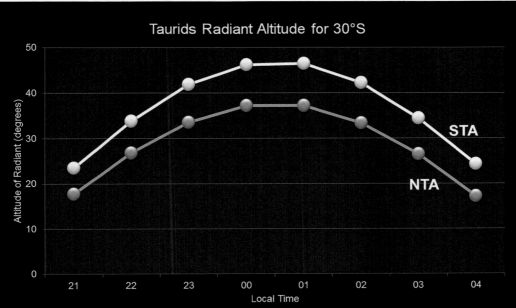

Durban, South Africa; Perth, Western Australia; Coffs Harbour, NSW, Australia; Coquimbo, Chile; Poto Alegre, Brazil.

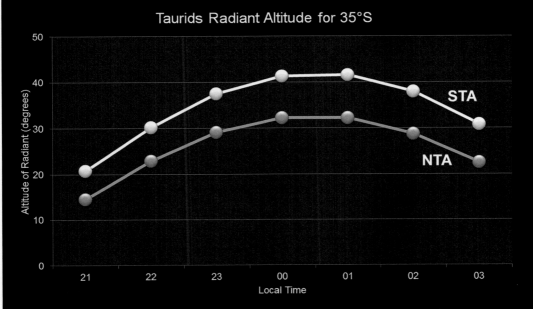

Curico, Chile; Buenos Aires, Argentina; Montevideo, Uruguay; Mangonui, New Zealand.

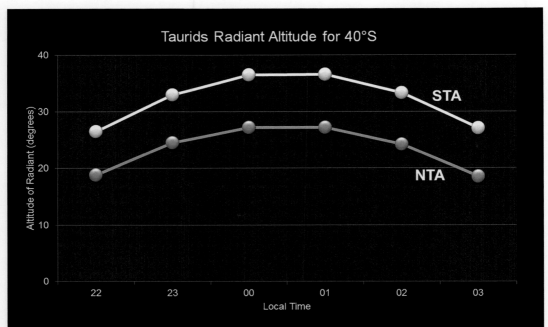

Valdivia, Chile; El Cuy, Argentina; Wanganui, New Zealand.

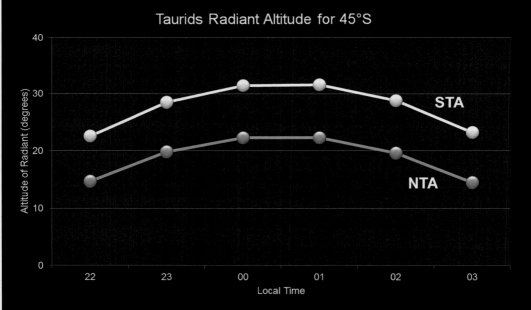

Camarones, Argentina; Oamaru, New Zealand.

The Taurids

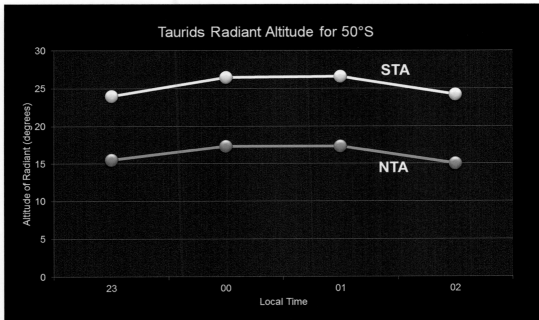

Puerto Santa Cruz, Argentina; Stanley, Falkland Islands.

The Leonids

Normally a quiet, docile cat, every 30-odd years this lion roars!

Observing Notes

Leonid storms are legendary. Occurring every 33 years or so, the 1999 event had an estimated ZHR of nearly 4000, while the previous storm in 1966 took place over the western United States and produced a ZHR believed to be a staggering 150,000! Some researchers have since suggested that it may have been less—a mere 70,000—but from an observer's point of view, once meteors start falling like rain, is it really possible to achieve an accurate estimate? Just enjoy the show!

But let's not get carried away. Leonid storms are a once or twice in a lifetime experience, and the next isn't due until 2034 when, if predictions are correct, the ZHR should be in the order of 5000 [23]. Unfortunately, 2034 is outside the lifespan of this *Atlas* so we won't dwell on it too much, other than to say, put the date in your diary.

During a 'normal' year, the Leonids are a much more modest shower. The ZHR is often quoted as being 10–15, but during 2010–2020, it has averaged a slightly higher 17, so it is worth making the effort to watch.

Activity usually starts around November 6 at $\lambda_\odot$ 224°, just a day after the Southern Taurids peak. Rates start to pick up about nine days later and reach maximum on November 18 at $\lambda_\odot$ 236°, after which they fall off just as rapidly as they arrived. In all, detectable activity lasts just under 3 weeks (Fig. 6.27). A few years either side of a storm, you might find rates somewhat enhanced, with a ZHR in the range of 30–60.

The earliest meteors radiate from close to ε Leonis. The radiant then move towards ζ Leonis, appearing to radiate from the lion's neck at maximum (Fig. 6.26). The radiant itself is usually quoted as being 5° in diameter, although the size has ranged from just under 4° to about 8°. There are a number of other showers active during the Leonids, none of which are likely to contaminate the shower, but do keep a look out for the α-Monocerotids, which can suddenly burst into life (Table 6.14).

Perhaps the most impressive aspect of the Leonids is just how swift they are. With a geocentric velocity of 69.7 km/s—just a tad off the maximum attainable of 72 km/s—they can start to ablate at an altitude of up to 200 km and can plunge more than 100 km before fizzling out. As a bonus, they tend to be fairly bright with an average magnitude of $\overline{m}_v$ +2.1 (Fig. 6.30).

The bad news is reserved for observers in the Southern Hemisphere. The Leonids are best observed after midnight. South of the Equator, the shower is only observable for a couple of hours from about 2 am LT, and anyone farther south than 25°S has little to no chance of seeing the display.

The Leonids

Table 6.12a: Leonid Activity Details

		Start	Maximum	End
Dates (approx.)		Nov 6	Nov 18	Nov 30
Solar longitude:	$\lambda_\odot$	224°	236°	248°
Right Ascension at max:	α		$10^h\ 16^m$ (154°)	
Declination at max:	δ		+21.4°	
Time of transit :			6.6^h L.T.	
ZHR (2010-2020 average):			17	
Radiant diameter:	$\varnothing$	5°	5°	5°

Table 6.12b: Leonid Additional Data

IAU Abbreviation: LEO		IAU Code: 013
Mean daily motion of radiant:	Δ	$\Delta\alpha$ = +0.59° $\Delta\delta$ = +0.29°
Geocentric velocity:	V_g	69.7 km/s (250,920 km/h)
Population index:	r	2.5
Mean magnitude:	$\bar{m}_v$	+2.1
Parent body: Comet 1P/Halley		
Best visibility: Post-midnight in the Northern Hemisphere		
The radiant is circumpolar from 79°N, i.e. north of Spitsbergen, Norway, Novosibirskiye Ostrova, Russia, Grise Fiord, Ellesmere Island, Canada and Qaanaaq (Thule), Greenland.		

Table 6.13: Leonid Maxima 2020-2030

Maximum ($\lambda_\odot$ 236°)				Moon
year	month	day	hour (UT)	Age (d)
2020	11	18	04:15	3
2021	11	18	10:26	14
2022	11	18	16:37	25
2023	11	18	22:40	5
2024	11	18	04:55	18
2025	11	18	10:59	28
2026	11	18	17:06	9
2027	11	18	23:21	19
2028	11	18	05:22	2
2029	11	18	11:34	13
2030	11	18	17:52	23

Table 6.14: Other Active Meteor Showers During the Leonids
Calendar dates are approximate. Use solar longitude $\lambda_\odot$

Shower	IAU Code	Start $\lambda_\odot$	Max $\lambda_\odot$	End $\lambda_\odot$	At Max. α	At Max. δ	V_g (km/s)	ZHR
ν-Eridanids	NUE	Aug 24 151°	Sep 24 181°	Nov 16 234°	$08^h\ 28^m$ 127°	+16	67	<1
Southern Taurids	STA	Sep 10 168°	Nov 5 223°	Nov 20 238°	$03^h\ 33^m$ 53°	+12.9°	26.1	5
o-Eridanids	OER	Oct 16 203°	Nov 5 223°	Nov 24 242°	$03^h\ 00^m$ 45°	-04°	29	1
Northern Taurids	NTA	Oct 20 207°	Nov 12 230°	Dec 10 258°	$03^h\ 57^m$ 59°	+22.3°	28.7	5
χ-Taurids	CTA	Oct 20 207°	Nov 4 222°	Nov 17 235°	$03^h\ 32^m$ 53°	+25°	41	1
Andromedids	AND	Oct 26 213°	Nov 6 224°	Nov 17 235°	$00^h\ 38^m$ 10°	+24°	19	<1
κ-Ursae Majorids	KUM	Nov 3 221°	Nov 7 225°	Nov 10 228°	$09^h\ 49^m$ 147°	+45°	66	<1
November Orionids	NOO	Nov 7 225°	Nov 29 247°	Dec 17 265°	$04^h\ 39^m$ 70°	+16	43	<1
ρ-Puppids	RPU	Nov 10 228°	Nov 14 232°	Nov 20 238°	$08^h\ 23^m$ 126°	-25°	58	<1
σ-Hydrids	HYD	Dec 3 251°	Dec 9 257°	Dec 21 269°	$08^h\ 07^m$ 121°	+4°	62	7
α-Monocerotids	AMO	Nov 15 233°	Nov 21 239°	Nov 25 243°	$07^h\ 48^m$ 117°	+0.9°	63	Var.
November θ-Aurigids	THA	Nov 17 235°	Nov 26 244°	Dec 1 249°	$06^h\ 13^m$ 93°	+35	33	<1

Discovery

The Leonids were first recorded in CE 855 by the Islamic scholar Imam Ibn al-Jawzī [24], who noted that in Yemen, "…many stars fell and they were many, uncounted. This stayed the whole night, from the late evening until the morning the stars were disorderly and arranged like locusts." Further events were recorded in mainly Chinese, Japanese and Korean annals throughout the 10th to the 17th centuries. The first scientists to observe a Leonid storm were the Prussian naturalist Alexander von Humboldt and French botanist Aimé Bonpland from Cumaná, Venezuela in 1799 [25]. That particular storm was seen from northern England to the Bahamas and south at least as far as northern Brazil. Despite its sudden and spectacular appearance, it failed to attract the attention of most scientists.

A further 34 years were to pass before the scientific community took note of the shower, in particular the Connecticut physicist and astronomer Denison Olmsted. The Great Leonid Storm of 1833, as it became known, showed for the first time that shower meteors radiated from a particular area of the sky. Olmsted set himself the task of estimating the position of the newfound radiant, which he located at α 150° ($10^h\ 00^m$), δ +20° (the currently accepted values are α 154° ($10^h\ 16^m$), δ +21.4°) [26].

It was left to the German physician and amateur astronomer Heinrich Olbers[3] to point out in 1837 that there is a periodicity to the Leonids of 33–34 years [27]. He predicted the next storm to be 1867. In 1864, Hubert Newton disagreed. He calculated the Leonids' period to be 33.25 years and predicted the storm would occur a year earlier. Newton turned out to be correct in part, and rates were variously estimated to be between 2000 and 5000 meteors per hour. However, 1867 also produced a storm, believed by some to be more intense than the 1866 display. Rates remained high in 1868—up to 1000 meteors per hour—and 1869, when the rate was about 200 meteors/h.

Rates fell to about a dozen meteors per hour in subsequent years, though several scientists predicted a storm in 1899. Things look promising in 1898, when a rate of about 100 meteors/h was achieved, but the following year saw, at best, just 40 meteors/h. It was subsequently realized that a close approach to Saturn in 1870 and Jupiter in 1898 had perturbed the meteoroid stream quite significantly so that it missed the Earth by more than twice the 1866 distance. Astronomers had to wait until 1901 to see another strong display, with maximum activity rising to about 400 meteors/h.

1932 was slated as being the next possible storm, but UK observers saw 'only' 240 meteors/h, and another cycle had to pass before a storm appeared—but what a storm! 1965 saw promising rates of 120 meteors/h, but what was more impressive was how bright the Leonids were: the average magnitude was $\overline{m}_v$ −3. Then 1966 arrived and with it, the strongest meteor storm ever witnessed. It is difficult to truly estimate the ZHR of such events, but as mentioned, observers in the western US reckoned that activity reached 150,000 meteors/h [28].

As 1999 approached, there was a flurry of predictions as to how active the Leonids would be. Significantly enhanced activity was predicted to occur between 1998 and 2002. Storms actually occurred in 1999 (ZHR 3700), 2001 (ZHR 4500) and 2002 (ZHR 2500), with the year 2000 producing a ZHR of 480.

Origins

Following the 1866 storm, one of the giants of French science, Urbain Le Verrier, calculated the orbit of the Leonids and pointed out that material was not spread evenly around the orbit but was concentrated in certain segments of the meteoroid stream [29]. At about the same time, Theodor von Oppolzer calculated the orbit of a comet that had been discovered in 1865–66 by Ernst Tempel in Marseilles and Horace Tuttle at Harvard. He noted that Comet Tempel-Tuttle 1866 I had an orbital period of 33.17 years. This prompted several astronomers to point to a possible link with the Leonids. The comet now has the IAU designation 55P/Tempel-Tuttle (Fig. 6.33 and Table 6.16).

---

[3] Better known for *Olbers' Paradox*: if there are stars in every direction, why isn't the sky uniformly bright? Olbers also discovered the asteroids (2) Pallas and (4) Vesta.

Table 6.15: Leonid Timeline

CE 855	Leonids first recorded appearance by Imam Ibn al-Jawzī.
CE 902	Leonid storm observed from Egypt and Italy.
10th-17th centuries	Leonids recorded in 10th, 11th, 13th, 14th, 16th and 17th centuries.
1799	Alexander von Humboldt and Aimé Bonpland witness the Leonid storm from Cumaná, Venezuela.
1832	High rates observed in Europe and the Middle East.
1833 Nov 13	The Great Leonid Storm indicated that the meteors appeared to radiate from a particular part of the sky.
1834	Denison Olmsted notes that the shower's storm is of short duration. He calculates the radiant to be $\alpha\ 150°, \delta\ +20°$.
1837	Heinrich Olbers suggests a period of 33-34 years and predicts the next storm to occur in 1867.
1864	Hubert Newton calculates the period of the Leonids to be 33.25 years and predicts the next storm to be 1866 November 13-14.
1865 Dec 19	Ernst Tempel in Marseilles discovers a m_v+6 comet near β UMa.
1866 Jan 6	Horace Tuttle, Harvard, independently discovers Tempel's comet. They share the discovery, and comet 1866 I, Tempel-Tuttle, enters the history books.
1866	Newton's predicted storm proves to be accurate, with estimated ZHRs in the region of 2,000 to 5,000.
1867	Le Verrier calculates the Leonid orbit and points out that the meteoroids are not spread evenly around the orbit, but rather, a clump of material exists that produces the periodic storms.
1867	Storm occurs, which may have been stronger than in 1866.
1867	Theodor von Oppolzer calculates the orbital period of Comet Tempel-Tuttle to be 33.17 years. Various astronomers realize the similarity to the Leonid orbit.
1868	Rates of up to 1,000 meteors/h.
1869 Nov 14	Rates about 200 per hour.
1870-1897	Rates of 10-15 per hour.
1898	Rates peak at about 100 meteors/h over the US.
1899	The predicted storm does not appear due to the stream's close encounter with Saturn in 1870 and Jupiter in 1898, which increased the Earth-stream distance to twice what it was in 1866. Rates are just 40 meteors/h.
1901 Nov 13-14	Rates of about 400 meteors/h recorded at Carlton College, Minnesota.
1902	Few Leonids recorded due to moonlight.
1903	Rates of 200 meteor/h recorded over Ireland.
1932	Rates of 240 meteors/h recorded over the UK.
1965	ZHR of 120 with an average magnitude of m_v -3.
1966 Nov 17	Storm with an estimated ZHR of 150,000 reported over the western United States.
1967-1969	ZHR 100-150
1971	ZHR 170
1998	ZHR 350
1999	ZHR 3,750
2000	ZHR 190
2001	ZHR 1,500
2002	ZHR 3,140
2003	ZHR 116

Fig. 6.24 The Leonids, 1868
Étienne Trouvelot's somewhat fanciful illustration of the Leonid storm of 1868. The meteor in the centre of the plate is particularly impressive!

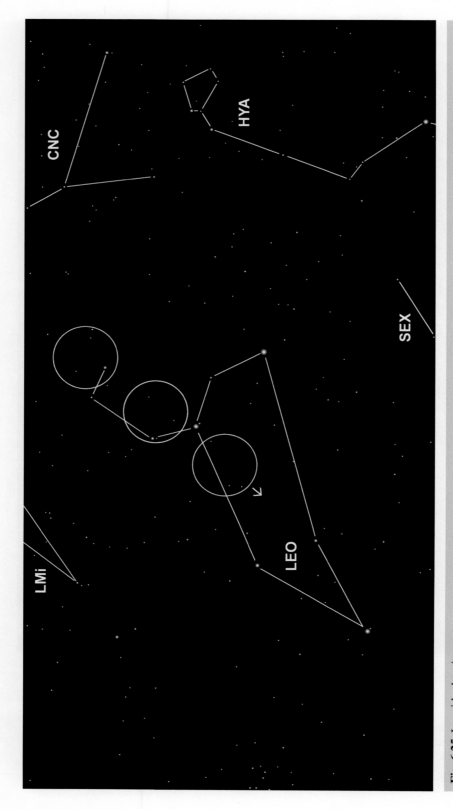

Fig. 6.25 Leonids sky view

The Leonids

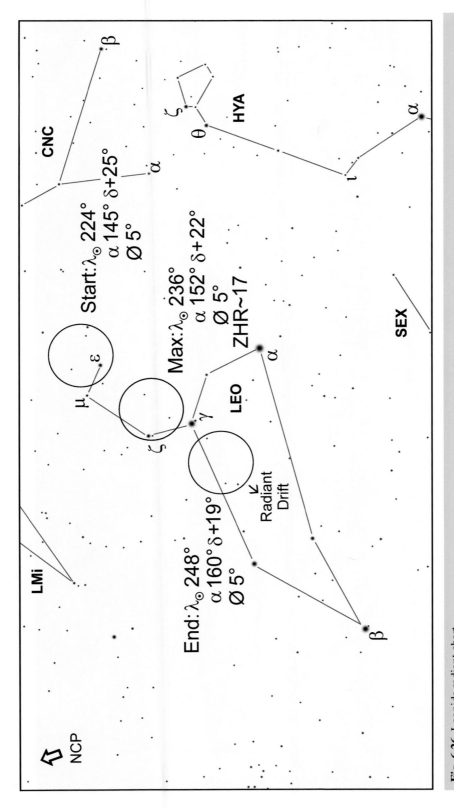

Fig. 6.26 Leonids radiant chart

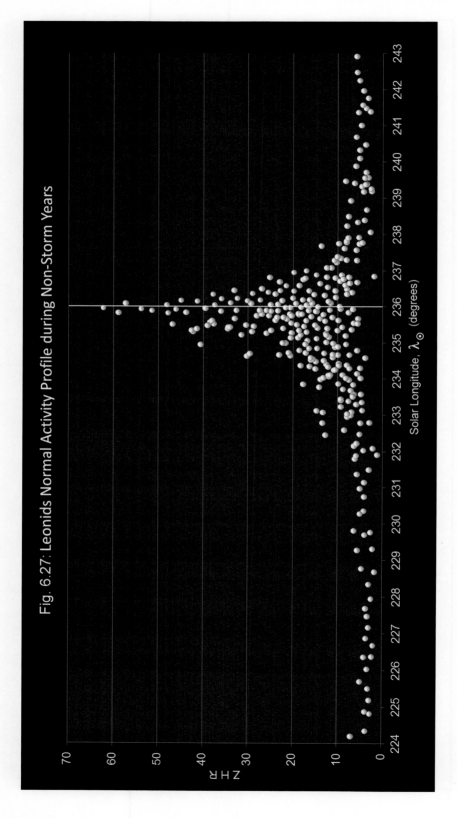

Fig. 6.27: Leonids Normal Activity Profile during Non-Storm Years

The Leonids

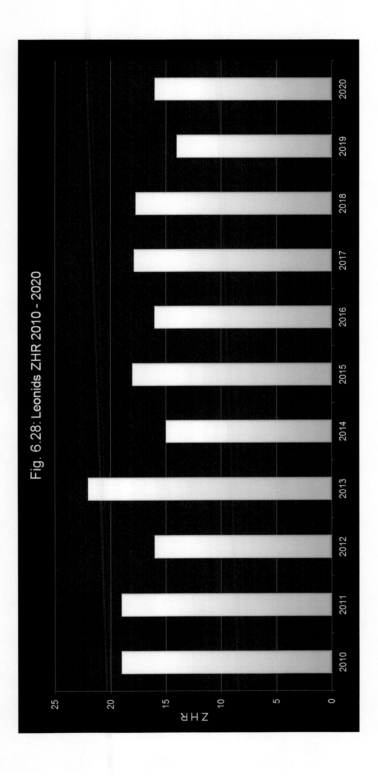

Fig. 6.28: Leonids ZHR 2010 - 2020

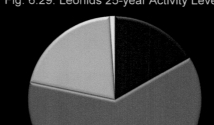

Fig. 6.29: Leonids 25-year Activity Levels

- 16% Low (<15)
- 63% Medium (15-18)
- 20% High (18-25)
- 1% Very High (>25)

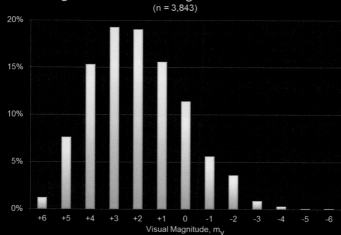

Fig. 6.30 Leonids Magnitude Distribution (n = 3,843)

Fig. 6.31: Leonids Geocentric Velocity

69.7 km/s
250,920 km/h

The Leonids

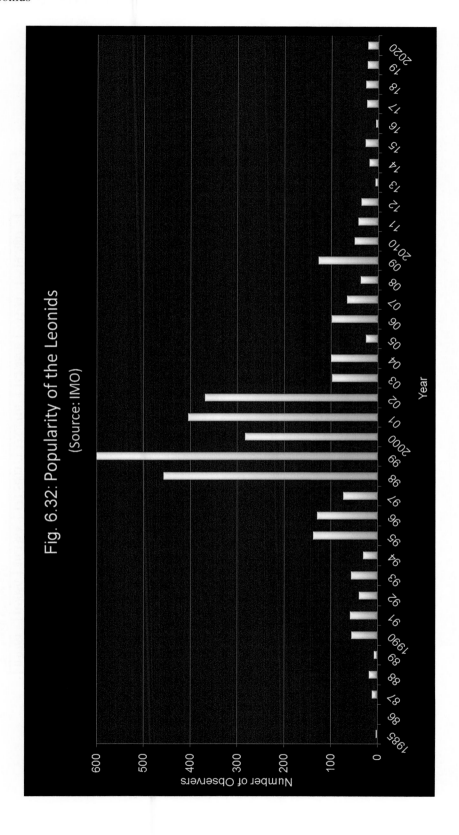

Fig. 6.32: Popularity of the Leonids
(Source: IMO)

236   6   The Major Meteor Showers: October to December

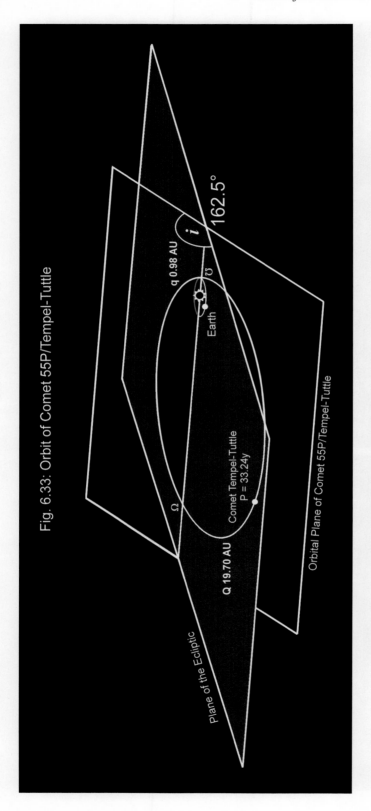

Fig. 6.33: Orbit of Comet 55P/Tempel-Tuttle

Table 6.16: Orbital Data for the Leonid Parent Body*

Comet 55P/Tempel-Tuttle

q AU	e	a AU	ω °	Ω °	i °	P y	Q AU	b AU
0.9764	0.9056	10.3383	172.5003	235.2710	162.4866	33.24	19.7007	4.3848

*For Epoch 1998-08-15.0

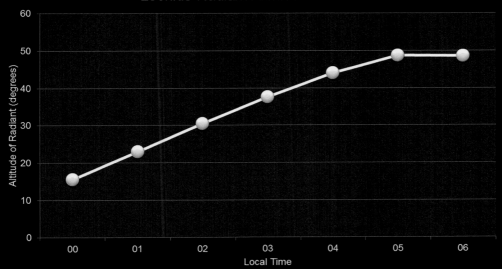

Lerwick, Scotland; Oslo, Norway; Stockholm, Sweden; Helsinki, Finland; St. Petersburg, Russia; Nanontalik, Greenland.

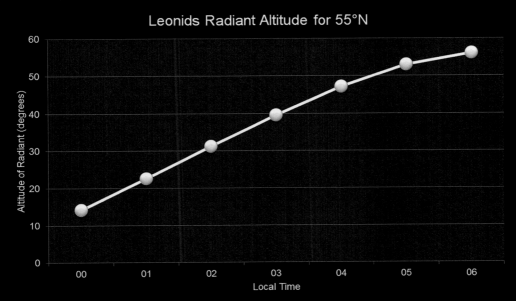

Newcastle upon Tyne, England; Copenhagen Denmark; Malmö, Sweden; Vilnius, Lithuania; Moscow, Russia; Chelyabinsk, Russia; Edmonton, Canada.

Adjust local time for daylight saving if applicable

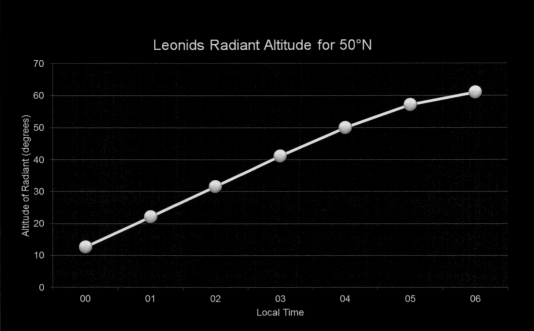

London, England; Frankfurt, Germany; Bern, Switzerland; Kiev, Ukraine; Qaraghandy, Kazakhstan, Qiqihar, China, Vancouver and Winnipeg, Canada.

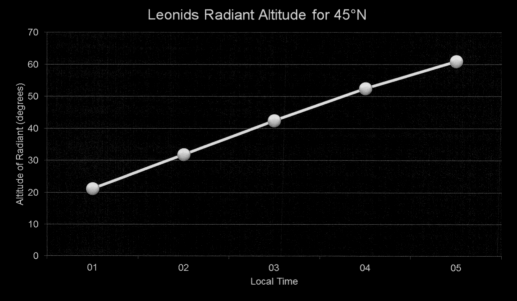

Bordeaux, France; Zagreb, Croatia; Ulaanbaatar, Mongolia; Harbin, China; Portland and Minneapolis, USA; Toronto and Montreal, Canada.

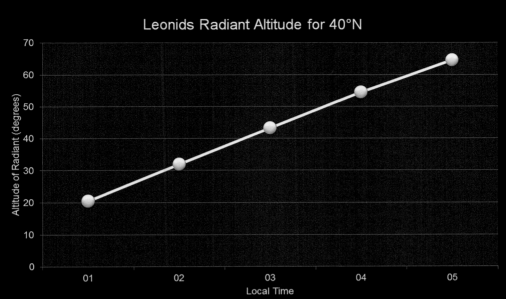

Porto, Portugal; Rome, Italy; Istanbul, Turkey; Toshkent, Uzbekistan; Beijing, China; Sapporo, Japan; Sacramento, Salt Lake City, Denver, Kansas City, Indianapolis, Pittsburgh and Philadelphia, USA.

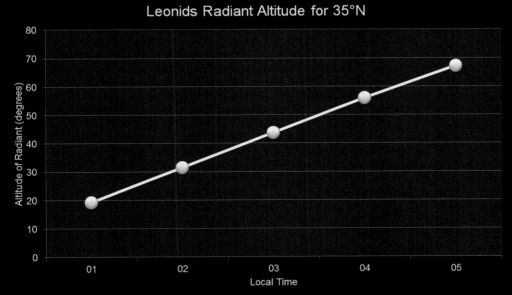

Casablanca, Morocco; Malaga, Spain; Athens, Greece; Tehran, Iran; Jinan, China; Soul, South Korea; Los Angeles, Memphis and Charlotte, USA.

240 6 The Major Meteor Showers: October to December

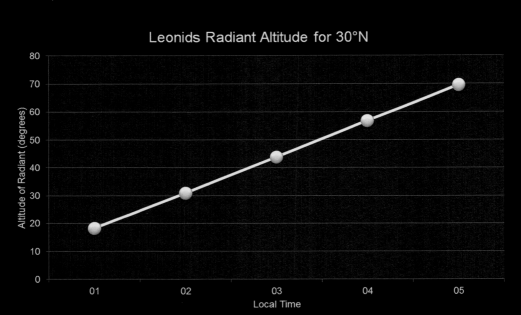

Madeira, Portugal; Tripoli, Libya; Cairo, Egypt; Basra, Iraq; Multan, Pakistan; Lahore, India; Shanghai, China; Tokyo, Japan; San Diego, Austin and Jacksonville, USA.

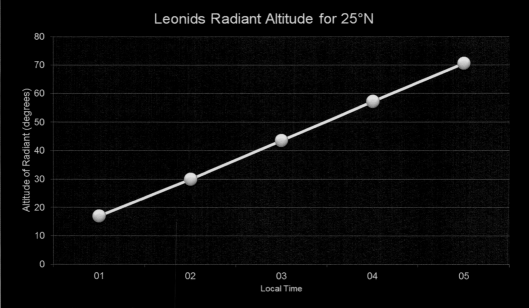

Las Palmas, Canary Islands; Aswan, Egypt; Karachi, Pakistan; Varanasi, India; Kunming, China; T'aipei, Taiwan; Torreon, Mexico; Miami, USA.

The Leonids

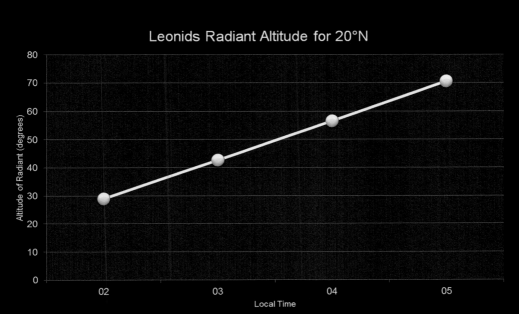

Jeddah, Saudi Arabia; Mumbai, India; Haikou, China; Campeche, Mexico; Santiago de Cuba, Cuba.

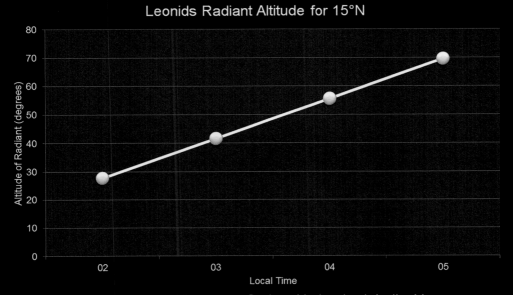

Dakar, Senegal; Khartoum, Sudan; Hyderabad, India; Yangon, Myanmar; Manila, Philippines; Guatemala City, Guatemala

242   6   The Major Meteor Showers: October to December

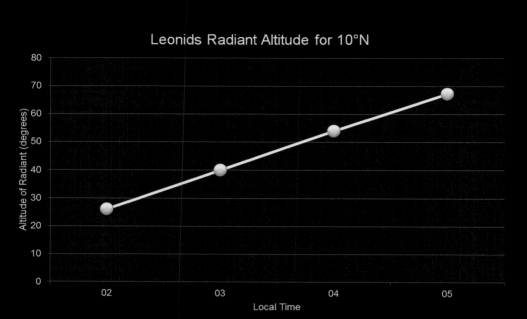

Conakry, Guinea; Kaduna, Nigeria; Addis Ababa, Ethiopia; Ho Chi Minh City, Vietnam; Manila, Philippines; San Jose, Costa Rica; Caracas, Venezuela.

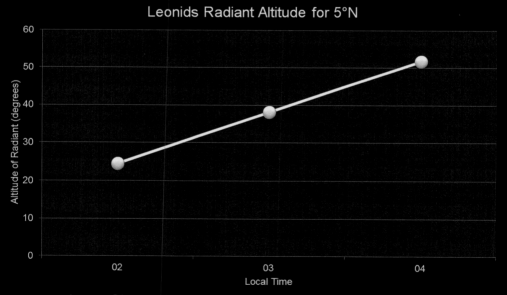

Accra, Ghana; Colombo, Sri Lanka; George Town, Malaysia; Bogota, Colombia; Cayenne, French Guiana.

The Leonids

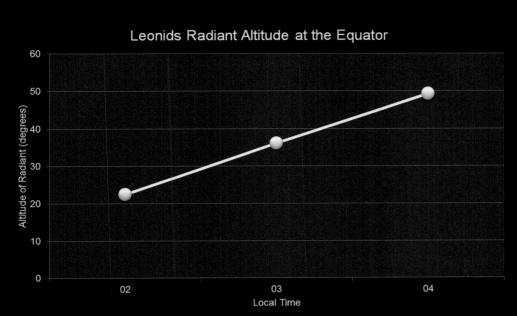

Libreville, Gabon; Kampala, Uganda; Singapore; Quito, Ecuuador; Belem, Brazil.

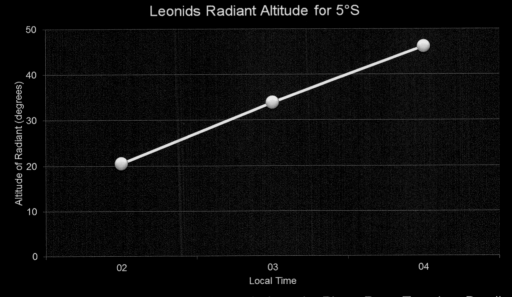

Kinshasa, D.R.Congo; Jakarta, Indonesia; Piura, Peru; Teresina, Brazil.

244 6 The Major Meteor Showers: October to December

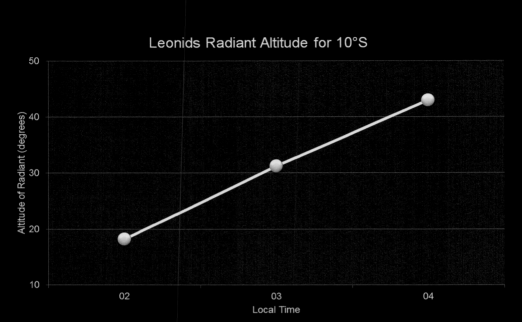
Luanda, Angola; Kolwetzi, D.R.Congo; Kupang, Indonesia, Maceio, Brazil.

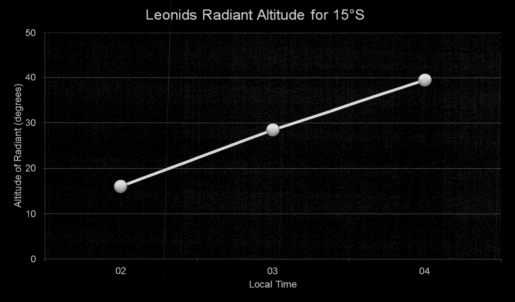
Lusaka, Zambia; Blantyre, Malawi; Coronation Island, Western Australia.

The Leonids

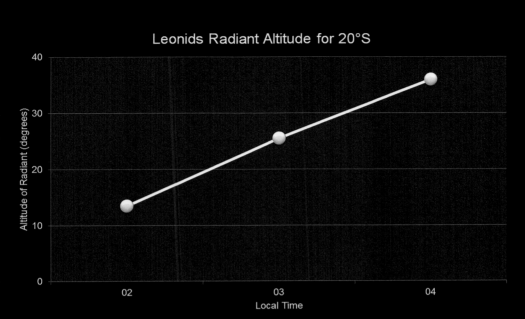

Bulawayo, Zimbabwe; Beira, Mozambique; Port Hedland, Western Australia; Bowen, Queensland, Australia; Iquique, Chile; Cariacica, Brazil.

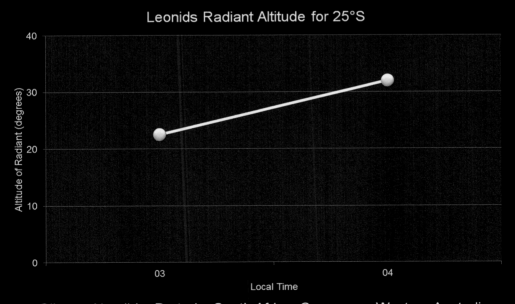

Gibeon, Namibia; Pretoria, South Africa; Carnarvon, Western Australia; Bundaberg Central, Queensland, Australia; Taltal, Chile; Asuncion, Paraguay.

The Geminids

Often the most prolific meteor shower of the year, the Geminids are probably close to their best ever performance. A century from now, they may be gone.

Observing Notes

The Geminids are a beautiful shower to watch. The cold, settled air of late autumn makes the meteors appear crisp and sharp. They plunge deep into the atmosphere, decelerating from 34 km/s to typically 15 km/s, leaving behind globules of material in their wake and occasionally a long-duration train. But trains are the exception rather than the rule. The brightest Geminids appear shortly after maximum, so don't be in a hurry to pack up and go to bed.

Observers in the Northern Hemisphere are in for a marathon as far as the Geminids are concerned. Providing you can put up with the cold night air of the mid- to high-latitudes, you can watch the shower from early evening through to dawn. Even for those living in the Tropics, the shower is visible from late evening—9 or 10 pm LT—and is only interrupted by the rising Sun. In fact, observers as far south as Buenos Aires in Argentina, Montevideo in Uruguay and Auckland, New Zealand will be able to watch the shower from about 11 pm to 3 am LT. The only problem for high Northern observers—apart from the Moon—is bad winter weather. However, if the weather is kind and the Moon is new, it would seem almost criminal not to make the effort to observe the Geminids. As we'll see in the next section, this shower isn't going to be around forever.

Compared to some showers, the Geminids have a fairly short duration of just about 2 weeks (Fig. 6.37). Early activity can usually be detected around December 4 ($\lambda_\odot$ 252°). Nothing much happens for about a week; the ZHR totters around 10, but starts to pick up fairly rapidly about 3 days before maximum, which falls on December 14 throughout the 2020s ($\lambda_\odot$ 262.2°). Maximum isn't pinpoint sharp but is stretched out a bit, and there's the prospect of seeing a double peak separated by perhaps a couple of hours, with the secondary peak appearing earlier than the primary. Then there is a fairly rapid fall-off in activity—but don't be tempted to call it a night. The individual meteoroids within the stream have become separated by mass due to a process called the Poynting-Robertson effect, discussed in Chap. 2. When a meteoroid absorbs energy from the Sun, it re-radiates some of that energy at a slightly different wavelength. This causes a tangential drag on the particle, slowing it down. The effect is most dominant on smaller particles but negligible for substantial objects. Because of the orientation of the Geminid orbit in relation to the Earth, we encounter the smaller particles first. The larger meteoroids—the ones that produce the brightest meteors—fall into our atmosphere last. Hang around until after maximum and you should see the best and brightest meteors, which are often reported to be bluish in colour.

Since the Geminids were discovered in the early 1800s, not only has activity steadily increased, but so has the average magnitude. In the 1930s, the average magnitude would seem to have been around $\overline{m}_v$ +3. Today it is about $\overline{m}_v$ +1.9 overall, and $\overline{m}_v$ +1.6 post-maximum. While most meteors generally last for between 0.4–0.8 s, the Geminids show an unusually high number of meteors with a duration of 1 s or longer. Rates are consistently high. If the past 25 years are anything to go by, there is a 76% chance of witnessing a ZHR in excess of 120 in any one year (Fig. 6.40).

The Geminids

Table 6.17a: Geminid Activity Details					
			Start	Maximum	End
Dates (approx.)			Dec 4	Dec 14	Dec 17
Solar longitude:		$\lambda_\odot$	252°	262.2°	265°
Right Ascension at max:		α		$07^h\ 27^m$ (112°)	
Declination at max:		δ		+32.3°	
Time of transit :				2.1^h L.T.	
ZHR (2010-2020 average):				139	
Radiant diameter:		$\varnothing$	5°	5°	5°

Table 6.17b: Geminid Additional Data		
IAU Abbreviation: GEM		IAU Code: 004
Mean daily motion of radiant:	Δ	$\Delta\alpha$ = +0.97° $\Delta\delta$ = -0.23°
Geocentric velocity:	V_g	33.6 km/s (120,960 km/h)
Population index:	r	2.6
Mean magnitude:	$\bar{m}_v$	+1.9
Parent body: (3200) Phaethon		
Best visibility: Late evening and post-midnight in the Northern Hemisphere		
The radiant is circumpolar from 58°N, i.e. Lewis, Scotland; Kristiansand, Norway; Linköping, Sweden; Pärnu, Estonia; St. Petersburg, Russia; Magadan, Russia; Anchorage, Alaska, USA; Watson Lake, Canada; Bellin, Canada and Uummannarsuaq (Nanortalik), Greenland.		

Table 6.18: Geminid Maxima 2020-2030				
Maximum($\lambda_\odot$ 262.2°)				Moon
year	month	day	hour (UT)	Age (d)
2020	12	14	00:49	1
2021	12	14	06:59	10
2022	12	14	13:14	21
2023	12	14	19:15	2
2024	12	14	01:26	14
2025	12	14	07:36	25
2026	12	14	13:41	5
2027	12	14	19:53	17
2028	12	14	01:57	27
2029	12	14	08:05	9
2030	12	14	14:23	20

Table 6.19: Other Active Meteor Showers During the Geminids
Calendar dates are approximate. Use solar longitude $\lambda_\odot$

Shower	IAU Code	Start $\lambda_\odot$	Max $\lambda_\odot$	End $\lambda_\odot$	At Max. α	At Max. δ	V_g (km/s)	ZHR
Northern Taurids	NTA	Oct 20 207°	Nov 12 230°	Dec 10 258°	$03^h 57^m$ 59°	+22.3°	28.7	5
November Orionids	NOO	Nov 7 225°	Nov 28 246°	Dec 17 265°	$06^h 04^m$ 91°	+16	43	3
Southern χ-Orionids[1]	ORS	Nov 13 231°	Dec 2 250°	Dec 21 269°	$05^h 29^m$ 82°	+18°	26	1
Northern χ-Orionids	ORN	?	Dec 12 260°	?	$05^h 52^m$ 88.1°	+25.7°	29	<1
December σ-Virginids	DSV	Nov 22 239°	Dec 21 250°	Jan 25 269°	$12^h 58^m$ 194°	+8°	66	1
December Monocerotids	MON	Nov 27 245°	Dec 9 257°	Dec 20 268°	$06^h 40^m$ 100°	+8°	41	3
December α-Draconids	DAD	Nov 30 248°	Dec 8 256°	Dec 15 263°	$13^h 29^m$ 202°	+59°	44	1
Puppid-Velids[2]	PUP	Dec 1 249°	Dec 7 255°	Dec 26 274°	$08^h 12^m$ 123°	-45	40	9
December φ-Cassiopeiids	DPC	Dec 1 249°	Dec 6 254°	Dec 8 256°	$01^h 36^m$ 24°	+50	16	Var.
December κ-Draconids	DKD	Dec 2 250°	Dec 4 252°	Dec 7 257°	$12^h 29^m$ 187°	+70°	41	2
ψ-Ursae Majorids	PSU	Dec 2 250°	Dec 5 253°	Dec 10 258°	$11^h 16^m$ 169°	+42	62	1
σ-Hydrids	HYD	Dec 3 251°	Dec 9 257°	Dec 21 269°	$08^h 20^m$ 125°	+2°	62	7
Phoenicids[3]	PHO	Nov 28 246°	Dec 2 250°	Dec 8 256°	$01^h 12^m$ 18°	-53°	18	Var.
December Leonis Minorids	DLM	Dec 6 254°	Dec 19 267°	Jan 18 297°	$10^h 40^m$ 160°	+30°	63	3
σ-Serpentids	SSE	Dec 7 255°	Dec 27-28 275°-276°	Jan 12 292°	$16^h 16^m$ 244°	-1.7°	45.5	<1
Comae Berenicids	COM	Dec 12 260°	Dec 16 264°	Dec 23 271°	$11^h 35^m$ 174°	+19°	65	3

[1]Occasionally produces fireballs.
[2]Not a single shower but a complex of several, which together produce a ZHR of about 9. Date of maximum and exact position of the radiants vary depending on which component is active.
[3]Outburst in 1956 with a ZHR of 100.

The Geminids

	Table 6.20: Geminid Timeline
1830 Dec 12/13	Ludvig F. Kämtz discovers the Geminids, recording 40 meteors.
1860	Rate of one meteor per minute, observed by F. Miller and students of Stanmore School, Maryland.
1862	Annual nature of the Geminids recognized by B.V. Marsh, citing G. Wood's observations of the previous year (Philadelphia), and independently by R.P. Greg of Prestwich, England.
1863	A.S. Herschel determines the radiant to be at $\alpha 105.5°(07^h02^m)$, $\delta +30.5°$.
1866	Strong activity with rates at about one per minute.
Late 1800s to early 1930s	Low levels of activity with only the occasional burst, e.g. 1920 and 1923.
1923	W.F. Denning publishes first radiant ephemeris for the Geminids.
1946	Jodrell Bank Experimental Station first detects Geminids.
1947	Shape and size of the meteoroid stream determined from Jodrell Bank data.
1960s-1970s	ZHR averages 80.
1983 Oct 11	Asteroid 1983 TB discovered in IRAS data by two British astronomers, S.F. Green and J.K. Davies. The object is later named (3200) Phaethon.
1983 Oct 25	Fred Whipple associates 1983 TB with the Geminids.
2009 June	NASA's STEREO-A spacecraft reveals Phaethon doubles in brightness as it approaches perihelion, possibly due to a cloud of dust being released. The phrase *rock comet* is introduced.
2017	R. Blaauw determines that most of the mass of the Geminid meteoroid stream is contained in larger particles. The stream's mass is estimated to be 1.6×10^{16} g.
2018 Nov 6	NASA's Parker Solar Probe images the Geminid meteoroid stream, revealing it to be about 100,000 km wide and following Phaethon for about 20 million km.
2020	Eric MacLennan, Mikael Gravnik and Athanasia Tolian publish evidence that thermal fracturing of (3200) Phaethon would produce enough dust to sustain the meteoroid stream.

A number of other showers are active during the Geminids, with the Monocerotids, the November Orionids and the Southern χ-Orionids waiting to catch out the unsuspecting observer, so a little care is needed to avoid contamination (Table 6.19).

If poor weather hampers observation, don't despair! The shower's activity level stays at around half the maximum value for about 20 h. With some luck, you'll get clear skies during that time.

Discovery

There is no mention of the Geminids prior to the 1800s. They do not appear in Chinese, Japanese or Korean annals, nor in Middle East records. The first apparent observation was made on 1830 December 12/13 by the Prussian physicist Ludvig F. Kämtz, who recorded 40 meteors, although it is not known how many were sporadics [30].

Over the next thirty years, activity gradually picked up. In 1860, a group of students from the Stanmore School, Maryland, under the supervision of F. Miller, recorded one meteor per minute. But it was left to B.V. Marsh [31] of Philadelphia and, independently, Robert Greg [32] of Prestwich (now part of Greater Manchester), England, to recognize the annual nature of the shower in 1862. The following year, Alexander Herschel determined the radiant to be at α 105.5° (07^h 02^m), δ +30.5°, not far from the currently accepted position of α 112° (07^h 27^m), δ +32.3°.

Activity remained high during the 1860s but seems to have fallen off from the late 1800s to the 1930s. The Jodrell Bank Experimental Station began monitoring the Geminids in 1946 and managed to determine the shape and size of the stream the following year.

Rates again began to increase so that by the 1960s–1970s, ZHRs of 60–80 were not uncommon. Since then, the shower has become steadily more active and the individual meteors brighter. Nowadays, the ZHR averages 139 (Fig. 6.39). There seems to be a consensus that this situation will not continue forever and that we are watching the Geminids at their best. One hundred years ago, the activity profile of the shower was the reverse of what it is today: a sharp rise to maximum and then a gradual decline. All these factors indicate a stream that is drifting away from the Earth and will perhaps disappear completely early in the twenty-second century. One hundred years from now, meteor observers may look back on the 2020s with envy.

Origins

Until the 1980s the Geminids were considered an orphan stream, as there was no sign of its parent comet. Then, in 1983, everything change. While studying data from the Infra-Red Astronomy Satellite (IRAS) two British astronomers, Simon Green and John Davies, spotted a new Apollo-type asteroid, which was given the designation 1983 TB. Just 2 weeks later, the father of meteor research, Fred Whipple (then well into his 70s), notice something that no one else had: 1983 TB was well and truly imbedded in the orbit of the Geminids [33]. The object was later named (3200) Phaethon, after one of the sons of Helios, the Sun god.

Finding an asteroid in the middle of a meteoroid stream put the proverbial cat amongst the pigeons. Meteoroid streams were born of comets, not asteroids. Everyone knew that! So what was Phaethon doing lurking inside the Geminids? Perhaps it isn't an asteroid? Perhaps it's a defunct comet? [34] Perhaps it is an asteroid that just so happens to be in the same orbit as the Geminids? Perhaps the asteroid and the stream resulted from the breakup of a larger asteroid? Or a comet? Or perhaps something in between? [35] Nearly four decades later, we still do not have a definitive answer, but a consensus is forming.

The orbit of the Geminids, and that of (3200) Phaethon, is the smallest of all the major meteoroid streams. At the farthest point in the orbit, the stream and the asteroid lie 2.4 AU from the Sun, slightly less than one and a half times the distance of Mars. The orbit then swings around towards perihelion and comes to within 0.14 AU, or 20 million km, of the Sun. That's less than half the distance of Mercury (Table 6.21 and Fig. 6.44).

The meteors themselves are somewhat unusual. They plunge deeper into the atmosphere than most major shower meteors and decelerate more strongly; they are generally brighter, they last longer and during breakup they visibly release globules of material. Their density is thought to be around 2.9 g/cm^3, or about the same as a typical stony meteorite. The brighter Geminids appear to be more cohesive than cometary fireballs. Spectral analyses reveal that the meteoroids are depleted in volatiles, no doubt due to their close encounters with the Sun.

Phaethon is also unusual. Technically, it is bluish in colour. Comets tend to be reddish. Belonging to the taxonomic Type B asteroids—'B' for bluish—they are a subgroup of C-type or carbonaceous asteroids [36]. This means that Phaethon is quite dark: just a little lighter than charcoal. Its dimensions are 5.7 × 4.7 km. That's a little bit larger than most Apollo asteroids, but a little bit smaller than most cometary nuclei.

Phaethon approaches the Sun much closer than any other known asteroid, with the exception of 2021PH$_{27}$, and as a result is subject to very high temperatures. At aphelion, the temperature is around minus 85 °C. At perihelion, the temperature is about 500 °C. But at perihelion, Phaethon's temperature itself is much higher, perhaps around 800 °C as the body retains some of its heat. Silicate minerals on Earth begin to melt at 600 °C under one atmosphere of pressure, and all are molten at 1200 °C. In the vacuum of space, minerals do not melt but vaporize, and at lower temperatures. The short orbital period of just 1.43 years means that Phaethon is continually heated and cooled without respite, and this never-ending thermal onslaught is thought to weaken and destroy its outermost layers. In 2009, NASA's STEREO-A spacecraft revealed that Phaethon doubles in brightness as it approaches perihelion, possibly due to a cloud of dust being released [37]. The phrase *rock comet* was coined to describe its behaviour, but more recently, the term *active asteroid* has come into use.

Particles that are released during the thermal destruction of Phaethon's surface behave in different ways. Those that are released pre-perihelion have markedly different orbits to those released post-perihelion. The result is two separate stream that cross one another, rather like a pair of scissors. In 2017, R. Blaauw was able to determine the total mass of the Geminid stream to be 1.6 × 10^{16}g, or just 5.16% of the mass of the Perseids. Blaauw noted that most of the mass was in the form of larger particles [38]. In 2020, Eric MacLennan, Mikael Gravnik and Athanasia Tolian published evidence that the thermal fracturing of (3200) Phaethon could produce enough dust to sustain the meteoroid stream [39].

Asteroids were not supposed to produce meteoroid streams; that was the job of comets as their ices sublimate when close to perihelion. But it would seem that, given the right circumstances, asteroids may also have a role to play in laying down meteoroid streams. Since the unusual nature of (3200) Phaethon was discovered, more than three dozen objects have been identified as displaying comet-like activity, including (1) Ceres and (101955) Bennu (see Chap. 7: Discovering Minor Showers).

252　　　　　　　　　　6　The Major Meteor Showers: October to December

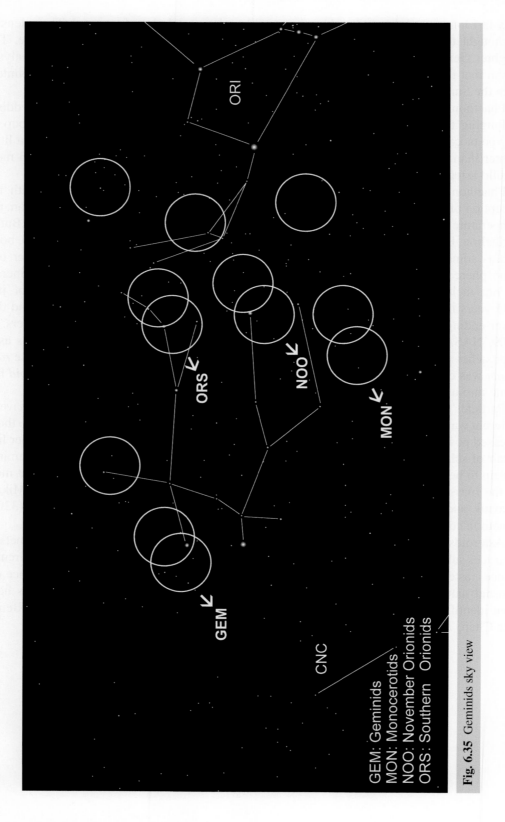

Fig. 6.35 Geminids sky view

The Geminids 253

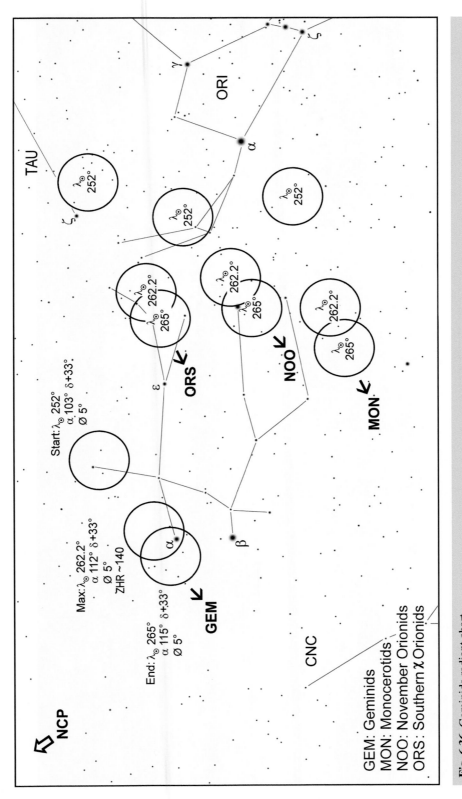

Fig. 6.36 Geminids radiant chart

GEM: Geminids
MON: Monocerotids
NOO: November Orionids
ORS: Southern χ Orionids

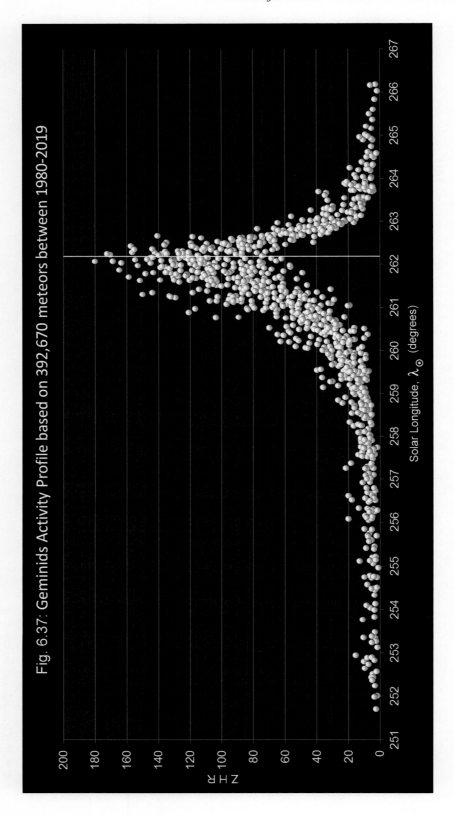

Fig. 6.37: Geminids Activity Profile based on 392,670 meteors between 1980-2019

The Geminids

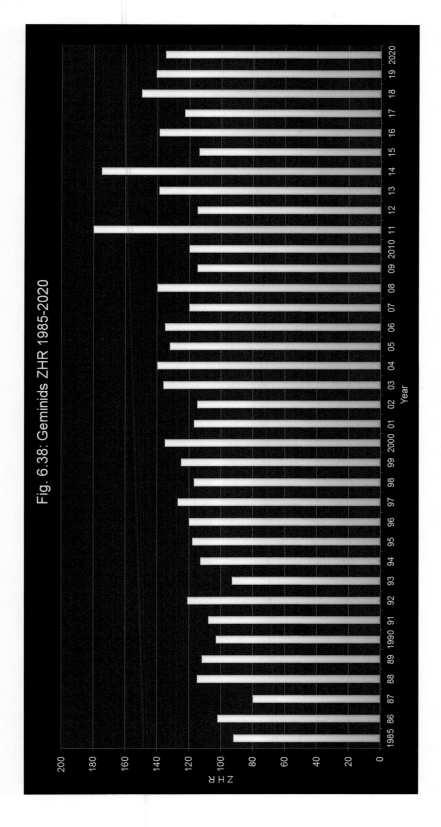

Fig. 6.38: Geminids ZHR 1985-2020

256 6 The Major Meteor Showers: October to December

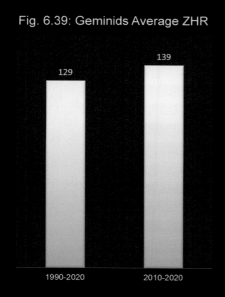

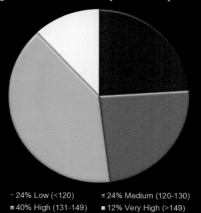

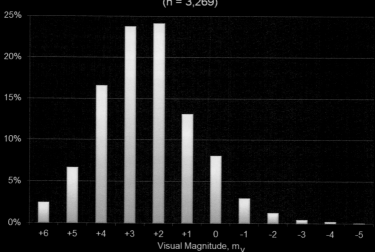

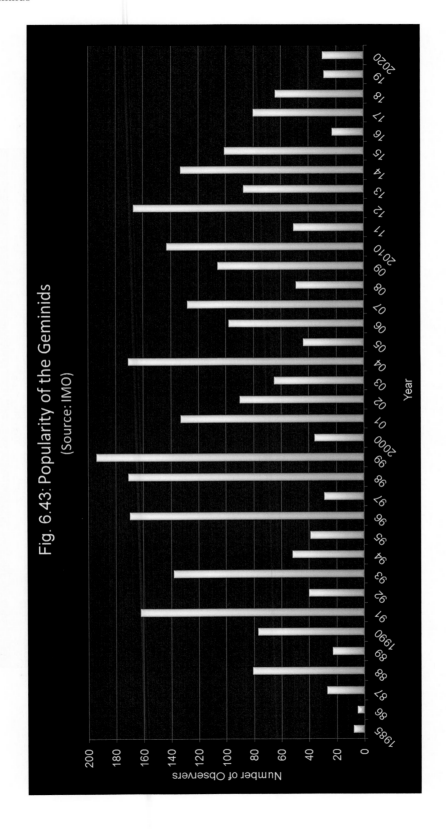
Fig. 6.43: Popularity of the Geminids (Source: IMO)

258 6 The Major Meteor Showers: October to December

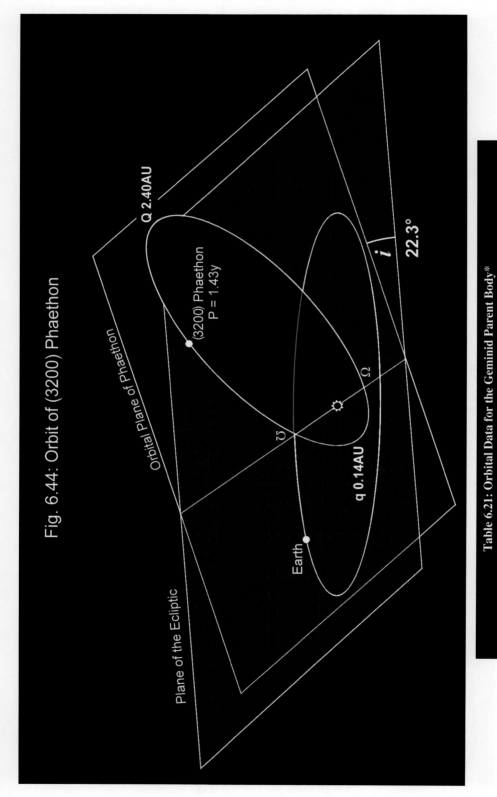

Fig. 6.44: Orbit of (3200) Phaethon

Table 6.21: Orbital Data for the Geminid Parent Body*

3200 Phaethon							
q	e	a	ω	Ω	i	P	Q	b
AU		AU	°	°	°	y	AU	AU
0.1401	0.8898	1.2714	322.1867	265.2177	22.2595	1.43	2.4027	0.5802

*For Epoch 202005-31.0

The Geminids

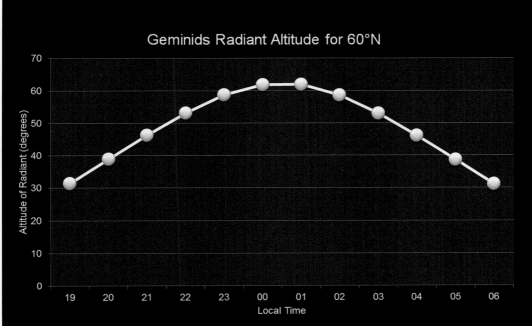

Lerwick, Scotland; Oslo, Norway; Stockholm, Sweden; Helsinki, Finland; St. Petersburg, Russia; Nanontalik, Greenland.

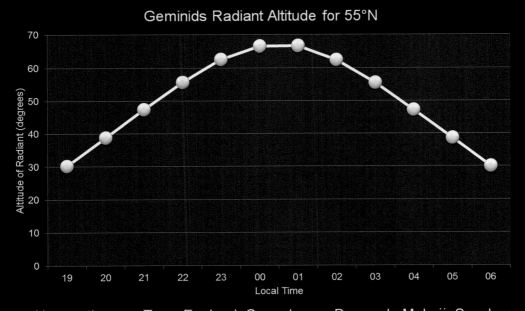

Newcastle upon Tyne, England; Copenhagen Denmark; Malmö, Sweden; Vilnius, Lithuania; Moscow, Russia; Chelyabinsk, Russia; Edmonton, Canada.

260   6   The Major Meteor Showers: October to December

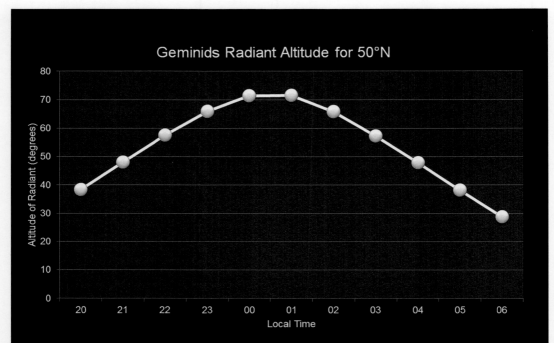

London, England; Frankfurt, Germany; Bern, Switzerland; Kiev, Ukraine; Qaraghandy, Kazakhstan, Qiqihar, China, Vancouver and Winnipeg, Canada.

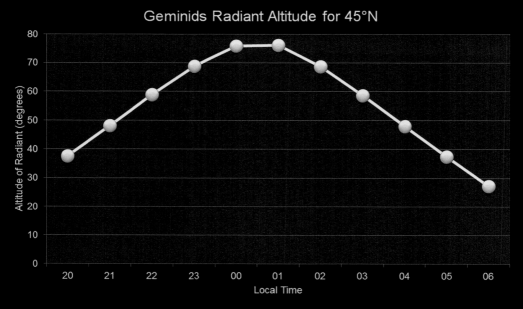

Bordeaux, France; Zagreb, Croatia; Ulaanbaatar, Mongolia; Harbin, China; Portland and Minneapolis, USA; Toronto and Montreal, Canada.

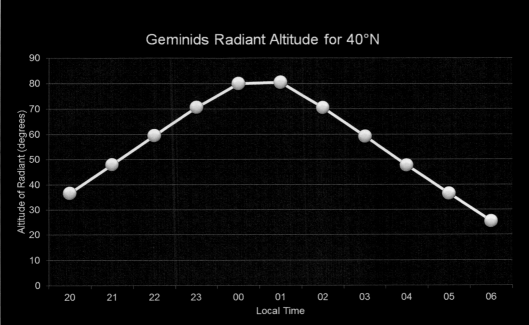

Porto, Portugal; Rome, Italy; Istanbul, Turkey; Toshkent, Uzbekistan; Beijing, China; Sapporo, Japan; Sacramento, Salt Lake City, Denver, Kansas City, Indianapolis, Pittsburgh and Philadelphia, USA.

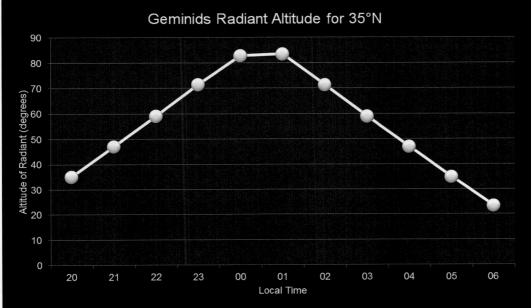

Casablanca, Morocco; Malaga, Spain; Athens, Greece; Tehran, Iran; Jinan, China; Soul, South Korea; Los Angeles, Memphis and Charlotte, USA.

262 6 The Major Meteor Showers: October to December

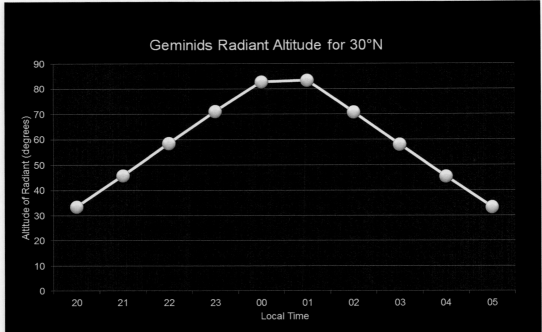

Madeira, Portugal; Tripoli, Libya; Cairo, Egypt; Basra, Iraq; Multan, Pakistan; Lahore, India; Shanghai, China; Tokyo, Japan; San Diego, Austin and Jacksonville, USA.

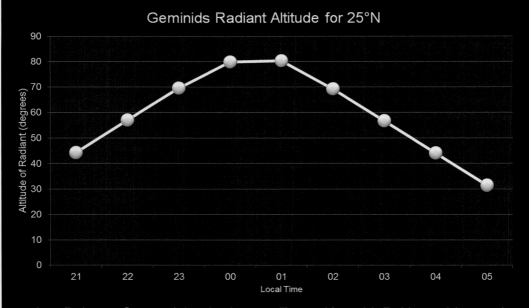

Las Palmas, Canary Islands; Aswan, Egypt; Karachi, Pakistan; Varanasi, India; Kunming, China; T'aipei, Taiwan; Torreon, Mexico; Miami, USA.

The Geminids

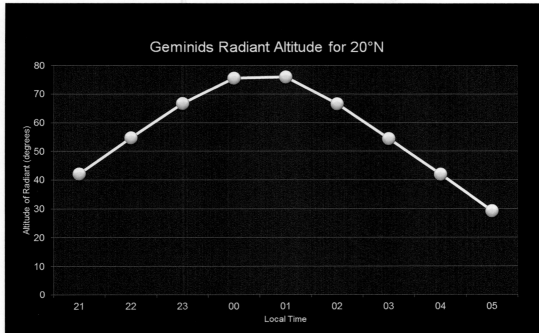

Jeddah, Saudi Arabia; Mumbai, India; Haikou, China; Campeche, Mexico; Santiago de Cuba, Cuba.

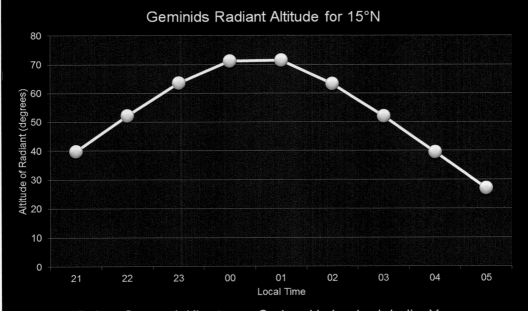

Dakar, Senegal; Khartoum, Sudan; Hyderabad, India; Yangon, Myanmar; Manila, Philippines; Guatemala City, Guatemala

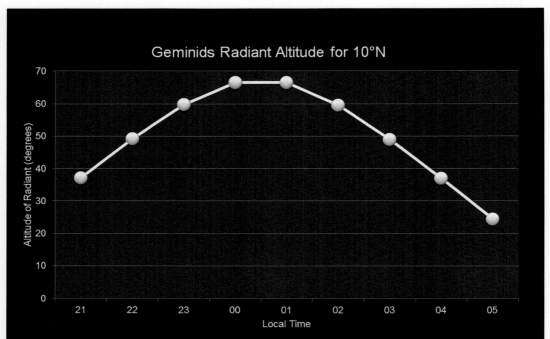

Conakry, Guinea; Kaduna, Nigeria; Addis Ababa, Ethiopia; Ho Chi Minh City, Vietnam; Manila, Philippines; San Jose, Costa Rica; Caracas, Venezuela.

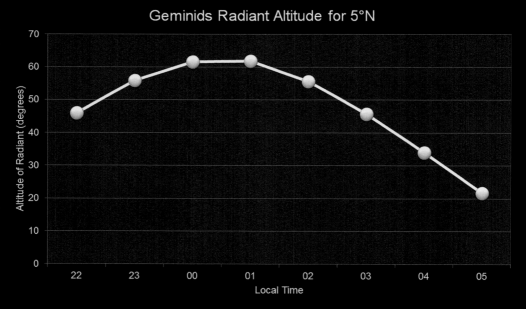

Accra, Ghana; Colombo, Sri Lanka; George Town, Malaysia; Bogota, Colombia; Cayenne, French Guiana.

The Geminids

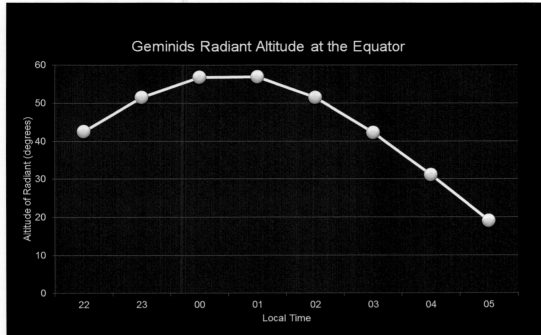

Libreville, Gabon; Kampala, Uganda; Singapore; Quito, Ecuuador; Belem, Brazil.

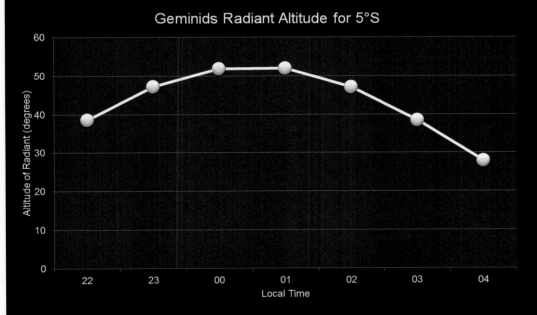

Kinshasa, D.R.Congo; Jakarta, Indonesia; Piura, Peru; Teresina, Brazil.

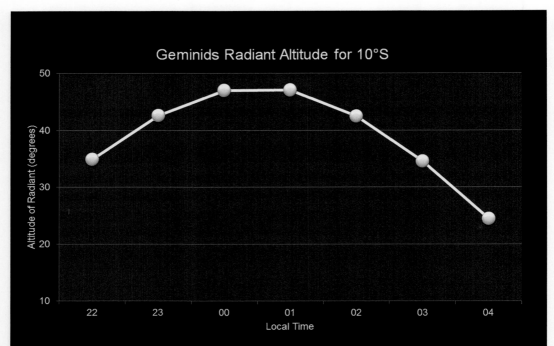

Luanda, Angola; Kolwetzi, D.R.Congo; Kupang, Indonesia, Maceio, Brazil.

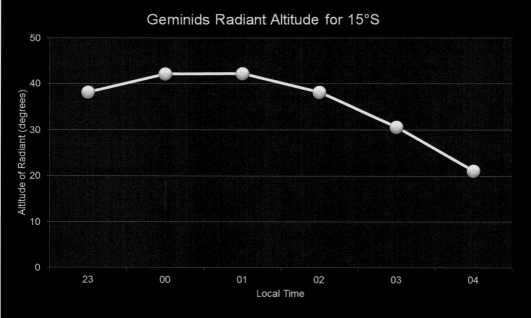

Lusaka, Zambia; Blantyre, Malawi; Coronation Island, Western Australia.

The Geminids

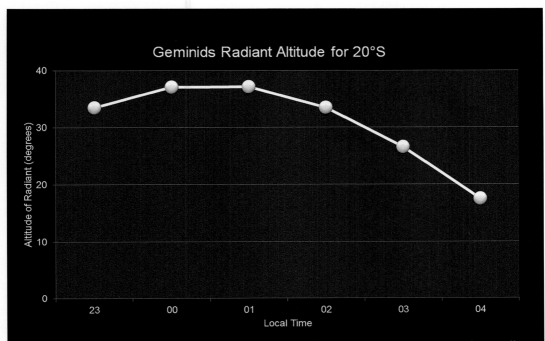

Bulawayo, Zimbabwe; Beira, Mozambique; Port Hedland, Western Australia; Bowen, Queensland, Australia; Iquique, Chile; Cariacica, Brazil.

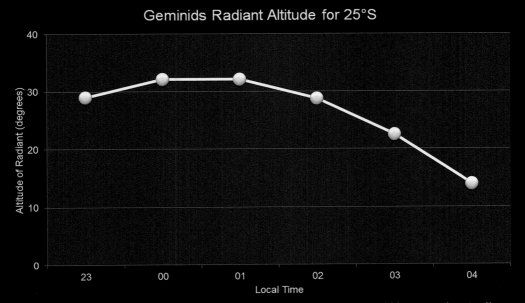

Gibeon, Namibia; Pretoria, South Africa; Carnarvon, Western Australia; Bundaberg Central, Queensland, Australia; Taltal, Chile; Asuncion, Paraguay.

268 6 The Major Meteor Showers: October to December

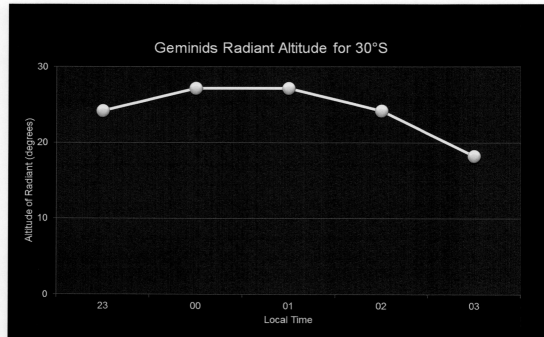

Durban, South Africa; Perth, Western Australia; Coffs Harbour, NSW, Australia; Coquimbo, Chile; Poto Alegre, Brazil.

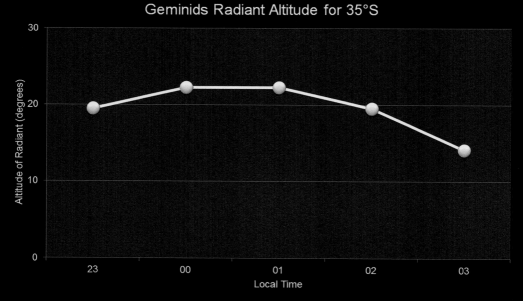

Curico, Chile; Buenos Aires, Argentina; Montevideo, Uruguay; Mangonui, New Zealand.

The Ursids

The year ends with the Ursids, a modest shower that can produce an occasional outburst. Like the Quadrantids, the Ursids can suffer from cold nights and poor weather but have the advantage of being easily observable from early evening in mid- to high-northern latitudes.

Observing Notes

The Ursids are a short-duration shower lasting about 10 days at most. They are only observable from the Northern Hemisphere—anyone south of 25°N won't have much luck spotting Ursid meteors—but for most observers, the real advantage is that the radiant is high in the sky for most of the night. This means that from 35°N northwards, the shower can be observed from early evening. And it is in desperate need of good quality observation.

The first meteoroids start hitting the atmosphere around December 17 ($\lambda_\odot$ 265°). Don't expect anything spectacular at first; it will take a few days for numbers to increase. About a day before maximum, activity can rapidly pick up, reaching a sharp peak around December 23 ($\lambda_\odot$ 265°) and then just as rapidly subside. A few days later you'll be lucky to spot any stragglers (Fig. 6.48).

When rates are high, the peak can be quite pronounced, but low activity can make it difficult to identify exactly when maximum occurs [40]. The ZHR is often quoted as 10–15, which is probably about right. The problem with quoting the shower's ZHR is that it often produces bursts of high activity—perhaps on half a dozen or so occasions over the last 40 years (Fig. 6.49)—and these can skew the average. Figure 6.50 shows the average ZHR for the periods 1990–2020 and 2010–2020. According to these figures, it would seem that activity has fallen from an average ZHR of 23 to 17. However, that's because of bursts in 1993, 2000 and 2014. If we look at the median, which minimizes the effects of the outlying bursts, the activity appears to have shown a modest increase from a ZHR of 13 to 15. Overall, there is a 3 in 4 chance that you'll see a ZHR in the range of 10 to 20 meteors/h, but there is a possibility of a ZHR in the high-20s in 2021. There are no other showers in the area to contaminate the results (Table 6.24).

The radiant is often described as very diffuse, perhaps 12° in diameter, perhaps elongated. Because the shower is of such short duration, and the radiant centres overlap during the whole of the active period, this can make the radiant appear to be smeared across the sky. During outbursts, some observers claim that the radiant shrinks, possibly to less than 1°, a feature that was first noted in 1947 by J.P.M. Prentice [41]. The average magnitude is nothing to write home about at $\overline{m}_v$ +3.2, but high activity brings brighter meteors including some fireballs.

The shower has been predicted to display bursts of activity at various times during the last twenty years, hardly any of which have materialized. This may be because the model on which these predictions were made requires adjustment, or the bursts have been so short-lived that they have gone completely unnoticed. This is a distinct possibility as the shower is not particularly popular with observers. According to the International Meteor Organization, only 8 people submitted observations of the shower in 2020, even though the ZHR was predicted to be 490. The Ursids are expected to put on strong displays in 2028 and 2030, with ZHRs of between 180–190 [42].

The Ursids

Table 6.22a: Ursid Activity Details

		Start	Maximum	End
Dates (approx.)		Dec 17	Dec 23	Dec 26
Solar longitude:	$\lambda_\odot$	265°	271°	274°
Right Ascension at max:	α		$14^h 28^m$ (217°)	
Declination at max:	δ		+75.4°	
Time of transit :			8.7^h L.T.	
ZHR (2010-2020 average):			15	
Radiant diameter:	$\varnothing$	12°	12°	12°

Table 6.22b: Ursid Additional Data

IAU Abbreviation: URS IAU Code: 015 AKA: Tuttleids, Umids	
Mean daily motion of radiant: Δ	$\Delta\alpha + 0.05°$ $\Delta\delta - 0.31°$
Geocentric velocity: V_g	32.9 km/s (118,440 km/h)
Population index: r	3.0
Mean magnitude: m_v	+3.2
Parent body: (3200) Phaethon	
Best visibility: All night above 35°N; post-midnight between 25°N and 30°N.	
The radiant is circumpolar from 15°N, i.e. Dakar, Senegal; Khartoum, Sudan; Hyderabad, India; Yangon, Myanmar; Manila, Philippines; Guatemala City, Guatemala.	

Table 6.23: Ursid Maxima 2020-2030

Maximum ($\lambda_\odot$ 271°)				Moon
year	month	day	hour (UT)	Age (d)
2020	12	22	16:15	8
2021	12	22	22:33	18
2022	12	23	04:42	1
2023	12	23	10:44	11
2024	12	22	16:59	22
2025	12	22	23:03	2
2026	12	23	05:13	14
2027	12	23	11:25	25
2028	12	22	17:23	6
2029	12	22	23:39	17
2030	12	23	05:52	28

Table 6.24: Other Active Meteor Showers During the Ursids
Calendar dates are approximate. Use solar longitude $\lambda_\odot$

Shower	IAU Code	Start $\lambda_\odot$	Max $\lambda_\odot$	End $\lambda_\odot$	At Max. α	At Max. δ	V_g (km/s)	ZHR
Southern χ-Orionids[1]	ORS	Nov 13 231°	Dec 2 250°	Dec 21 269°	$05^h\,29^m$ 82°	+18°	26	1
December σ-Virginids	DSV	Nov 22 239°	Dec 21 250°	Jan 25 269°	$12^h\,58^m$ 194°	+8°	66	1
December Monocerotids	MON	Nov 27 245°	Dec 9 257°	Dec 20 268°	$06^h\,40^m$ 100°	+8°	41	3
σ-Hydrids	HYD	Dec 3 251°	Dec 9 257°	Dec 21 269°	$08^h\,07^m$ 121°	+4°	62	7
December Leonis Minorids	DLM	Dec 6 254°	Dec 20 268°	Feb 4 315°	$10^h\,40^m$ 160°	+30°	63	5
σ-Serpentids	SSE	Dec 7 255°	Dec 27-28 275°-276°	Jan 12 292°	$16^h\,16^m$ 244°	-1.7°	45.5	<1
Comae Berenicids	COM	Dec 12 260°	Dec 16 264°	Dec 23 271°	$11^h\,40^m$ 175°	+18°	63.5	3
α-Hydrids[2]	AHY	Dec 17 266°	Jan 1 280°	Jan 17 296°	$08^h\,19^m$ 124.9°	-7.0°	43.2	1

[1] Occasionally produces fireballs.
[2] Radiant position quoted for $\lambda_\odot\,271°$ (Ursids max).

Discovery

There is perhaps only one record of Ursid activity prior to the twentieth century: 1795 December 20, recorded in Japanese annals as 'Stars fell like a shower.' [43] The next mention was not until 1916, when William F. Denning reported in *The Observatory* that meteors radiated from a point near β Ursae Minoris [44]. In 1921, Denning had sufficient observations to relate the shower with Comet Mechain-Tuttle (now 8P/Tuttle). And then, nothing! The Ursids became a forgotten shower.

Quite by accident, M. Dzubák of the Skalnaté Pleso Observatory, Slovakia, rediscovered the shower in 1945 after the end of the Second World War. Initially, Bečvář gave the activity rate as 169 meteors per hour [45], but four years later in 1951, Zdeněk Ceplecha reevaluated the observations—he called the shower the Umids—using a method that was to eventually become the zenithal hourly rate. His estimate was 48 meteors/h, taking into account meteors that were counted more than once by the small team of observers [46].

Activity appeared to decline over the next few decades. During the 1960s, it was much the same as it is today—a ZHR of 10 to 15—but that fell to just a few meteors in the 1970s, except for 1973, when radio observations picked up a daylight burst with a ZHR of 30. Towards the end of the 70s (1979 to be precise), the ZHR reached 26 for Norwegian observers [47], then 43 for Japanese observers in 1981, and 122 for Europeans in 1986. Other bursts have occurred since then.

There was, however, something wrong with the Ursid observations. The strong bursts of activity consistently occurred *six years after* the comet had reached perihelion, by which time the comet was half an orbit away! It was Peter Jenniskens and Esko Lyytinen who finally worked out what was happening. [48] They discovered that the comet's orbit was too far from the Earth to produce any direct activity. Modelling the orbits of the dust particles revealed that they became trapped in a 7:6 resonance with Jupiter. Over a period of

The Ursids

600 years, the trapped dust would eventually reach the Earth but, by that time, it had drifted halfway around the orbit, lagging behind the comet by 6.67 years.

	Table 6.25: Ursid Timeline
1795 Dec 20	Japanese annals record that 'Stars fell like a shower'
1916	First modern record of activity from the Ursids, reported in *The Observatory* by W.F. Denning. He notes that the meteors appeared to radiate from near β Ursae Minoris.
1921	Denning associates the Ursids with Comet Mechain-Tuttle (now 8P/Tuttle).
1920s - 1945	The Ursids become a forgotten shower.
1945 Dec 22	M. Dzubák of the Skalnaté Pleso Observatory, Slovakia, rediscovers the shower. Antonín Bečvár[1] estimates the hourly rate to be 169 meteors/hour, six years after the comet passed through perihelion.
1947	J.P.M.Prentice is the first person to estimate that the radiant is <1° across at the time of maximum activity.
1947	First radio observations of the Ursids by Jodrell Bank. The shower is studied until 1953.
1951	Zdeněk Ceplecha re-evaluates Bečvár's observations (which he calls the Umids) using a new method that is to become the *zenithal hourly rate* (ZHR). His estimate is 48 meteors/hour with a burst of 108 at 18:00 UT. Ceplecha notes that high activity occurs when the comet is at the opposite side of its orbit.
1960s	ZHR around 10-15.
1967 Mar 31	Comet Tuttle reaches perihelion.
1970s	ZHR of just a few meteors per hour during the 1970s.
1973 Dec 22	Radio observations detect a daylight burst equivalent to a ZHR of 30. Again, 6 years after the comet's perihelion passage.
1979	Norwegian astronomers observe a short outburst with an estimated ZHR of about 26.
1980 Dec 14	Comet Tuttle reaches perihelion.
1981	Japanese astronomers observe 5 bright fireballs over 6 hours. ZHR 43.
1986 Dec 22	Strong activity (ZHR 122) detected by observers in Europe, 6 years after perihelion passage of the comet. Several fireballs were also observed.
1993	ZHR 105
1994	ZHR 50, Japan
2000	ZHR 90. Meteors are mostly faint, m_v +3 to +5.
2006	P. Jenniskens predicts a ZHR of around 35, but the shower attains only 15. Many meteors are unusually bright.
2007	ZHR 22
2020	ZHR predicted to be 420–490 but only reaches 19.
2028	Predicted ZHR 190.
2030	Predicted ZHR 180.
	[1] Antonín Becvár has a minor planet named in his honour. Its number is easy to remember: 4567.

Origins

It was Denning who initially linked the meteoroid stream with Comet Mechain-Tuttle, now called 8P/Tuttle [44]. Bečvár came to the same conclusion and the shower was occasionally called the Tuttleids, though the name never really caught on (Figs. 6.55 and Table 6.26).

The comet was originally discovered by the French astronomer Pierre Mechain from Paris on 1790 January 9 (designated as 1790 II) and was rediscovered by Horace Tuttle in 1858 January 5 (1858 I).

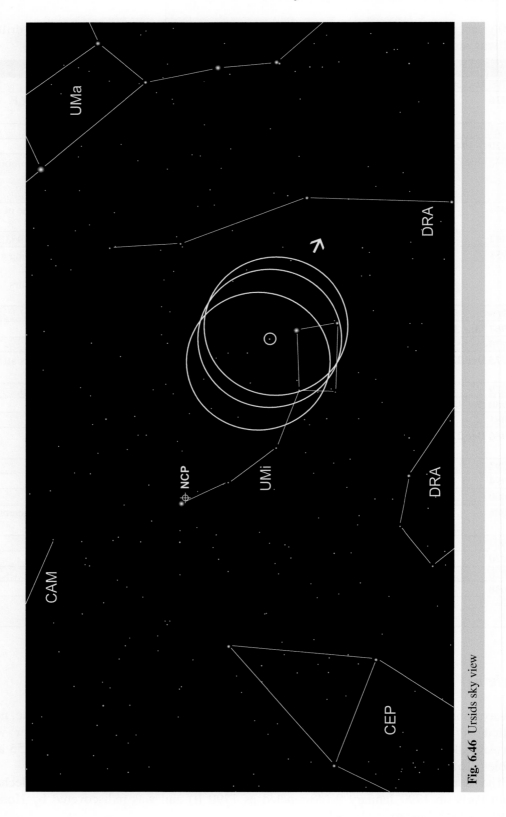

Fig. 6.46 Ursids sky view

The Ursids

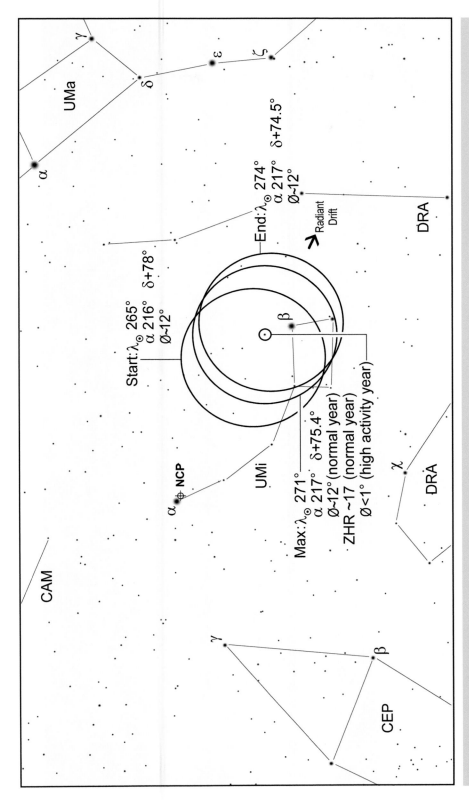

Fig. 6.47 Ursids radiant chart

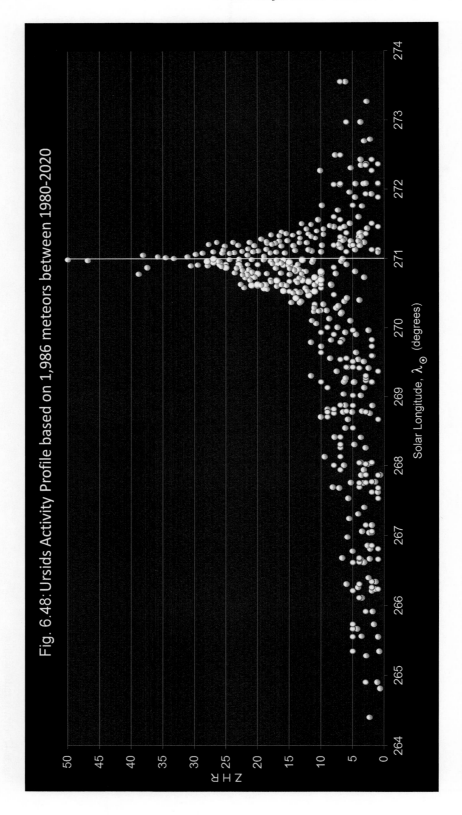

Fig. 6.48: Ursids Activity Profile based on 1,986 meteors between 1980-2020

The Ursids

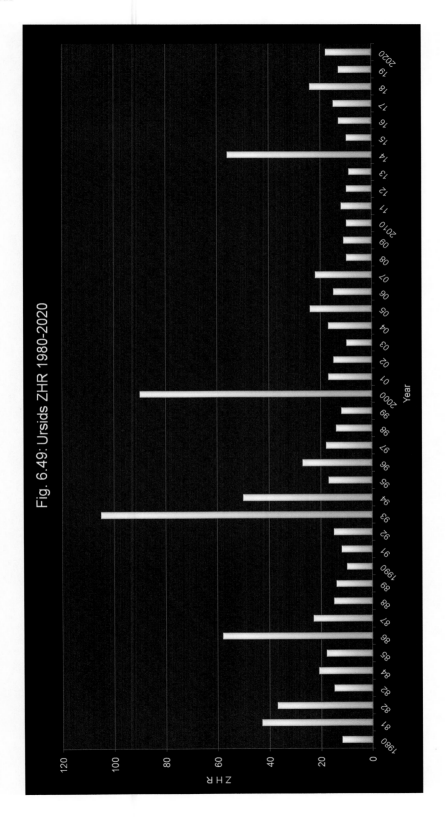

Fig. 6.49: Ursids ZHR 1980-2020

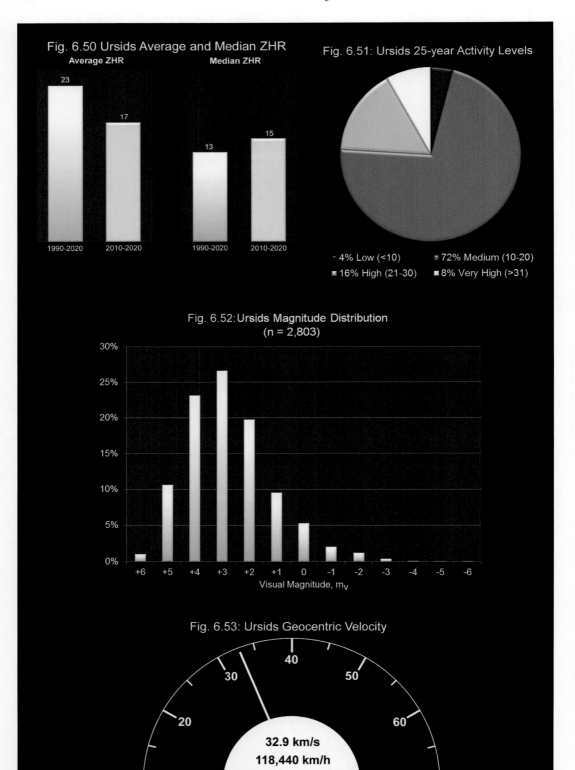

The Ursids

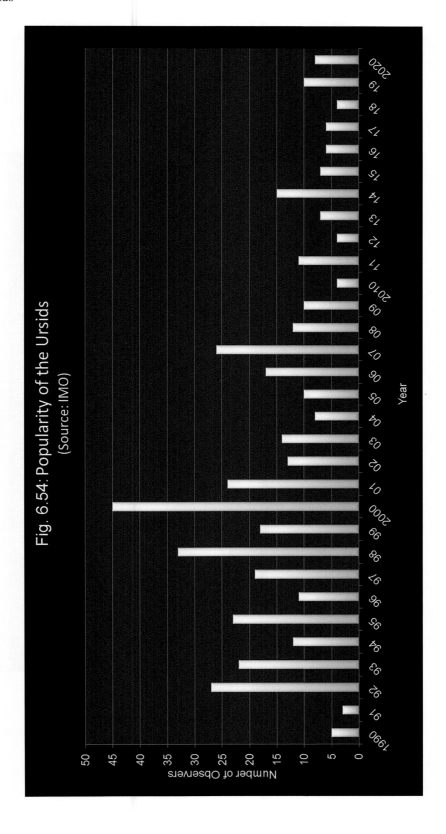

Fig. 6.54: Popularity of the Ursids
(Source: IMO)

280 6 The Major Meteor Showers: October to December

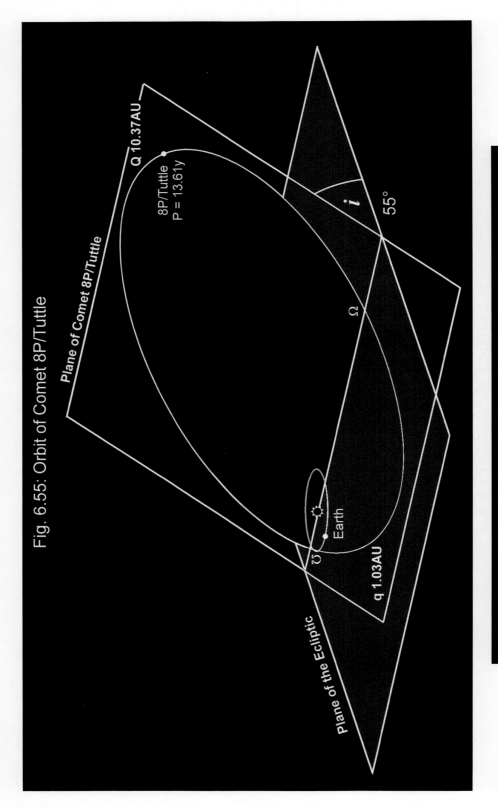

Fig. 6.55: Orbit of Comet 8P/Tuttle

Table 6.26 : Orbital Data for the Ursid Parent Body*

Comet 8P/Tuttle						
q AU	e	a AU	ω °	Ω °	i °	P y	Q AU	b AU
1.0271	0.8198	5.7007	207.5074	270.3416	54.9832	13.6078	10.3741	3.2645

*For Epoch 2007-12-19.0

The Ursids

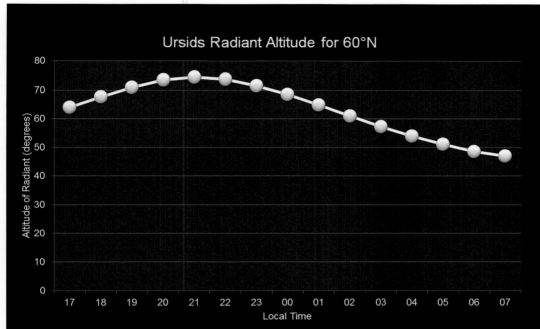

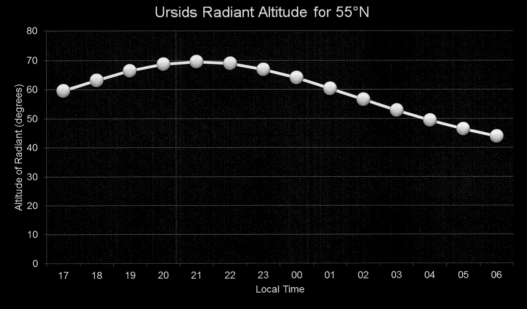

Fig. 6.56a–h Ursids radiant altitude graphs

282 6 The Major Meteor Showers: October to December

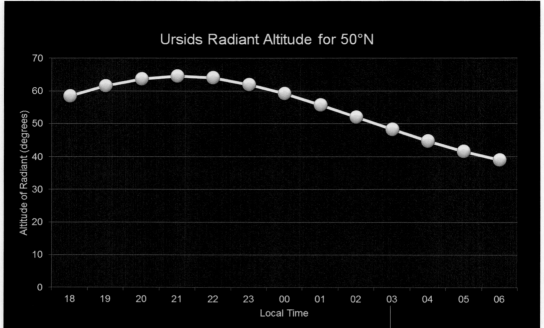

London, England; Frankfurt, Germany; Bern, Switzerland; Kiev, Ukraine; Qaraghandy, Kazakhstan, Qiqihar, China, Vancouver and Winnipeg, Canada.

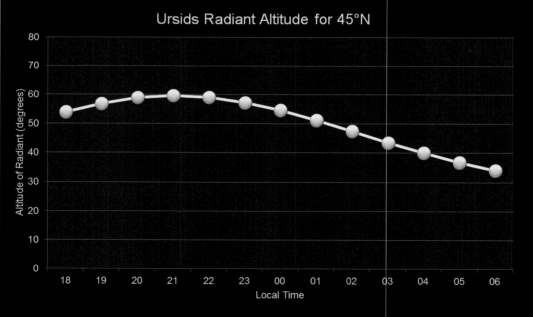

Bordeaux, France; Zagreb, Croatia; Ulaanbaatar, Mongolia; Harbin, China; Portland and Minneapolis, USA; Toronto and Montreal, Canada.

The Ursids

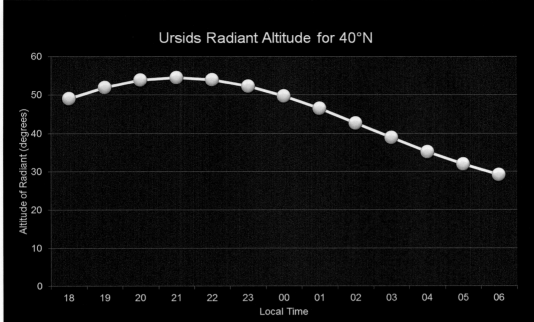

Porto, Portugal; Rome, Italy; Istanbul, Turkey; Toshkent, Uzbekistan; Beijing, China; Sapporo, Japan; Sacramento, Salt Lake City, Denver, Kansas City, Indianapolis, Pittsburgh and Philadelphia, USA.

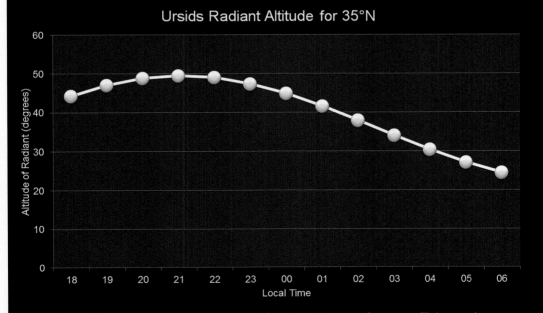

Casablanca, Morocco; Malaga, Spain; Athens, Greece; Tehran, Iran; Jinan, China; Soul, South Korea; Los Angeles, Memphis and Charlotte, USA.

284 6 The Major Meteor Showers: October to December

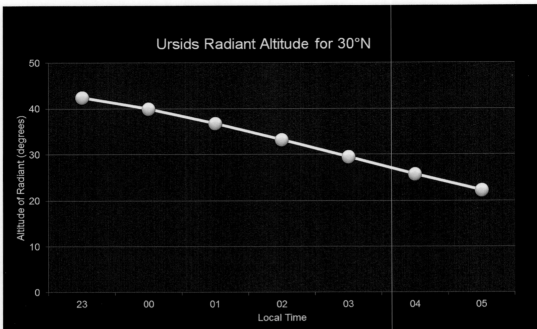

Madeira, Portugal; Tripoli, Libya; Cairo, Egypt; Basra, Iraq; Multan, Pakistan; Lahore, India; Shanghai, China; Tokyo, Japan; San Diego, Austin and Jacksonville, USA.

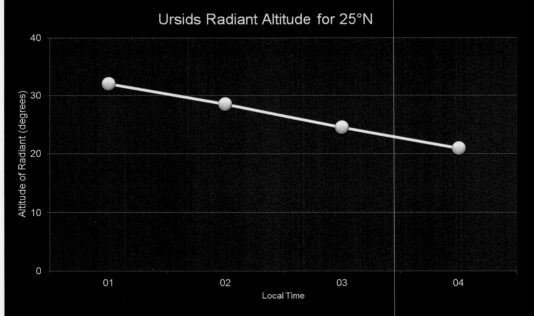

Las Palmas, Canary Islands; Aswan, Egypt; Karachi, Pakistan; Varanasi, India; Kunming, China; T'aipei, Taiwan; Torreon, Mexico; Miami, USA.

References

1. Denning, F. W. (1918). Meteoric showers of October. *The Observatory, 41*, 60.
2. Prentice, J. P. M. (1939). The radiants of the orionid meteor shower. *Journal of the British Astronomical Association, 49*, 148.
3. Znojil, V. (1968). Frequency occurrence of small particles in meteor showers. II. Orionids, ε-Geminids. *Bulletin of the Astronomical Institute of Czechoslovakia, 19*, 311.
4. Benzenberg, J. F., & Brandes, H. W. (1800). An attempt to determine the distance, speed and orbit of shooting stars. *Annalen der Physik, 6*, 224.
5. Quetelet, L. A. J. (1845). Catalogue of the main appearances of shooting stars. *Nouveaux Mémoires de l'Académie Royale des Sciences et Belles-Lettres de Bruxelles, 12*, 23.
6. Herschel, A. S. (1865). Radiant points of shooting-stars. *Monthly Notices of the Royal Astronomical Society, 26*, 51.
7. Falb, R. (1868). The comet halley and its meteors. *Astronomische Nachrichten, 72*, 361.
8. Montanari, G. (1676). *The great meteor flying flame seen over Italy on the evening of March 31st.* Bologna.
9. Denning, W. F. (1878). Suspected repetition, or second outbursts from radiant points, and on the long duration of meteor showers. *Monthly Notices of the Royal Astronomical Society, 38*, 111.
10. Olivier, C. P. (1925). *Meteors* (p. 93). Williams and Wilkins.
11. Hindley, K. B. (1972). Taurid meteor stream fireballs. *Journal of the British Astronomical Association, 82*, 287.
12. Asher, D. J., & Clube, S. V. M. (1993). An extraterrestrial influence during the current glacial-interglacial. *Quarterly Journal of the Royal Astronomical Society, 34*, 481.
13. Heis, E. (1867). The radiation points of the falling stars. *Astronomische Nachrichten., 69*, 157.
14. Kronk, G. W. (2014). *Meteor showers: An annotated catalogue* (p. 291). Springer.
15. Kronk, G. W. (2014). *Meteor showers: An annotated catalogue* (p. 292). Springer.
16. Almond, M. (1951). The summer daytime meteor streams of 1949 and 1950. III. Computation of the orbits. *Monthly Notices of the Royal Astronomical Society, 111*, 37.
17. Kresák, L. (1978). The Tunguska object: A fragment of comet encke? *Bulletin of the Astronomical Institute of Czechoslovakia, 29*, 129.
18. Brown, P., Marchenko, V., Moser, D. E., Weryk, R., & Cooke, W. (2013). Meteorites from meteor showers: A case study of the Taurids. *Meteoritics & Planetary Science, 48*, 270.
19. Štohl, J. (1986) *On meteor contribution by short-period comets. The Exploration of Halley's Comet.* Volume 2: Dust and Nucleus. p. 225.
20. Knopf, O. (1931). About the origin of the meteors. *Astronomische Nachrichten., 242*, 161.
21. Hoffmeister, C. (1937). *The meteors, their cosmic and terrestrial relationships.* Leipzig.
22. Clube, S. V. M., & Napier, W. M. (1984). The microstructure of terrestrial catastrophism. *Monthly Notices of the Royal Astronomical Society, 211*, 953.
23. Jenniskens, P. (2006). *Meteor showers and their parent comets* (p. 631). Cambridge University Press.
24. Cook, D. (1991). A survey of muslim material on comets and meteors. *Journal for the History of Astronomy., 30*, 131.
25. von Humboldt, A. (1818). [Williams, H.M. (transl.)]: *Personal narrative of travels to the equinoctial region of the new continent (Vol.3)* p. 331. London.
26. Olmsted, D. (1834). Observations on the meteors of 13 Nov. 1833. *American Journal of Science and the Arts., 25*, 363.
27. Olbers, H.W.M. (1837). *Yearbook for 1837.* Stuttgart.
28. Milon, D. (1967). Observing the 1966 Leonids. *Journal of the British Astronomical Association., 77*, 89.
29. LeVerrier, U. J. J. (1867). *Weekly Reports of the Sessions of the Academy of Sciences., 64*, 94.
30. Kamtz, L. (1836). *Lehrbuch der Meteorologie (Vol 3)*, 297. Halle.
31. Marsh, B. V., & Wood, G. (1863). *American Journal of Science and the Arts (2nd Series), 35*, 302.
32. Greg, R. P. (1864). *Report of the Annual Meeting of the British Association for the Advancement of Science., 33*, 149.

33. Marsden, B. G. (1983). *IAUC 3881: 1983 TB and the Geminid Meteors; 1983 SA; KR Aur (Circular No. 3881)*. Central Bureau for Astronomical Telegrams.
34. Cochran, A. L., & Barker, E. S. (1984). Minor planet 1983TB: A dead comet? *Icarus, 59*, 296.
35. Green, S. F., Meadows, A. J., & Davies, J. K. (1985). Infrared observations of the extinct cometary candidate minor planet (3200) 1983TB. *Monthly Notices of the Royal Astronomical Society, 214*, 29.
36. JPL-CalTech Small Body Database: 3200 Phaethon (1983 TB). Retrieved from https://ssd.jpl.nasa.gov. Accessed 27 Sept 2020.
37. NASA Marshall Space Flight Center.Retrieved from www.nasa.gov/centers/marshall/news/lunar/phaethon.html. Accessed 02 Oct 2020.
38. Blaauw, R. (2017). The mass index and mass of the Geminid meteoroid stream as determined with radar, optical and lunar impact data. *Planetary and Space Science., 143,* 83.
39. MacLennan, E.M., Toliou, A. and Granvik, M. (2020). *Dynamical evolution and thermal history of asteroids (3200) phaethon and (155140) 2005 UD*. arXiv:2010.10633.
40. Collins, H. (1983). *Ursid Report*. Meteor! 25, 10.
41. Prentice, J. P. M. (1948). Visual observation of bečvář's meteor stream. *Journal of the British Astronomical Association, 58*, 140.
42. Jenniskens, P. (2006). *Meteor showers and their parent comets* (p. 641). Cambridge University Press.
43. Jenniskens, P. (2006). *Meteor showers and their parent comets* (p. 609). Cambridge University Press.
44. Denning, W. F. (1916). Mechain-Tuttle's comet of 1790-1858 and a meteoric shower. *The Observatory., 39*, 466.
45. Bečvář, A. (1946). *IAU circular no. 1026*. Central Bureau for Astronomical Telegrams.
46. Ceplecha, Z. (1951). Umids-Bečvář's meteor stream. *Bulletin of the Astronomical Institutes of Czechoslovakia, 2*, 156.
47. Frossard, F. (1979). Activité météore d'hiver. *Le Ciel Déchu., 16*, 86.
48. Jenniskens, P. (2006). *Meteor showers and their parent comets* (p. 263). Cambridge University Press.

Chapter 7

Discovering Minor Showers

The IAU Meteor Data Center lists more than one thousand meteor showers. Nearly all of these are minor showers display very limited activity, and some probably no longer exist. For a vast majority of the showers, information is sparse to say the least, and more observation is needed. But with most minor showers having a ZHR of just 1 or 2, is it really worth making the effort to monitor their activity? Perhaps after reading this chapter, you will be encouraged to spend at least some time hunting down these elusive showers.

It is unfortunate that so many researchers incorrectly use the terms *meteor shower* and *meteoroid stream,* for when it comes to minor activity, the distinction is important. One golden rule to bear in mind is that *a minor meteor shower does not necessarily indicate a minor meteoroid stream.* There are various scenarios in which a meteoroid stream could be just as densely populated as the Geminids, Perseids or Quadrantids, but the way in which the Earth encounters the stream results in only minor activity.

Minor showers can be categorised as belonging to one of six different types, at least in theory. In practice, it is not always obvious which category applies—an issue that can only be resolved by continuous, methodical observation.

Type 1: Remnant Shower

Meteoroid streams have a limited lifespan tied to that of the parent comet. Once a meteoroid stream is laid down by a passing comet, it immediately begins to degrade. Both gravitational and non-gravitational forces, such as the Poynting-Robertson effect, cause the individual dust particles to spread out. Depending on the size, shape and location of their orbits, some streams will last a few centuries while others may last millennia. In terms of the lifetime of the Solar System, meteoroid streams are transient: they appear and disappear in the blink of a cosmic eye.

If the parent comet is in a short period orbit, then it is likely to replenish the stream at regular intervals, adding to its longevity. Comets do not last forever, of course, and once the comet becomes an inert body, having developed a crust that prevents outgassing and the further loss of particles as the comet approaches the Sun, the meteoroid stream's days are numbered. Dying comets are not the only source of dying meteoroid streams, however. A comet can be perturbed into a new orbit that is completely different from its original orbit, and lay down an entirely new stream, or it can be ejected from the Solar System altogether. Sometimes comets become completely disrupted, as in the case of Biela's Comet.

In 1772, a comet was discovered by Jacques Montaigne, and independently by Charles Messier, which had an orbital period of just 6.6 years. It was the third periodic comet to be discovered—after Halley and Encke—and was named Comet Biela after Wilhelm von Biela, who was the first to calculate its orbit. The comet was observed at several returns, but in 1845 it appeared to have split in two. This was the first stage in what became its total disintegration. Nearly 30 years later, in 1872, the Earth passed through the comet's remnants, resulting in a storm of around 16,000 meteors/h (Fig. 7.1). Nothing was seen in subsequent years until 1885, 13 years later, when rates reached 75,000 meteors/h. It was a further 14 years before the Bielids, as they became known, put on another strong display, but this time it was 'only' a few hundred meteors per hour. The shower almost disappeared completely after 1899, although a double peak was detected in 1940 with a rate of about 30 meteors/h. Having been renamed the Andromedids, the shower attracted renewed interest in the 1970s when observations showed that, rather than being dead and forgotten, the shower was still producing a few meteors every hour. Nowadays, with a ZHR <1, the Andromedids are fast fading away but at the same time replenishing the zodiacal dust cloud. The stream has now gyrated away from the Earth's orbit—but only temporarily. 100 years from now, the Earth will again encounter the stream and, if there is anything left, a new meteor shower may occur.

The total disruption of a meteoroid stream's parent body marks the beginning of the end of the shower. Often, this is a gradual process that can take centuries to complete, but in the case of the Andromedids, dispersion appears to have been quite rapid. Monitoring how the shower performs will not only provide more insight into how streams die out, but may also reveal whether there are clusters of material still in orbit around the Sun. Enhanced Andromedid activity was detected in 2008 and 2011 by the Canadian Meteor Orbit Radar (CMOR) [1]. These events were caused by the release of material by Comet Biela more than three and a half centuries ago in 1649. Only observation will reveal if there are any larger particles still in the stream that can produce significant visual activity, or whether all that is left is fine dust.

It is undoubtedly true that many of the minor meteor showers we see each year are the skeletal remnants of once substantial streams, perhaps rivalling the Geminids, the Perseids and the Quadrantids.

Type 1: Remnant Shower

Fig 7.1: Biela's Comet (top) and the Andromedids (bottom)

Angelo Secchi's depiction of the breakup of Biela's Comet (top) and Amédée Guillemin's portrayal of the resulting Andromedid storm of 1872 November 27, as seen from France. (Images: Angelo Secchi in Guillemin's *World of Comets*, and the Andromedids in Guillemin's *Le Ciel*, both 1877)

Type 2: Close Approach

There are probably countless meteoroid streams crisscrossing the Solar System. Unless they directly cross the Earth's orbit though, we will know nothing of their existence. However, there must be streams that just skirt by the Earth's gravitational pull. As the Earth sweeps by, our planet captures the relatively few meteoroids at the very edge of the stream each year, producing regular but limited displays. The problem with a close approach is that it impossible to tell what sort of meteoroid stream has been encountered. Is it a remnant stream or the edge of a more massive stream that has the potential to rival the major meteor showers?

Because everything in the Solar System is gyrating, a close approach may sometimes result in a...

Type 3: One-Off Display

There are numerous reports of showers that are seen just once and never again. This could be because the Earth just brushes the edge of a meteoroid stream but, a year later, the separation is once again too great to produce a shower. Any observer noticing unusual activity from a particular location should inform the IAU's Meteor Data Centre. This could include, for example, activity noticed over several nights or an outburst of a significant number of meteors during a short period of time. If the event is not a one-off, then you may have discovered a...

Type 4: New Shower

Sometimes a new shower appears and becomes an annual event. This could be due to the Earth's orbit and that of the stream gradually drifting into one another, or it could be due to the Earth encountering a stream that has only recently been laid down, perhaps by a comet with an orbital period of thousands of years recently arriving at perihelion.

In 2019, Quanzhi Ye predicted that a new shower would be found in the constellation Sculptor, caused by the disintegration not of a comet, but of an active asteroid, (101955) Bennu (Fig. 7.2a). The asteroid was discovered on 1999 September 11 as part of a near-Earth asteroid survey by the Lincoln Near-Earth Asteroid Research (LINEAR) project. It was named after a mythological Egyptian bird by 9-year-old Michael Puzio from North Carolina [3]. Bennu belongs to a group of small bodies that are essentially rocky, or rock-piles (Fig. 7.2b), but which release solid particles and volatiles rather like a comet. Because of the characteristics of its orbit, Bennu is classed as an Apollo asteroid.

Fig. 7.2a Asteroid (101955) Bennu
The half-a-kilometre active asteroid (101955) Bennu was discovered in 1999 and was the target of the OSIRIS-REx sample return mission in 2020. Credit: NASA/Goddard/University of Arizona

Fig. 7.2b The surface of Bennu
The surface of Bennu is littered with loose rock fragments. Credit: NASA/Goddard/University of Arizona

Ye modelled the dust release from Bennu back to CE 1500 and forward to 2100. Bennu's ejecta takes only about 450 years to disperse into the zodiacal cloud. Observations of the object by the OSIRIS-REx space probe indicate that its ejecta consists of centimetre-sized objects rather than fine dust. Calculations showed that the Earth would encounter the dust trails released between 1523 and 1791 on 2019 September 25 at 12:26 UT, and a meteor display would last for 6 h. Although the debris is large, the low geocentric velocity meant that the resulting meteors would be quite dim. Tim Cooper and four members of the Astronomical Society of South Africa spotted five possible Bennu meteors on September 24/25 from a radiant at α 5° (00^h 20^m), δ −34° in the constellation of Sculptor, in line with Ye's prediction [4]. Further displays up to 2028 are given in Table 7.1. The shower is visible all night from any point in the Southern Hemisphere (Fig. 7.3).

Table 7.1: Predicted Meteor Activity from Bennu		
Date	Time (U.T.)	Duration (h)
2021 Sep 24	21:16	11
2022 Sep 25	02:47	8
2025 Sep 24	23:20	8
2027 Sep 25	10:54	8
2028 Sep 24	12:19	9

A month after Tim Cooper and his members observed the Bennu shower, the OSIRIS-REx probe managed to scoop material from the surface of the asteroid as part of the project's Touch-and-Go sample return mission. At the time of writing, the material is heading back to Earth for analysis.

The **η-Lyrids** (Figs. 7.4a and 7.4b) is one of those cases where the discovery of a comet subsequently led to the discovery of the associated meteor shower.

On 1983 April 25, the Infra-Red Astronomy Satellite recorded a comet that was just 0.413 AU from the Earth. The comet had an orbital period of 970.5 years. On May 3, two amateur astronomers, schoolteacher Genichi Araki in Japan and retired schoolteacher, the legendary George Alcock in England, also spotted the comet and both notified the IAU's Central Bureau for Astronomical Telegrams. On May 4, the discovery of Comet IRAS-Araki-Alcock 1983 VII (C/1983 H$_1$) was announced to the world [5]. It was quickly realized that the comet would become a rare, naked-eye object of m_v +2.4 by May 11. The announcement captured the attention of Jack Drummond of the Steward Observatory in Tucson, Arizona, who pointed out that the comet could produce a meteor shower from the constellation Lyra. He successfully observed the shower on May 9 when the ZHR reached 5.1. IRAS-Araki-Alcock passed with 0.031 AU (4.65 million km) of the Earth on May 11. Although it was technically a 'new' shower, further investigation revealed a number of images of η-Lyrid meteors in various photographic surveys dating back to the 1930s.

At maxima, the η-Lyrids radiant is located at α 287° (19^h 08^m), δ +44°. It is best observed from the Northern Hemisphere from late evening until dawn. South of the Equator, the shower is a post-midnight spectacle with useful observation only possible down to about 30°S. Peak ZHR is likely to be 3.

Type 4: New Shower 293

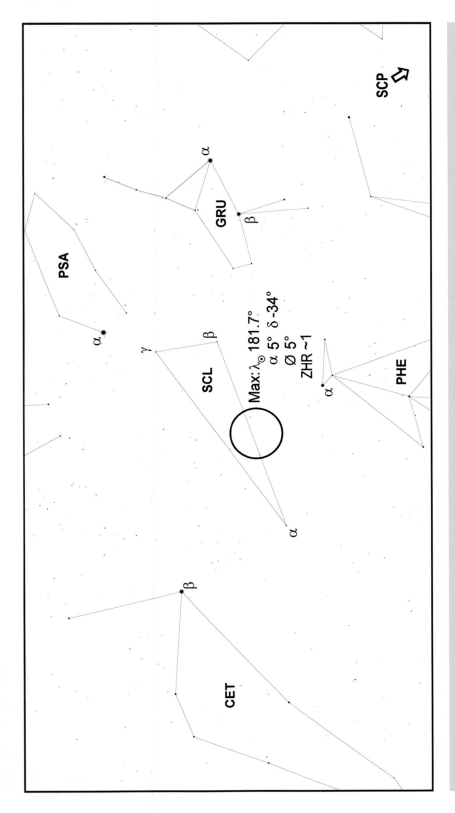

Fig. 7.3 The radiant for debris from asteroid Bennu is in the constellation sculptor

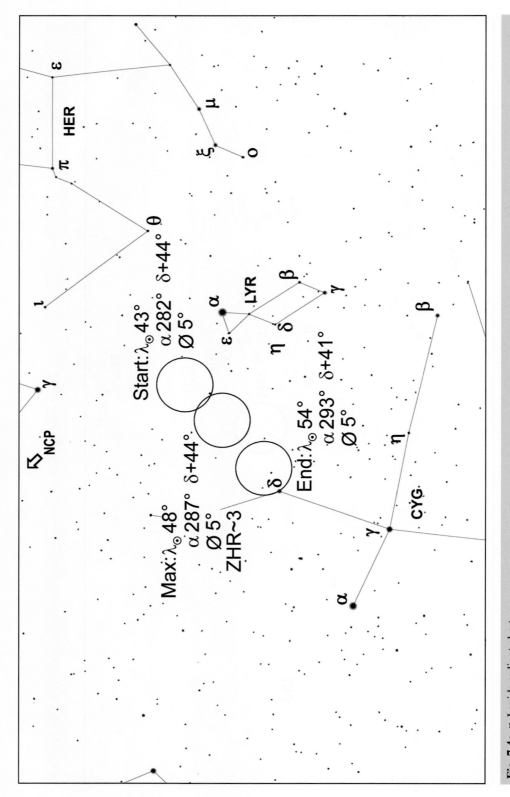

Fig. 7.4a η-Lyrids radiant chart

Type 4: New Shower

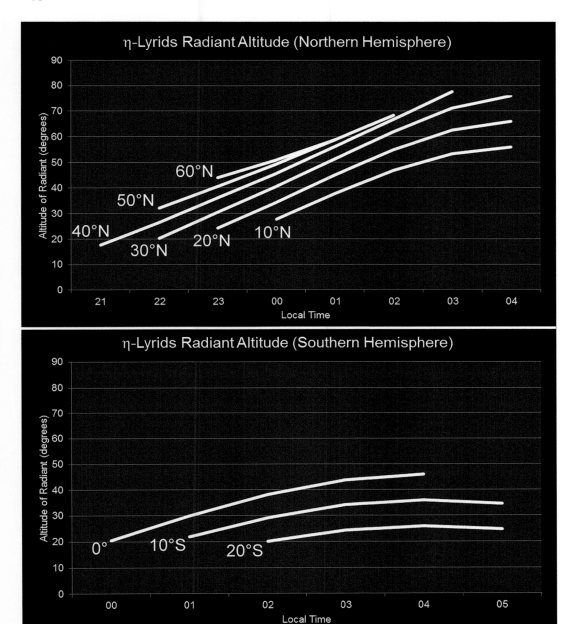

Fig. 7.4b η–Lyrids radiant altitude

A new minor shower was discovered radiating from Volans on 2015 December 31 by the Cameras for All-sky Meteor Surveillance (CAMS) technology, on New Zealand's South Island, which utilizes video cameras. Over a period of 6.5 h, a total of 21 **Volantids** were captured. This promising new shower, however, failed to put on a display the following year, or the year after that, and observers were beginning to think it was just a one-off display. In 2020, the shower suddenly reappeared with a peak ZHR of 11.2 ± 6.5 meteors/h [6]. Definitely a shower to watch out for.

Type 5: Developing Shower

Similar to Type 4, a Type 5 shower initially increases in strength from one year to the next. The **Geminids** are a good example. The stream is slowly sliding across the Earth's orbit. A couple of hundred years ago the shower did not exist, the stream having not quite reached the Earth's orbit at that point. Then, in the 1830s, minor activity sprang from a new radiant, which gradually increased in intensity over the years. Nowadays, the Geminids are often the most prolific shower of the year—but that won't last. During the course of the twenty-first century, activity from the Geminids is expected to subside as the stream continues its relentless drift across the Earth orbit, leaving us behind. In a little more than one hundred years from now, the Geminids will be just a distant memory (see Chap. 6 for more details).

Two other developing showers may be the α-Capricornids and the κ-Cygnids. The **α-Capricornids** (Figs. 7.5a and 7.5b) were first recorded in 1871 by the Hungarian astronomer Miklós de Konkoly-Thege [6]. The shower was studied in some depth by Michael Buhagiar and the Western Australian Meteor Society between 1969 and 1980 but attracts relatively little interest today. The α-Capricornids run from about July 3 to August 15 and can have several peaks from July 25 into the first week of August. The ZHR is around 5; the highest it has ever been was 10 in 1995. Estimates of the shower's radiant range from 5° by some observers in the south to a more diffuse 15°+ by some northern astronomers and overlaps the Anithelion radiant. It is not clear whether the α-Capricornids radiant varies in diameter over the period of activity or by how much [7]. At maxima, the radiant is located at about α 307° ($20^h\ 28^m$), δ −10° with the incoming meteors travelling at 22.7 km/h, somewhat slower than the Antihelion background. The meteors are often bright, colourful and fragment into several pieces—typical of ecliptic-bound showers. The shower displays the occasional fireball. Best time to observe is around midnight.

It was once believed that the α-Capricornids was a remnant shower, its parent comet long gone. A study by Peter Jenniskens and Jérémie Vaubaillon seems to indicate that this is not the case. The two researchers believe that the shower is relatively young, perhaps 4500–5000 years old, and contains 9×10^{13} kg of material. In their model, the Earth is only just encountering the outer edges of the meteoroid stream, and between 2220 and 2420, it will produce a shower that is stronger than any current stream.

Type 5: Developing Shower

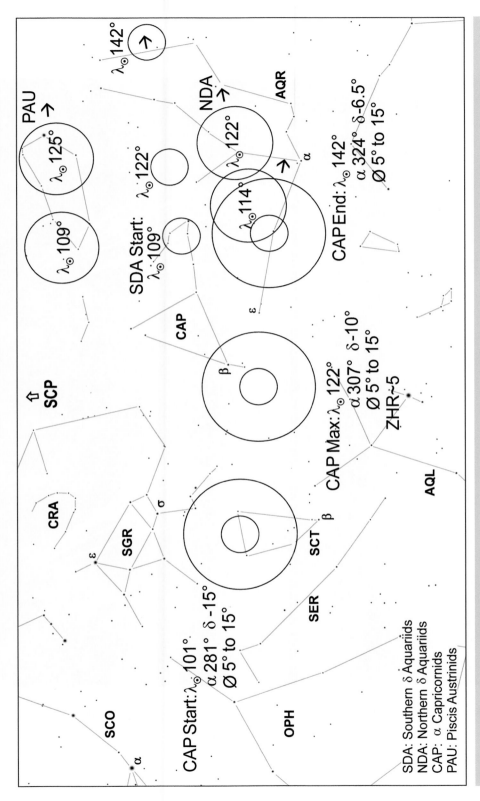

Fig. 7.5a α-Capricornids radiant chart

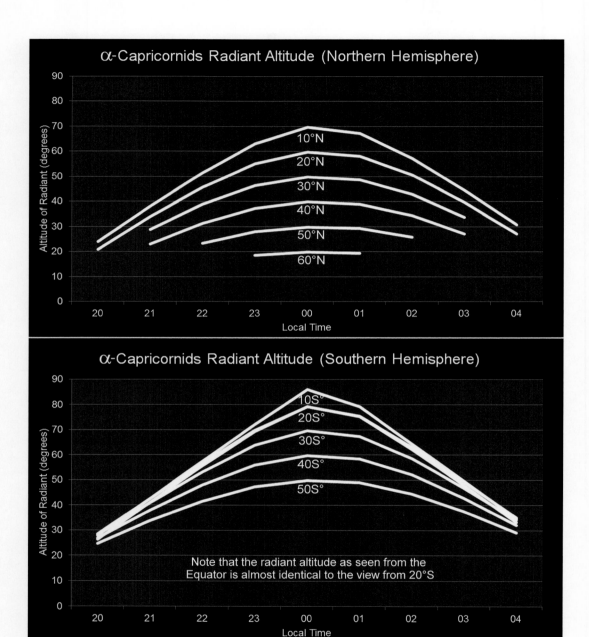

Fig. 7.5b α-Capricornids radiant altitude

Various comets and asteroids have been proposed as the stream's parent body, but the most likely candidate is 169P/NEAT, a 5.7-km-diameter comet that was originally designated an asteroid, 2002 EX_{12}. It is now believed that 169P/NEAT is a fragment of a larger comet that was disrupted 4000–5000 years ago.

Meteors from the **κ-Cygnids** (Figs. 7.6a and 7.6b) share some of the properties of the α Capricornids—bright; a tendency to fragment—but the shower has no obvious parent body. Visual activity occurs from August 3 to August 24 with maximum at $\lambda_\odot$ 145°, around August 17, from a 6°-diameter radiant at α 284.7° (18^h 59^m), δ +59°. Other forms of observation have detected the shower as early as July 26 and as late as September 1. The relatively slow 23 km/s meteors usually produce a ZHR of 3, but occasionally there is an outburst, as in 1993, when the ZHR reached 14 (there was also a Perseid outburst at the same time!). There have been suggestions of a 7-year cycle, but the evidence is weak. Unlike the Perseids, the shower can be observed from dusk for pretty much all night in the Northern Hemisphere.

The first observations of the κ-Cygnids were by Francesco Denza in Italy, who inadvertently discovered the shower while observing the Perseids on 1869 August 11, later reported by George Tupman [9]. The low ZHR failed to capture the interest of observers, especially as there was the more prolific Perseids on offer. Only in 1954 did Fred Whipple manage to establish the shower [10]. He noted that the long duration of the shower suggested the breakup of a large comet, so the κ-Cygnids has long been regarded as a remnant shower. Jenniskens disagrees, citing orbital characteristics that point to it being an evolving meteoroid stream. As yet, the parent body has not been confirmed [11].

Type 6: Variable Activity Shower

There is a special category of meteor shower that displays minor activity for most of the time, but occasionally its rates will increase significantly, though there can be several reasons for this. The Draconid meteor shower is an interesting example.

It was Martin Davidson, Director of the BAA's Meteor Section, who first suggested that debris from Comet Giacobini-Zinner 1900 III (21P/Giacobini-Zinner) could encounter the Earth on 1915 October 10, two years after the comet had reached perihelion [12]. He calculated the radiant to be in the head of Draco. Sure enough, William Denning confirmed the presence of the 'Giacobinids', although activity levels seem to have been quite low [13]. Today, we call the shower the **October Draconids** (Figs. 7.7a and 7.7b), though it has also been called the γ-Draconids or simply the Draconids. Throughout the 1920s, the Draconids were either completely absent or barely detectable, with the exception of 1926. As the Earth crossed the orbit of the comet on October 9/10, 70 days before Giacobini-Zinner, rates started to pick up. The first indication was a brilliant fireball at 22:16 GMT, which was seen by hundreds of people in Britain. It left behind a persistent train that lasted for about half an hour. After that, meteors were seen at a rate of between 14 and 17 per hour.

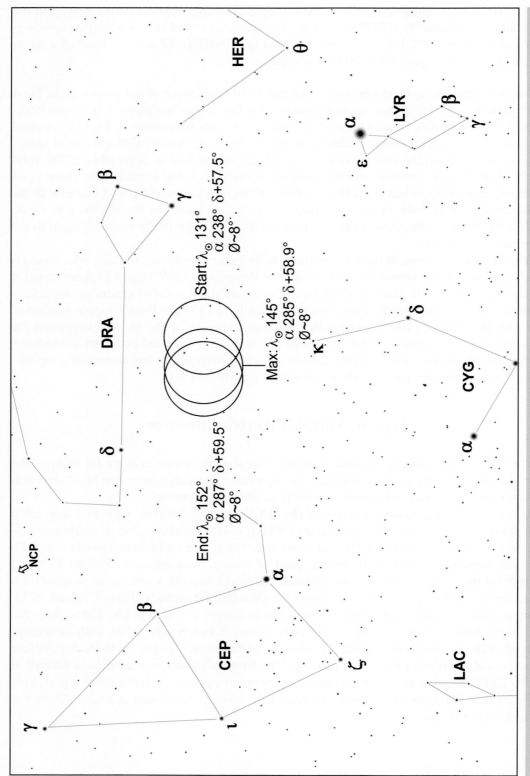

Fig. 7.6a κ-Cygnids radiant chart

Type 6: Variable Activity Shower

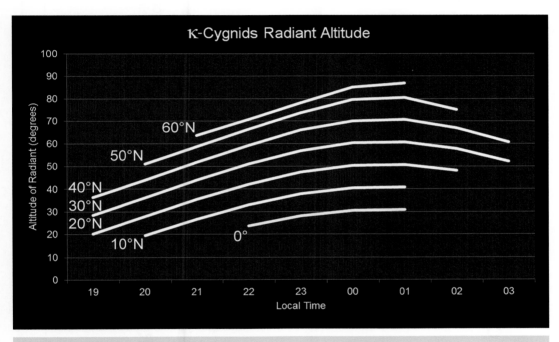

Fig. 7.6b κ-Cygnids radiant altitude

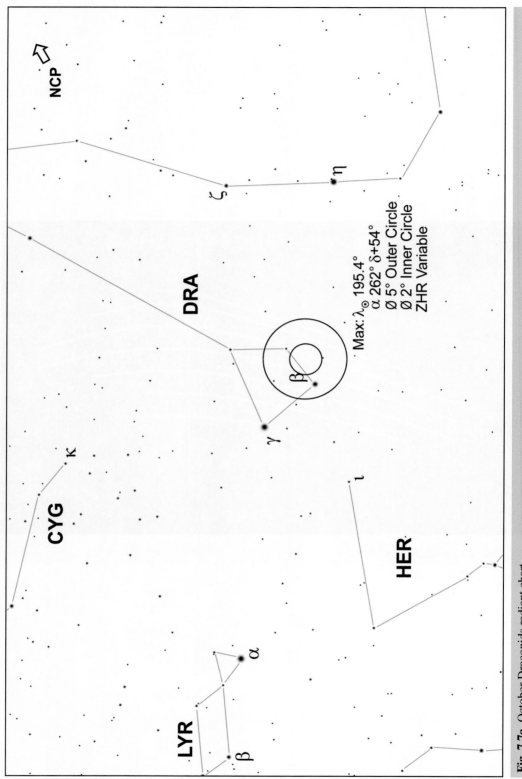

Fig. 7.7a October Draconids radiant chart

Type 6: Variable Activity Shower

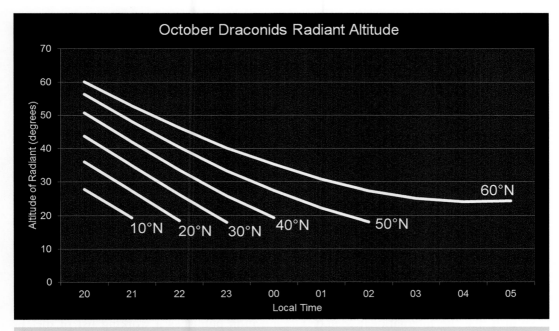

Fig. 7.7b October Draconids radiant altitude

From 1927 to 1932, few, if any Draconids were seen. As 1933 arrived, Giacobini-Zinner reached perihelion on July 15, and the Earth crossed the descending node just 80 days after the comet. Rates sky rocketed with an estimated ZHR of 10,000. Most of the meteors were quite faint—m_v +3 to +5—slow and yellowish. However, it was not the recent passage of the comet that produced this particular storm, but rather dust trails released during 1900 and 1907.

Following the 1933 storm, rates returned to their normal level. There was a prediction that 1939 would see another increase in rates, but Earth crossed the comet's orbit 136 days ahead of comet, so nothing was seen. The shower was virtually non-existent until 1946 when a second storm occurred, again with a ZHR of 10,000.

From 1947 to 1984, the shower, when it did appear, produced only very minor activity. There was a radar burst in 1952, caused by dust that was too small to be detected visually [14], and 1959 was slated to be a good year but disappointed everyone; Jupiter had perturbed the comet so it didn't come any closer than 0.058 AU (8.7 million km) to the Earth's orbit. 1972 saw a ZHR of 10–15, but rates were generally about 1 or 2 throughout most of the 1970s and early 1980s [15].

1985 saw a brief burst over Japan with a ZHR of 700, but a predicted display in 1988 simply did not materialize. 1998 was almost a repeat of 1985 with a slightly higher ZHR of 720, again over Japan, with rates of 5 to 10 for a few days either side of maximum. A few other years saw high rates—1990 ZHR 20–30; 2005 ZHR 150; 2011 ZHR 300 and 2018 ZHR 110—but activity has been quite muted generally. Enhanced activity is predicted for 2024, 2025 and 2030 [16].

The October Draconids are best seen from mid- to high-northern latitudes from early evening onwards. The farther south you go, the shorter the night at this time of year. From 10°N, observers are limited to just 1 h of observation. The shower cannot effectively be observed from the Southern Hemisphere. With the shower lasting just five days, the radiant barely moves across the sky. Estimates of the average radiant size range from 5° to a very compact 2°.

Ones to Watch

There are a number of established minor showers that can be observed each year. However, because most observers ignore minor showers, there is a dearth of reliable information to such an extent that even basic details, such as start, maximum and end dates, position of the radiant and drift, often contain a large degree of error. While technology can help to constrain certain parameters, all technology has its limitations and own level of error, so visual observations can help provide a fuller understanding of activity and the physical nature of a shower.

The following showers are all worth pursuing. Shower details are given in Appendix 9. Note also that the International Meteor Organization publishes a working list of meteor showers, which highlights minor showers in need of more observation.

α-Centaurids (ACE)

The α-Centaurids (Figs. 7.8a and 7.8b) are a Southern Hemisphere shower and are difficult for any observers north of the Equator. The shower begins around January 31 and ends February 20, with maximum activity at 319.2° (about Feb 8) from α 210° ($14^h 00^m$), δ −59°.

The first probable observation of the α-Centaurids was by V. Williams [17] from Sydney, Australia on 1889 February 10/11. Nothing more was recorded until 1938, when Cuno Hoffmeister probably picked up activity from his site in South West Africa. However, the shower was only formally recognized during the 1970s, when Michael Buhagiar and members of the Western Australia Meteor Society undertook systematic observations [18]. The ZHR is normally around 6 but can occasionally be higher; 1974 saw rates of 12–15 meteors per hour and 169 α-Centaurids were recorded by three Australian observers in 1980 (ZHR 30). This particular outburst produced many yellow meteors [19]. The average magnitude was an impressive $\overline{m}_v$ +0.54 (usually it is $\overline{m}_v$ +2.45) and almost half the meteors had trains. The geocentric velocity is 58 km/s, making them quite swift. The parent body of the stream is unknown.

α-Centaurids (ACE)

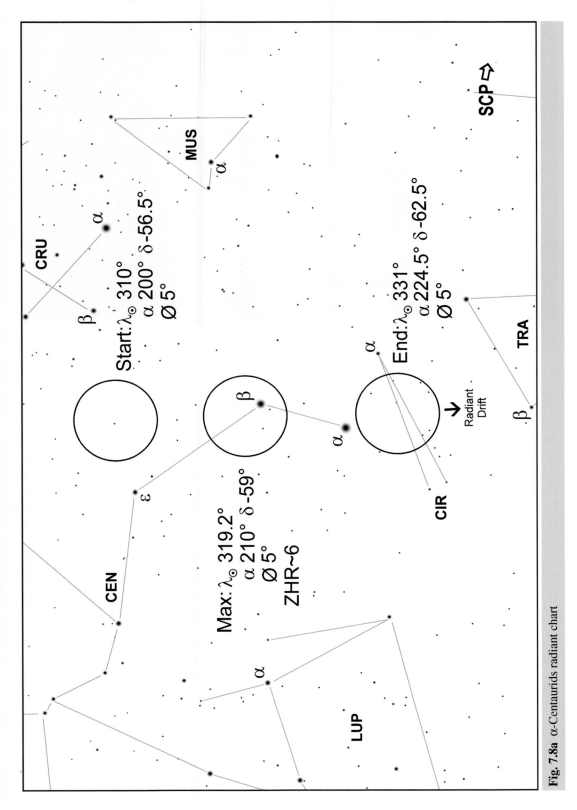

Fig. 7.8a α-Centaurids radiant chart

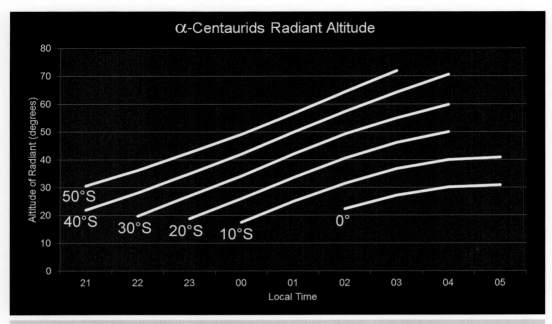

Fig. 7.8b α-Centaurids radiant altitude

June Boötids (JBO)

The June Boötids (Figs. 7.9a and 7.9b) is one of those showers that has the ability to suddenly surprise. Most of the time its ZHR is just a few meteors per hour, but every now and then there will be an outburst of activity, which can last from perhaps 30 min to several hours. The shower's normal limits are June 22 to July 2, with maximum around June 27 at $\lambda_\odot$ 95.7° from an average radiant at α 224° (14^h 56^m), δ +48°, sandwiched between Boötes, Hercules, Draco and Ursa Major. The radiant is quite large: somewhere between 12° and 15°. At 18.7 km/s, the meteors appear slow.

The shower favours observers in the Northern Hemisphere, particularly those in mid-northern latitudes around 45°N. Anyone much father north is hampered by short nights and twilight. Activity can be seen from 10°S prior to midnight, but observers farther south will find the radiant hugging the horizon, with the chances of spotting any meteors being significantly reduced.

William Denning probably discovered the shower on 1916 June 28, although to his credit, he later pointed out that Edward Lowe had spotted "many meteors" in 1860 and 1861 on June 30 [20]. Initially called the ι-Draconids, the shower later came to be known as the Pons-Winneckids, after its association with Comet Pons-Winnecke was uncovered by both Denning and Charles Olivier.

June Boötids (JBO)

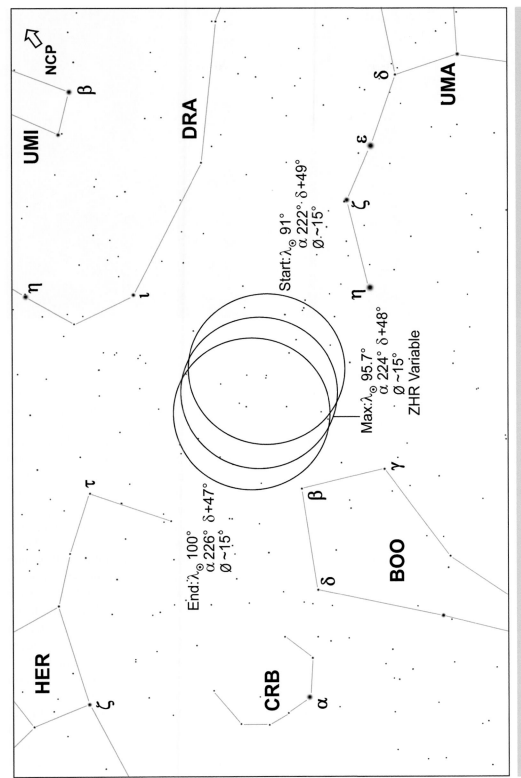

Fig. 7.9a June Bootids radiant chart

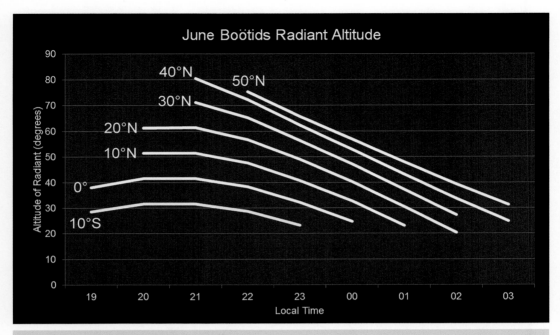

Fig. 7.9b June Bootids radiant altitude

Following Denning's discovery, or rediscovery, of the shower, activity slumped, with few if any meteors recorded. There may have been an outburst witnessed by K. Nakamura in Japan on 1921 July 3, who noted 135 meteors in just 35 min, although several researchers have questioned Nakamura's account [21]. Other observations from around the world at that time suggested a ZHR of no more than 10, but it is entirely possible that Nakamura did observe a short-period burst. Observers belonging to the Russian Mïrovédénië (World Studies) Society observed 500 meteors/h on 1927 June 27 from their site in Uzbekistan [22]. European observers saw at best a ZHR of about 30. Again, this may have been due to a short burst of activity, but another explanation is that the skies over Uzbekistan were exceptionally clear, with a limited magnitude perhaps as low as m_v +7.5. The observers noted that 90% of the meteors were of 5th magnitude of less.

Seventy years passed without any noteworthy appearances of the June Boötids. Then, without any warning, the shower put on a command performance on the night of 1998 June 27, with a maximum ZHR of 250 ± 50 for observers in Japan. This was no short-lived burst. The shower's strong display lasted for a total of seven hours, with observers in both Europe and the USA reporting high rates, bright greenish meteors and almost 50% displaying long-duration trains that lasted for more than 1 s. The Ondrejov Observatory in the Czech Republic captured a m_v −7.9 fireball with a terminal mass of about 0.14 kg [23]. Subsequent investigations revealed the debris was from a trail laid down by Pons-Winnecke in 1825.

Not much has happened since then. 2004 saw a respectable ZHR of 18, but a prediction of enhanced activity in 2010 didn't really materialize, although a near-full Moon kept many observers in their beds. There was a 20-min outburst during daylight hours on 2018 June 27 at 16^h UT, detected by the RAMBo radio telescope in Bologna, Italy, and a superbolide, said to be as bright as the Sun, crossed the skies of Tajikistan on 2008 July 23, which had an orbit similar to 7P/Pons-Winnecke [24]. There is a possibility the bolide resulted in an as-yet unrecovered meteorite fall.

June Boötids (JBO)

Since 1819, the orbit of 7P/Pons-Winnecke has changed significantly, with the inclination increasing from 10.8° to 22.4° and the perihelion distance increasing from 0.77 AU to 1.26 AU. The orbit now appears more stable (Fig. 7.10).

The June Boötids is one of those showers that observers should definitely keep an eye on.

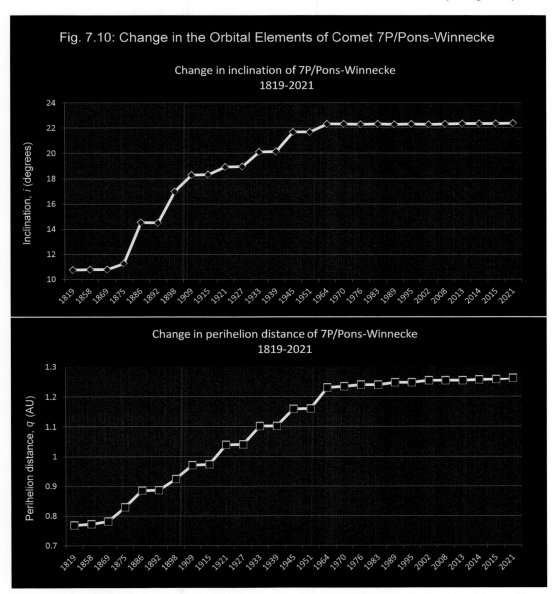

Fig. 7.10: Change in the Orbital Elements of Comet 7P/Pons-Winnecke

The orbit of comet Pons-Winnecke has changed significantly from 1819 but has stabilized since 1964

Aurigids (AUR)

The Aurigids (previously called the α-Aurigids) (Figs. 7.11a and 7.11b) were probably first observed in 1873 by J.E. Clark of Street in Somerset, England, when he observed 9 meteors on September 1. No other observations took place until Cuno Hoffmeister witnessed the shower in 1911, 1929 and 1930, though he didn't realize it at the time. The official discovery of the Aurigids had to wait until 1935, when Hoffmeister and colleague Arthur Teichgraeber at Hoffmeister's Sonneberg Observatory in Germany, as well as A. Vratnik, J. Vlcek and J. Stepanek of the Czech Astronomical Society at the Stefanik Observatory in Prague, independently saw a short outburst with a ZHR of somewhere between 30–50 [25]. More than half a century passed before the shower put on another good display in 1986, when Istvan Tepliczky, a member of the Hungarian Meteor and Fireball Network in Tata, witnessed a display of fast, bright and often yellow meteors, many with persistent trains [26]. Tepliczky estimated the average magnitude to be $\overline{m}_v$ +0.54, with the brightest meteor being m_v −4. The ZHR was calculated to be around 40.

The year 1994 saw rates climb to a ZHR of between 37–55 depending on where the observer was located. The first predicted outburst was in 2007. Again, bright meteors were the order of the day, and although the outburst lasted only about 20 min, it would have scaled up to a ZHR of about 130. The next outburst is predicted to be in 2021 with a ZHR of 50–100 [27].

The Aurigids are best observed from mid- to high-northern latitudes. At 50°N—London, Frankfurt, Vancouver and Winnipeg—the radiant is visible for about four hours after local midnight. At 10°S, barely an hour of observation is possible before dawn.

Vladimir Guth linked the Aurigid meteoroid stream with Comet Keiss 1911 II (C/1911N$_1$), an object with an orbital period of about 2500 years and an aphelion of 367.5 AU [28]. Only one other stream is associated with a long-period comet: the April Lyrids with Comet Thatcher (C/1801G$_1$) period of 415 years.

Aurigids (AUR)

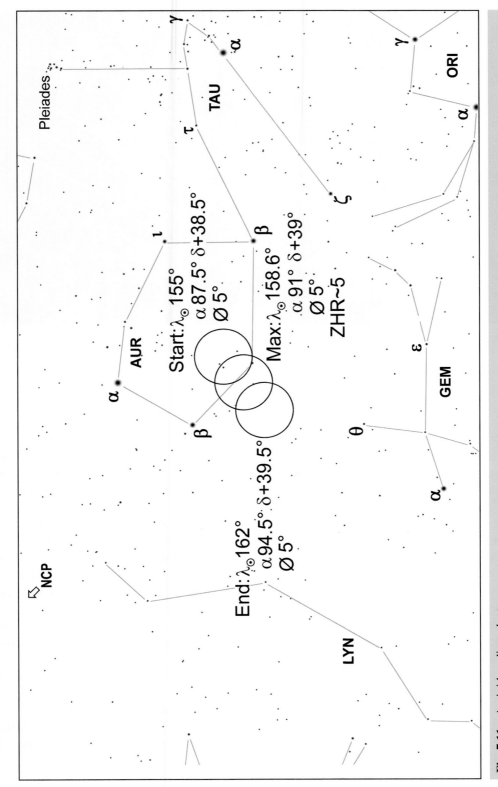

Fig. 7.11a Aurigids radiant chart

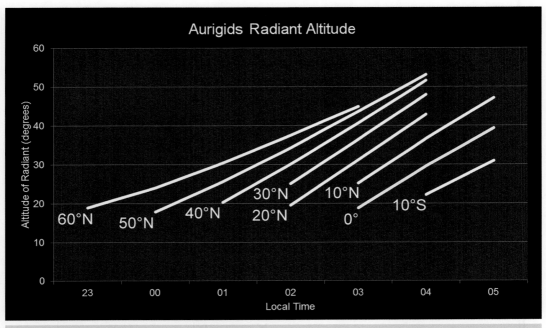

Fig. 7.11b Aurigids radiant altitude

Piscis Austrinids (PAU)

A sadly overlooked annual shower—because of its proximity to the Southern δ-Aquariids, or perhaps because the two are often confused—is the Piscis Austrinids (Figs. 7.12a and 7.12b). This shower is best viewed from the Southern Hemisphere, particularly from between about 20°S and 30°S where the radiant is a cat's whisker away from the zenith at about 2:30 LT on August 1.

The shower was probably first observed by Alexander Herschel [29] in 1865, and Denning also recorded it in 1898. For these observers, the shower would be only about 15° above the horizon at most, which perhaps is an indication of the absence of light pollution in the late-nineteenth century. New Zealand astronomer R.A. Mackintosh identified no fewer than seven radiants in the area in the 1930s (Table 7.2). Apart from a strong radar signature picked up from Christchurch, New Zealand in 1955, the shower more or less disappeared until 1965, when it was recovered by E.F. Turco, a Regional Director of the American Meteor Society, from Rhode Island [30]. He reported that most of the meteors were faint, although a couple of fireballs did occur.

Activity begins in mid-July with the radiant near the star θ Piscis Austrini, reaches maximum around August 1 ($\lambda_\odot$ 129°) with a ZHR of about 5, just about centred on α PsA, and fizzles out about nine days later as the radiant drifts over Sculptor. Careful observation is needed to avoid contamination with the Southern δ-Aquariids, and vice versa, made all the more difficult by the fact that meteors from both showers have almost identical velocities of around 43 km/s.

Piscis Austrinids (PAU)

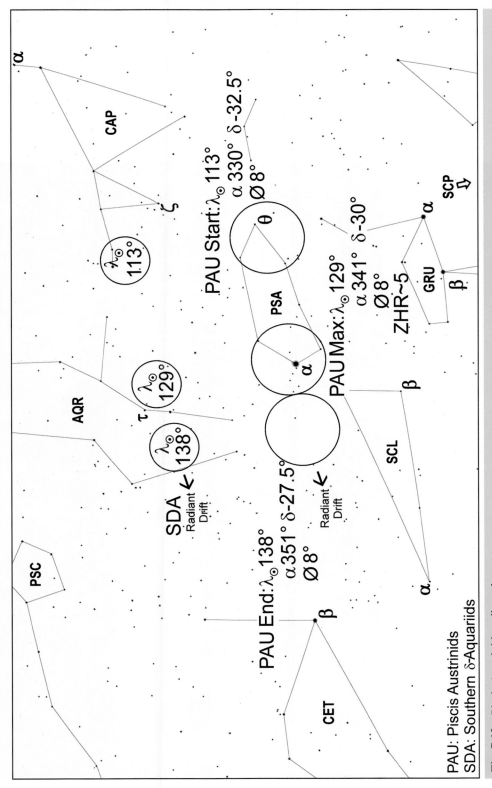

Fig. 7.12a Piscis Austrinids radiant chart

Table 7.2: Piscis Austrinids Radiants (R.A.Mackintosh)					
Radiant	Duration		α		δ
β-Pisces Australids	July 14-22	Start:	330.5°	$22^h 02^m$	-30°
		End:	339°	$22^h 36^m$	-30°
α-Pisces Australids	July 26 – August 8	Start:	337°	$22^h 28^m$	-33°
		End:	350°	$23^h 20^m$	-30°
Pisces Australids	July 28 – August 3		326°	$21^h 44^m$	-26°
λ-Pisces Australids	August 5-14	Start:	334°	$22^h 16^m$	-27.5°
		End:	339°	$22^h 36^m$	-26°
20 Pisces Australids	August 8-9		340.5°	$22^h 42^m$	-27°
θ-Pisces Australids	August 12-14		327°	$21^h 48^m$	-32°
ε-Pisces Australids	August 13-14		338°	$22^h 32^m$	-24°

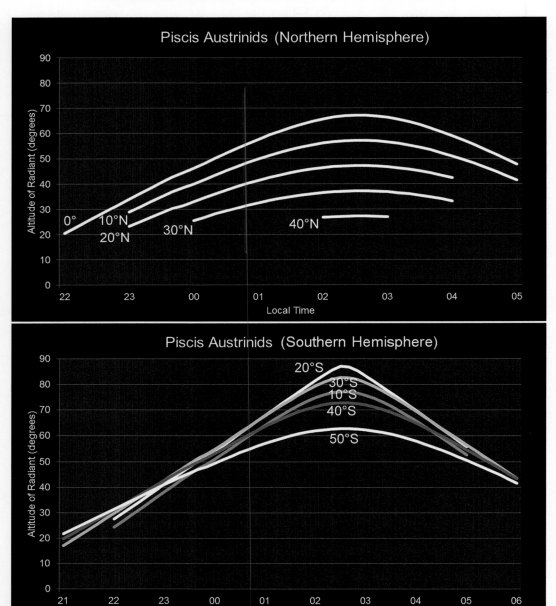

Fig. 7.12b Piscids Austrinids radiant altitude

September ε-Perseids (SPE)

Running from September 5–21, the September ε-Perseids (Figs. 7.13a and 7.13b) offer a great opportunity to hone your plotting skills during the time of year when temperatures are still quite mild. In the Northern Hemisphere, the shower can be observed just about all night, from dusk till dawn. At 40°N, the radiant almost touches the zenith at sunrise. Once you reach the Equator, it becomes a post-midnight event with useful observation taking place down to about 30°S.

The radiant first makes an appearance just north of 16 Persei, then charges eastwards through the constellation, hitting maximum activity as it just about reaches Algol, or β Persei, the so-called 'Demon Star' or 'Winking Demon' that fluctuates in brightness over about three days. It then heads towards ε Persei, where it dies out. The ZHR is about 5 at maximum, although there were outbursts in 2008 and 2013. Obviously, it would have made more sense to name the shower the September β-Perseids—there's already a β-Perseid shower—but the powers that be thought otherwise. That's not the only odd thing about the shower.

There have been numerous radiants reported to be active during this period in the east of Perseus. It was probably Cuno Hoffmeister [31] who named this activity the September Perseids in his *Meteorströme (Meteor Streams)* in 1936, which Robert Lunsford [32] also used in his *Meteors and How to Observe Them*, adding the IAU Code SPE (2009). Terentjeva referred to the shower as the ξ-Perseids during a study of photographic images taken by US and Canadian networks during the period of 1963–1984 September 9–17 [33]. The name 'September ε-Perseids' was used by the International Meteor Organization during its analysis of video images for 1999–2008. In all, there are at least 18 radiant points, many of which make little sense (Fig. 7.13b and Table 7.3). Carefully plotted meteors could reveal whether the currently accepted radiant is correct and determine how many of the other radiant points are still active.

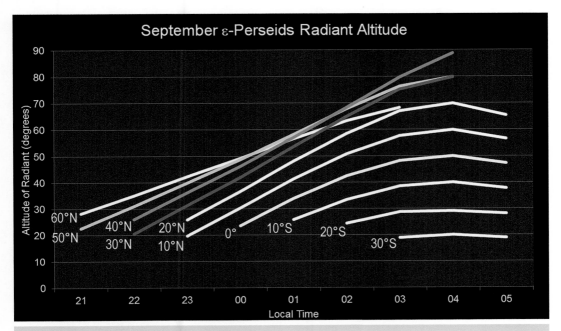

Fig. 7.13a September ε-Perseids radiant altitude

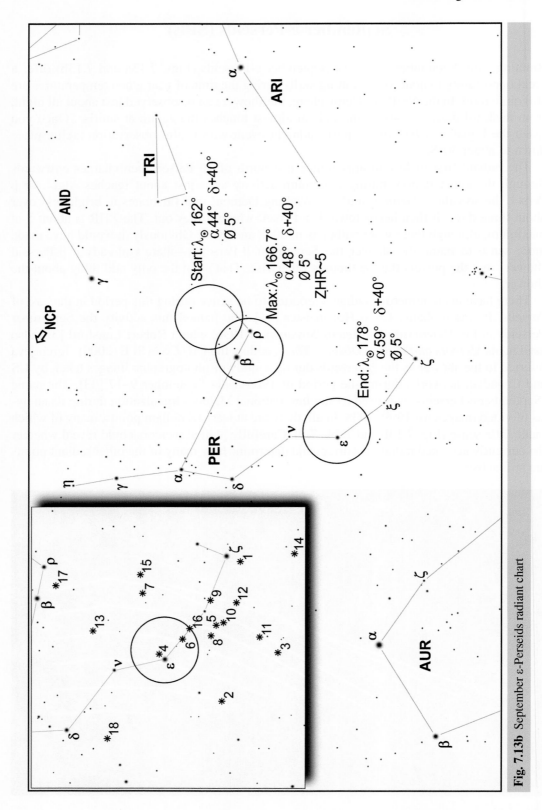

Fig. 7.13b September ε-Perseids radiant chart

September ε-Perseids (SPE)

Table 7.3: Reported Radiants in Perseus from Late-August to Mid-September

Ref.	Date	Observer	α	δ	Notes
1	1869 September 6	Zezioli	60°	+32°	per Denning
2	1871 September 7-15	Tupman	66°	+40°	per Denning
3	1872 August 24 to September 14	Italian Meteoric Association	67°	+35°	per Denning. Radiant determined from a total of 10 meteors.
4	1874 September 6	Denning	59°	+40°	Radiant near ε Persei
5	1877 September 4-16	Denning	61°	+36°	238 meteors. Swift, brighter than average, mostly trained.
6	1877 September 7	Denning	60°	+38°	5 meteors
5	1880 September 2-6	Denning	61°	+36°	4 meteors
7	?	Heiss	55°	+37°	
8	1885 September 3	Denning	62°	+37°	7 meteors
8	1885 September 8-10	Denning	62°	+37°	4 meteors
9	1877-1885 September 4-5	Denning	60°	+35°	5 meteors
10	1886 August 28 and September 7	Denning	62°	+36°	6 meteors
11	1898 September 15	Denning / Herschel	65°	+35°	Fireball
12	1911 September 18	Unknown	62°	+34°	per Denning. One meteor brighter than m_v +1
13	1936 September 16	Hoffmeister	52.9°	+40.8°	'September Perseids' 20 meteors in 3.7h.
14	1936 September 16	Hoffmeister	63°	+29°	4 meteors
15	1963-1984 September 9-17	Terentjeva	54°	+36°	US/Canada photographed fireballs. 'ξ-Perseids'
16	1988 September 3-20	Spanish Meteor Society	60°	+37°	41 meteors, mostly faint. V_g = 50 km/s
17	1999-2008	IMO	48°	+39.5°	Video Network 'September ε-Perseids'
18	2009	Lunsford	60°	+47	'September Perseids' (SPE)

November Orionids (NOO)

Running from about November 13 ($\lambda_\odot$ 231°) to December 6 ($\lambda_\odot$ 254°), this shower is sandwiched between the Geminids and the December Monocerotids and suffers as a result, with few observers being able to distinguish between the various showers (the Southern χ-Orionids are also active at the same time in the same area of sky!). However, whereas the Geminids have a velocity of about 24 km/s, both the Monocerotids and the November Orionids are almost twice as fast at 41 km/s. That makes it easier to separate them from the Geminids but devilishly difficult to extricate them from the Monocerotids. Maximum is said to be at $\lambda_\odot$ 246°—about November 28—with a ZHR of 3. See the section on the Geminids for a chart and visibility.

Southern χ-Orionids (ORS)

Of all the showers active in the Geminid region in November, the Southern χ-Orionids is the weakest with a ZHR of just 1. Activity starts about November 13 ($\lambda_\odot$ 231°)—the same date as the November Orionids—peaks about December 2 ($\lambda_\odot$ 250°), but hangs around for almost three weeks with the ZHR at less than 1. There is a northern component which, not surprisingly, is called the **Northern χ-Orionids** (ORN), beginning somewhat later on November 23 ($\lambda_\odot$ 241°) and peaking on December 12 ($\lambda_\odot$ 260°) before disappearing on December 18 ($\lambda_\odot$ 266°). With a ZHR of less than 1, it will hardly get observers excited, but for completeness, identifying meteors from this component is useful.

The stream has been tentatively linked to a couple of parent bodies, 2010LU$_{108}$ and 2002 XM$_{35}$. For details of the radiant position and altitude, please refer to the Geminids section.

σ-Hydrids (HYD)

Despite being observable from much of the globe, relatively little is known about the σ-Hydrids. It is thought to commence activity around December 3 ($\lambda_\odot$ 251°) and peters out around December 21 ($\lambda_\odot$ 269°) but no one is absolutely sure. Maximum occurs on December 9 ($\lambda_\odot$ 257°) although there have been reports of a second peak on December 14 ($\lambda_\odot$ 262°); this may be due to contamination by the Geminids. The shower can also be confused with the December Monocerotids, which are never far behind, but the σ-Hydrids are about 20 km/s faster (Figs. 7.14a and 7.14b).

The shower was probably first observed by Cuno Hoffmeisterin 1937 from present-day Namibia (then South West Africa) [31]. It was not recognized as a distinct shower until R.E. McCrosky and A. Posen published the results of the Harvard Meteor Survey 1952–54, which showed seven meteors emanating from the same radiant [34]. In 1978, members of the National Association of Planetary Observers' Meteor Section in Australia undertook a study of the shower and estimated the average magnitude to be $\bar{m}_v$ +3.06, with a little over 6% of meteors leaving behind persistent trains [35]. About one in five meteors were said to be yellow. Generally, the shower displays faint meteors with only the occasional bright members, although a magnitude m_v −8 fireball was seen over Japan in 1993.

The σ-Hydrids are best after local midnight in any part of the world. Those living within the Tropics will see the radiant peak around 3 am LT when its altitude is greater than 70°, allowing more than 90% of activity to be seen.

σ-Hydrids (HYD)

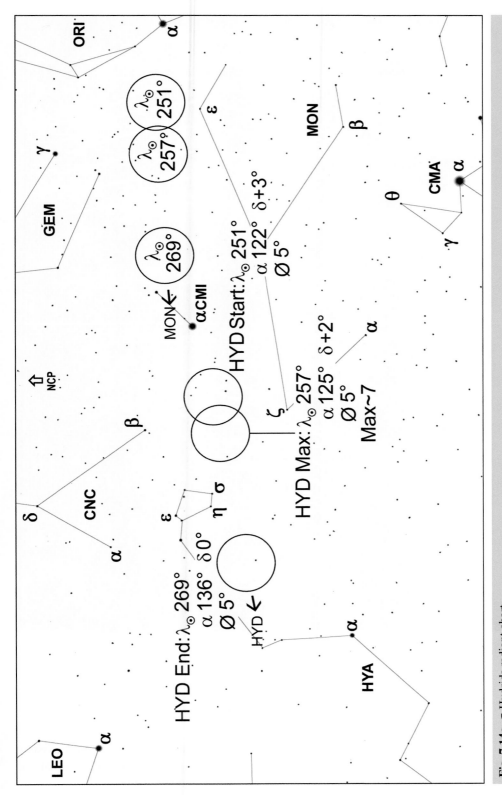

Fig. 7.14a σ-Hydrids radiant chart

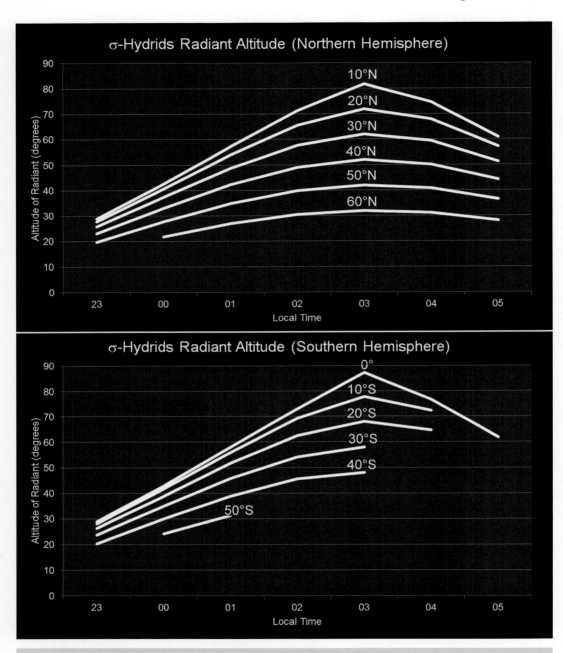

Fig. 7.14b σ-Hydrids radiant altitude

Phoenicids (PHO)

This Southern Hemisphere shower is rarely observed because of its low to non-existent ZHR, yet its rich history begs for more observation to help clarify the demise of the Phoenicid meteoroid stream and its parent comet (Figs. 7.15a and 7.15b).

Phoenicids (PHO)

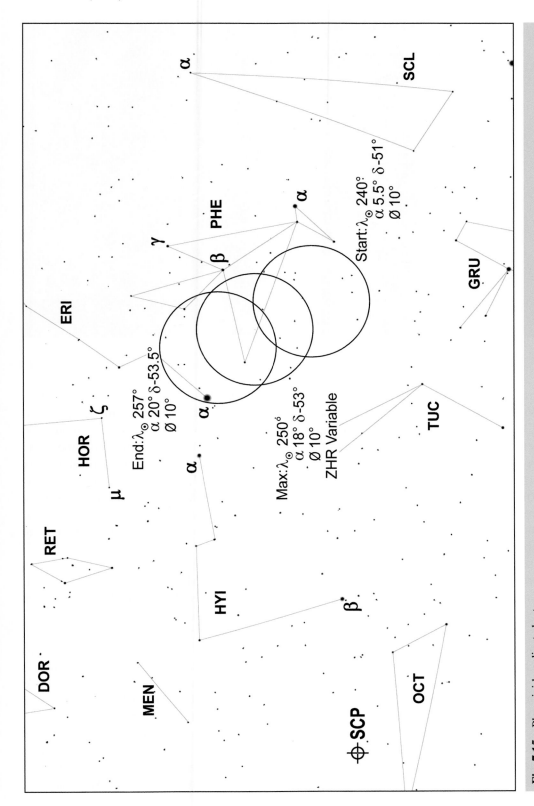

Fig. 7.15a Phoenicids radiant chart

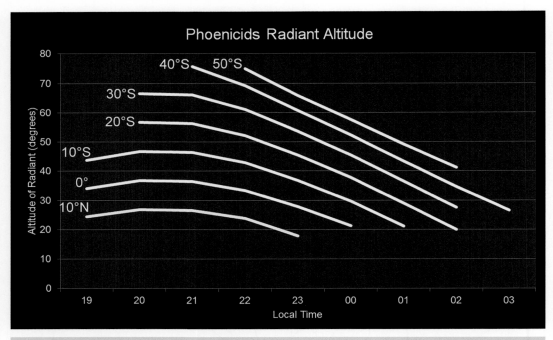

Fig. 7.15b Phoenicids radiant altitude

Probably the first record of Phoenicid activity dates back to 1887 when, on December 3, V. Williams of Sydney, Australia witnessed a short outburst of more than 1 meteor per minute [36]. Then nothing was recorded for 69 years, until a second outburst occurred on 1956 December 5, which was so impressive that it caught the attention of both astronomers and the wider public alike. S.C. Venter, the Meteor Section Director of the Astronomical Society of South Africa in Pretoria, estimated the ZHR to be 100 [37]. Half a world away in New South Wales, Australia, Charles Shain saw the outburst while taking a break from his work at the Radiophysics Laboratory [38]. He noted more than one meteor per minute, many of them slow, yellow and very bright, typically m_v −2. Other observers commented on the large number of exploding fireballs, said to be reddish, orange and yellow.

Following the 1956 outburst, the shower all but disappeared, with few observations recorded. The 1970s saw the rise of the Western Australia Meteor Society (WAMS) and annual, systematic observation of the shower between 1977 and 1986. In 1972, observers were treated to an outburst of 20 meteors/h. Their study of the shower in 1983 yielded 97 meteors over 63 h, with an average magnitude of $\overline{m}_v$ +3.27. Only 2% left trains. 1985 saw a slight improvement, $\overline{m}_v$ +2.38, 4.8% with trains [39], and this continued into 1986 with $\overline{m}_v$ 2.88 and 5.3% trained [40]. WAMS largely died out from the late 1980s and the Phoenicids were mostly ignored, but not entirely. There were modest enhancements in 2008, when the Earth passed within 0.00012 AU (18,000 km) of dust trails laid down in 1861 and 1866, and in 2014, when the ZHR reached 16 thanks to an encounter with the 1956 dust trail.

It was Harold B. Ridley, one-time Director of the British Astronomical Association's Meteor Section, who first made the connection between the Phoenicid meteoroid stream and Comet Blanpain 1819 IV (289P/Blanpain) [41]. The comet was discovered by the French astronomer Jean-Jacques Blanpain on 1819 November 28. He noted it had "a very small and confused nucleus". The radio astronomer A.A. Weiss confirmed the link with the comet in 1958 but also noted something unusual: the number of radio echoes was only about one third of the visual ZHR [42]. Usually, they outnumber the visual count by a significant factor. A number of possible explanations were forthcoming, but the most likely is that Weiss' data covered a period 6 h prior to the visual outburst, so most of the activity was simply missed.

The comet went missing, believed to have fragmented, and was given the designation D/1819 W_1, the D standing for 'Disappeared'. However, in 2003, a small object was discovered and, believed to be an asteroid, received the designation 2003 WY_{25}. It soon became apparent that 2003 WY_{25} was in an orbit very similar to Blanpain's Comet. Two years later, the British-American astronomer, David Jewitt, detected a faint coma and in July 2013 the Panoramic Survey Telescope and Rapid Response System (Pan-STARRS), located at Haleakala Observatory, Hawaii, recorded a significant outburst [43]. The cometary nature of 2003 WY_{25} was in no doubt and it was redesignated as Comet 289P/Blanpain.

Piecing all the evidence together, it would seem that the comet was disrupted by a close approach to Jupiter shortly before being discovered by Jean-Jacques Blanpain; his comment about the unusual appearance of the nucleus then begins to make sense. At perihelion, the comet was subjected to further gravitational forces, which fragmented the nucleus, and the comet was too small to be detected by nineteenth century instruments. The nucleus is now tiny—just 160 m across—and too small to produce the amount of dust in the meteoroid stream, which is estimated to be about 10^8 kg, so it is likely that the event that disrupted the nucleus in 1819 also led to the production of much of the dust that currently exists within the stream. The comet's orbit attains perihelion at 0.96 AU and reaches 5.13 AU at its farthest point (Jupiter lies between 4.95 and 5.46 AU). With an inclination of a little under 6°, it has an orbital period of 5.31 years.

There are a lot of unknowns attached to the Phoenicids and disagreement as to if or when the stream will again encounter the Earth and produce a strong display. Regular observation of the shower may yield clues as to its progress.

December Monocerotids (MON)

The December Monocerotids (originally just the Monocerotids) are a bit of an enigma. There are fireball reports for this period in Chinese annals from the eleventh to the sixteenth century [44], and a couple of accounts of fireballs in CE 381 November 20 and December 13, the latter apparently being audible. Originally, it was thought that the fireballs belonged to the Geminids, but we now know that the Geminid stream did not come anywhere near the Earth until just a couple of centuries ago, so it is at least possible that the Monocerotids were responsible for the activity that the Chinese so diligently recorded.

The shower was discovered by Fred Whipple in 1954 from photographs taken by the Harvard Observatory [10]. Two meteors captured on 1950 December 13 and 15 had orbits that were very similar to Comet Mellish 1917 I (C/1917 F_1), a Halley-type comet discovered by the American amateur astronomer John E. Mellish with an orbital period of 145 years and an inclination of 33°. A number of radar studies, however, show that the orbit for radar meteors, which tend to be smaller and fainter than photographic and visual meteors, has a lower inclination of just 22°.

The December Monocerotids have never been systematically studied by visual observers. The shower clashes with the Geminids, and what observations do exist tend to show that they peak at the same time as the Geminids. The problem is that the Monocerotid radiant is fairly close to the Geminid radiant and, to complicate matters further, there are also the November Orionids and the Southern χ-Orionids active about the same time. It takes some pretty skilful observing to correctly identify which meteors belong to which shower; most observers are likely to class every meteor as a Geminid. The shower can also be confused with the σ-Hydrids, though at 62 km/s, they are somewhat faster than the Monocerotids' leisurely 41 km/s.

The shower is active from about November 27 ($\lambda_\odot$ 245°) to December 20 ($\lambda_\odot$ 268°), with maximum around December 9 ($\lambda_\odot$ 257°) when the ZHR reaches 3. The shower can be seen from most places on Earth. For a radiant chart and visibility, please see the chapter on the Geminids.

Puppid-Velids (PUP)

The Southern Hemisphere is bombarded by a succession of minor meteoroid streams from late November to late March. Most of the time they never amount to much—a combined ZHR so low that visual observers find the showers are just about impossible to separate from the sporadic background, except in December. Between about December 1 through to December 26, the rates from the Puppid-Velid Complex (Figs. 7.16a and 7.16b) pick up and reach a combined maximum of 8 or 9 in a good year. When is a little more difficult to pin down. Current thinking is December 7, but take that with a large pinch of salt.

With a declination of δ -45°, the shower is very much in the domain of those observers in the Southern Hemisphere, although useful observations can be made from as far north as the Tropic of Cancer, 23°N. South of the Equator, the shower can be observed from mid-evening through till dawn. The complex displays the occasional fireball.

This is an under-observed period that desperately needs more good observation. Cuno Hoffmeister published the first account of activity from the complex in 1948, following his time in South West Africa in the 1930s. Astronomers in Christchurch, New Zealand studied the complex in the 1950s using radar [34], but since then there has been a scarcity of observations and consequently relatively little published—a good opportunity to add to the body of knowledge!

Puppid-Velids (PUP) 325

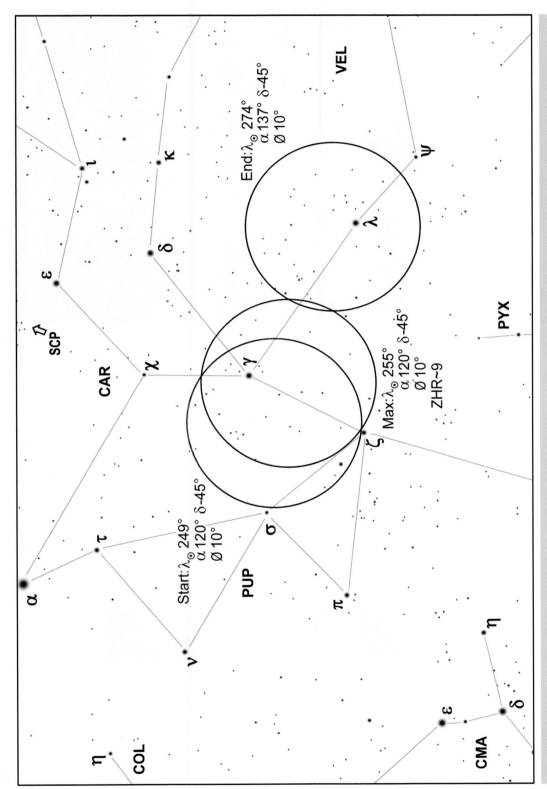

Fig. 7.16a Puppid-Velids radiant chart

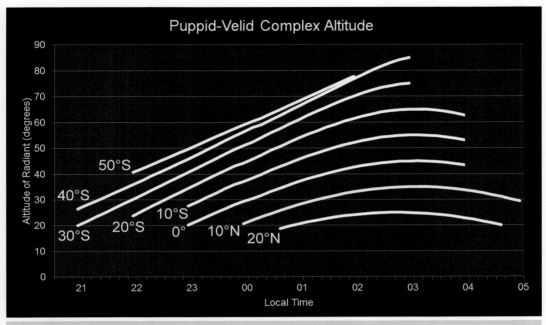

Fig. 7.16b Puppid-Velid complex altitude

Observing Minor Showers

Minor showers are observed in much the same way as major showers, with one difference. There is an argument that, when plotting meteors from minor showers, an observer should artificially increase the size of the radiant as they look farther away from the radiant centre to compensate for errors in backtracking meteors to their source. For example, regardless of the published size of the radiant, if an observer is looking 30° from the radiant centre, they should regard the radiant as being 17° in diameter. This *optimum radiant diameter* is calculated to capture true shower members while minimizing contamination by non-shower, or sporadic, meteors. Table 7.4 and Fig. 7.17 give the optimum radiant diameter for various distances.

Table 7.4: Optimum Radiant Diameter for Minor Showers			
Distance from Radiant Centre	Optimum Diameter	Distance from Radiant Centre	Optimum Diameter
15°	14.0°	45°	19.5°
20°	14.5°	50°	20.0°
25°	16.0°	55°	21.0°
30°	17.0°	60°	21.8°
35°	17.8°	65°	22.5°
40°	18.5°	70°	23.0°

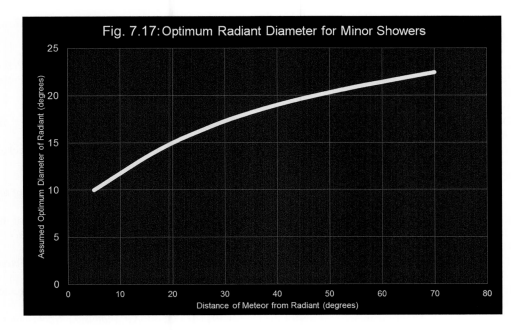

Fig. 7.17: Optimum Radiant Diameter for Minor Showers

New shower claims: If you think you have found a new minor shower, then details should be reported to the IAU Meteor Data Centre.

References

1. Wiegert, P. A., Brown, P. G., Weryk, R. J., & Wong, D. K. (2013). The return of the andromedids meteor shower. *The Astronomical Journal., 145*, 70.
2. Ye, Q. (2019). Prediction of meteor activities from (101955) Bennu. *Research Notes of the American Astronomical Society, 3*, 56.
3. Morton, E. & Neal Jone, N. (2019). *Asteroid's features to be named after mythical birds.* Retrieved from www.nasa.gov/feature/goddard/2019/asteroid-features-to-be-named-after-mythical-birds
4. Cooper, T. (2020). Observations of potential meteors from asteroid 101955 Bennu. *Monthly Notes of the Astronomical Society of Southern Africa, 78*, 142.
5. Marsden, B. G. (1983). *IAUC 3796: 1983d.* Central Bureau for Astronomical Telegrams.
6. de Konkoly, N. (1880). A list of 410 radiation-points of shooting stars, deduced from observations in Hungary in the years 1871 to 1878, and of 80 probable radiants deduced from the 410 radiation-points. *Monthly Notices of the Royal Astronomical Society, 40*, 349.
7. Warne, K. (1981). Observations of the capricornids down under. *Meteor!, 20*, 24.
8. Jenniskens, P. (2006). *Meteor showers and their parent comets* (p. 438). Cambridge University Press.
9. Tupman, G. L. (1869). Results of the observations of shooting stars, made in the Mediterranean in the years 1869, 1870 and 1871. *Monthly Notices of the Royal Astronomical Society, 33*, 298.
10. Whipple, F. L. (1954). Photographic meteor orbits and their distribution in space. *Astronomical Journal., 59*, 201.
11. Jenniskens, P. (2006). *Meteor showers and their parent comets* (p. 442). Cambridge University Press.
12. Davidson, M. (1915). Meteor radiants from Débris of comets. *Journal of the British Astronomical Association., 25*, 272.
13. Denning, W. F. (1926). A new cometary meteor shower (1926 October 9). *Monthly Notices of the Royal Astronomical Society, 87*, 104.

14. Davies, J. G., & Lovell, A. C. B. (1955). The Giacobinid meteor stream. *Monthly Notices of the Royal Astronomical Society, 115*, 23.
15. Corn, E. J. (1979). October Showers. *Northern Meteor Network Bulletin, 6*, 76.
16. Jenniskens, P. (2006). *Meteor showers and their parent comets* (p. 679). Cambridge University Press.
17. Jenniskens, P. (2006). *Meteor showers and their parent comets* (p. 347). Cambridge University Press.
18. Buhagiar, M. (1980). *Southern hemisphere meteor stream list.* Western Australia Meteor Society.
19. Wood, J. (1980). *Western Australia Meteor Society Bulletin* 157.
20. Denning, W. F. (1916). Meteors: Remarkable meteoric shower on June 28. *Monthly Notices of the Royal Astronomical Society, 76*, 740.
21. Yamamoto, I. (1922). Observations in Japan of meteors probably connected with Pons Winnecke's comet. *The Observatory, 45*, 81.
22. Kronk, G. W. (2014). *Meteor showers: an annotated catalogue* (p. 111). Springer.
23. Spurný, P., & Borovička, J. (1998). Photographic observation of a June Boötid fireball. *Journal of the International Meteor Organization., 26*, 177.
24. Konovalova, N. A., Madiedo, J. M., & Trigo-Rodriguez, J. M. (2011). The physical properties of the June Bootids and the July 23, 2008 Superbolide. *42nd Lunar and Planetary Science Conference.*
25. Hoffmeister, C. (1936). Unexpected meteor shower. *Astronomische Nachrichten, 258*, 25.
26. Tepliczky, I. (1987). The maximum of the Aurigids in 1986. *Journal of the International Meteor Organization, 15*, 28.
27. Rendtel, J., Lyytinen, E., & Vaubaillon, J. (2020). Enhanced activity of the Aurigids 2019 and predictions for 2021. *Journal of the International Meteor Organization., 48*, 158.
28. Guth, V. (1936). About the meteor shower of the Comet 1911 II (Kiess). *Astronomische Nachrichten., 258*, 27.
29. Kronk, G. W. (2014). *Meteor showers: an annotated catalogue* (p. 136). Springer.
30. Turco, E. F. (1968). *Review of Popular Astronomy, 62*, 7.
31. Hoffmeister, C. (1936). Meteorströme. Leipzig.
32. Lunsford, R. (2009). *Meteors and how to observe them.* Springer.
33. Terentjeva, A. K. (1989). Fireball streams. *Journal of the International Meteor Organization, 17*, 242.
34. McCrosky, R. E., & Posen, A. (1961). Orbital elements of photographic meteors. *Smithsonian Contributions to Astrophysics, 4*, 78.
35. Wood, J. (1979). *Meteor News, 45*, 11.
36. Jenniskens, P. (2006). *Meteor showers and their parent comets* (p. 387). Cambridge University Press.
37. Venter, S. C. (1957). The great meteor shower of 1956 December 5. *Monthly Notes of the Astronomical Society of Southern Africa, 16*, 6.
38. Shain, C. A. (1957). A remarkable southern meteor shower. *The Observatory, 77*, 27.
39. Wood, J. (1986). The phoenicid meteor shower 1985. *Journal of the British Meteor Society., 16*, 69.
40. Wood, J. (1987). Meteor News. 70, 6.
41. Ridley, H. B. (1962). *BAA circular No. 382.* British Astronomical Association.
42. Weiss, A. A. (1958). The 1956 phoenicid meteor shower. *Australian Journal of Physics, 11*, 113.
43. Jewitt, D. (2006). Comet D/1819 W1 (Blanpain): Not dead yet. *Astronomical Journal, 131*, 2327.
44. Astapovič, I. S. & Terentjeva, A. K. (1968). Fireball radiants of the 1st-15th Centuries in physics and dynamics of meteors. Symposium No. 33 (Ed. Kresák, L. and Millman, P.M.). Reidel: Dordrecht.

Appendix A: Greek Alphabet

Table A1: Greek Alphabet							
α	alpha	ι	iota	ρ	rho		
β	beta	κ	kappa	σ	sigma		
γ	gamma	λ	lambda	τ	tau		
δ	delta	μ	mu	υ	upsilon		
ε	epsilon	ν	nu	φ	phi		
ζ	zeta	ξ	xi	χ	chi		
η	eta	ο	omicron	ψ	psi		
θ	theta	π	pi	ω, Ω	omega*		

*All lowercase except for omega, which shows lower and uppercase.

Appendix B: Constellation Abbreviations and Meteor Shower Names

The *Abbreviations (abbrv.)* are the three-letter IAU codes for each of the 88 constellations. Some abbreviations, such as UMA and UMI, are sometimes published as UMa and UMi.

The *Shower Names* are the preferred titles for showers that radiate from a particular constellation, e.g. the radiant in *Aquarius* near the star η *Aquarii* has the shower name *η-Aquariids*. However, note that some meteor showers also have 3-letter codes that may be the same as the constellation code, e.g. *CHA* is the abbreviation for the constellation *Chamaeleon* and also the shower *χ-Andromedids*, while *LYR* is the abbreviation for both the constellation *Lyra* and the code for the April Lyrids meteor shower. These abbreviations are indicated with an asterisk.

Appendix B: Constellation Abbreviations and Meteor Shower Names

Table B.1: Constellation Abbreviations and Meteor Shower Names

Abbrv.	Constellation	Shower Names	Abbrv.	Constellation	Shower Names
AND*	Andromeda	Andromedids	LAC	Lacerta	Lacertids
ANT	Antlia	Antliids	LEO*	Leo	Leonids
APS*	Apus	Apusids	LMI*	Leo Minor	Leonis Minorids
AQR	Aquarius	Aquariids	LEP*	Lepus	Leporids
AQL	Aquila	Aquilids	LIB	Libra	Librids
ARA	Ara	Araids	LUP	Lupus	Lupids
ARI*	Aries	Arietids	LYN	Lynx	Lyncids
AUR*	Auriga	Aurigids	LYR*	Lyra	Lyrids
BOO	Boötes	Boötids	MEN	Mensa	Mensids
CAE*	Caelum	Caelids	MIC*	Microscopium	Microscopiids
CAM*	Camelopardalis	Camelopardalids	MON*	Monoceros	Monocerotids
CNC	Cancer	Cancrids	MUS*	Musca	Muscids
CVN*	Canes Venatici	Canum Venaticids	NOR	Norma	Normids
CMA	Canis Major	Canis Majorids	OCT*	Octans	Octanids
CMI*	Canis Minor	Canis Minorids	OPH*	Ophiuchus	Ophiuchids
CAP*	Capricornus	Capricornids	ORI*	Orion	Orionids
CAR*	Carina	Carinids	PAV	Pavo	Pavonids
CAS	Cassiopeia	Cassiopeiids	PEG*	Pegasus	Pegasids
CEN	Centaurus	Centaurids	PER*	Perseus	Perseids
CEP	Cepheus	Cepheids	PHE*	Phoenix	Phoenicids
CET*	Cetus	Cetids	PIC*	Pictor	Pictorids
CHA*	Chamaeleon	Chamaeleontids	PSC*	Pisces	Piscids
CIR	Circinus	Circinids	PSA*	Piscis Austrinus	Piscis Austrinids
COL*	Columba	Columbids	PUP*	Puppis	Puppids
COM*	Coma Berenices	Comae Berenicids	PYX	Pyxis	Pyxidids
CRA	Corona Australis	Coronae Australids	RET	Reticulum	Reticulids
CRB	Corona Borealis	Coronae Borealids	SGE*	Sagitta	Sagittids
CRV	Corvus	Corvids	SGR	Sagittarius	Sagittariids
CRT	Crater	Craterids	SCO*	Scorpius	Scorpiids
CRU	Crux	Crucids	SCL	Sculptor	Sculptorids
CYG	Cygnus	Cygnids	SCT*	Scutum	Scutids
DEL*	Delphinus	Delphinids	SER*	Serpens	Serpentids
DOR	Dorado	Doradids	SEX	Sextans	Sextantids
DRA*	Draco	Draconids	TAU*	Taurus	Taurids
EQU	Equuleus	Equuleids	TEL*	Telescopium	Telescopiids
ERI*	Eridanus	Eridanids	TRI*	Triangulum	Triangulids
FOR	Fornax	Fornacids	TRA	Triangulum Australe	Triangulum Australids
GEM*	Gemini	Geminids	TUC	Tucana	Tucanids
GRU	Grus	Gruids	UMA	Ursa Major	Ursae Majorids
HER	Hercules	Herculids	UMI	Ursa Minor	Ursae Minorids
HOR	Horologium	Horologiids	VEL*	Vela	Velids
HYA	Hydra	Hydrids	VIR*	Virgo	Virginids
HYI	Hydrus	Hydrusids	VOL*	Volans	Volantids
IND	Indus	Indids	VUL	Vulpecula	Vulpeculids

Appendix B: Constellation Abbreviations and Meteor Shower Names

*Note that these Constellation Codes are also used as Shower Codes:

AND:	Andromedids	MIC:	Microscopiids
APS:	Daytime April Piscids	MON:	December Monocerotids
AQR:	Aquariids	MUS:	78 Ursae Majorids
ARI:	Daytime Arietids	OCT:	October Camelopardalids
AUR:	Aurigids	OPH:	May Ophiuchid Complex
CAE:	Caelids	ORI:	Orionids
CAM:	Camelopardalids	PEG:	μ-Pegasids
CVN:	Canum Venaticids	PER:	Perseids
CMI:	December Canis Minorids	PHE:	July Phoenicids
CAP:	Capricornids	PIC:	π^1-Cancrids
CAR:	Carinid Complex	PSC:	Piscid Complex
CET:	π-Cetids	PSA:	ψ-Aurigids
CHA:	χ-Andromedids	PUP:	γ-Puppids
COL:	Columbids	SGE:	March δ-Geminids
COM:	Comae Berenicids	SCO:	σ-Columbids
DEL:	December Lyncids	SCT:	σ-Cetids
DRA:	October Draconids	SER:	September Eridanids
ERI:	η-Eridanids	TAU:	Taurid Complex
GEM:	Geminids	TEL:	Telescopiids
LEO:	Leonids	TRI:	August Triangulids
LMI:	Leonis Minorids	VEL:	Puppid-Velid II Complex
LEP:	Leporids	VIR:	March Virginid Complex
LYR:	April Lyrids	VOL:	Volantids

Appendix C: Right Ascension to Degrees Conversion Table

Table C.1: Right Ascension to Degrees Conversion Table

R.A.	0ᵐ	4ᵐ	8ᵐ	12ᵐ	16ᵐ	20ᵐ	24ᵐ	28ᵐ	32ᵐ	36ᵐ	40ᵐ	44ᵐ	48ᵐ	52ᵐ	56ᵐ
0ʰ	0°	1°	2°	3°	4°	5°	6°	7°	8°	9°	10°	11°	12°	13°	14°
1ʰ	15°	16°	17°	18°	19°	20°	21°	22°	23°	24°	25°	26°	27°	28°	29°
2ʰ	30°	31°	32°	33°	34°	35°	36°	37°	38°	39°	40°	41°	42°	43°	44°
3ʰ	45°	46°	47°	48°	49°	50°	51°	52°	53°	54°	55°	56°	57°	58°	59°
4ʰ	60°	61°	62°	63°	64°	65°	66°	67°	68°	69°	70°	71°	72°	73°	74°
5ʰ	75°	76°	77°	78°	79°	80°	81°	82°	83°	84°	85°	86°	87°	88°	89°
6ʰ	90°	91°	92°	93°	94°	95°	96°	97°	98°	99°	100°	101°	102°	103°	104°
7ʰ	105°	106°	107°	108°	109°	110°	111°	112°	113°	114°	115°	116°	117°	118°	119°
8ʰ	120°	121°	122°	123°	124°	125°	126°	127°	128°	129°	130°	131°	132°	133°	134°
9ʰ	135°	136°	137°	138°	139°	140°	141°	142°	143°	144°	145°	146°	147°	148°	149°
10ʰ	150°	151°	152°	153°	154°	155°	156°	157°	158°	159°	160°	161°	162°	163°	164°
11ʰ	165°	166°	167°	168°	169°	170°	171°	172°	173°	174°	175°	176°	177°	178°	179°
12ʰ	180°	181°	182°	183°	184°	185°	186°	187°	188°	189°	190°	191°	192°	193°	194°
13ʰ	195°	196°	197°	198°	199°	200°	201°	202°	203°	204°	205°	206°	207°	208°	209°
14ʰ	210°	211°	212°	213°	214°	215°	216°	217°	218°	219°	220°	221°	222°	223°	224°
15ʰ	225°	226°	227°	228°	229°	230°	231°	232°	233°	234°	235°	236°	237°	238°	239°
16ʰ	240°	241°	242°	243°	244°	245°	246°	247°	248°	249°	250°	251°	252°	253°	254°
17ʰ	255°	256°	257°	258°	259°	260°	261°	262°	263°	264°	265°	266°	267°	268°	269°
18ʰ	270°	271°	272°	273°	274°	275°	276°	277°	278°	279°	280°	281°	282°	283°	284°
19ʰ	285°	286°	287°	288°	289°	290°	291°	292°	293°	294°	295°	296°	297°	298°	299°
20ʰ	300°	301°	302°	303°	304°	305°	306°	307°	308°	309°	310°	311°	312°	313°	314°
21ʰ	315°	316°	317°	318°	319°	320°	321°	322°	323°	324°	325°	326°	327°	328°	329°
22ʰ	330°	331°	332°	333°	334°	335°	336°	337°	338°	339°	340°	341°	342°	343°	344°
23ʰ	345°	346°	347°	348°	349°	350°	351°	352°	353°	354°	355°	356°	357°	358°	359°

© The Author(s), under exclusive license to Springer Nature Switzerland AG 2021
P. M. Bagnall, *Atlas of Meteor Showers*, The Patrick Moore Practical Astronomy Series,
https://doi.org/10.1007/978-3-030-76643-6

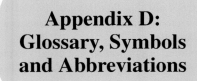

Appendix D: Glossary, Symbols and Abbreviations

Ablation	The removal of surface material from a meteoroid by vaporization, melting, chipping or sputtering as it passes through the atmosphere at high speed.
Active asteroid	An object that appears to be a rocky asteroid but which sometimes displays the appearance of a comet, releasing dust and volatiles into space.
Aërolite	A term used in older literature for a *meteorite*.
Amor asteroids	Amor-type asteroids have orbits that lie outside of the Earth's orbit. Named after (1221) Amor discovered in 1932, they have semi-major axes of >1 AU and perihelion distances of >1.017 AU. There are over 1800 known Amors. See also *Apollo asteroids, Aten asteroids* and *Atira asteroids*.
Antihelion radiant	The area of sky directly opposite the Sun from which some *meteors* appear to radiate.
Aphelion (Q)	The furthest point in an orbit from the Sun.
Apollo asteroids	Apollo-type asteroids have orbits that cross that of the Earth and are therefore potentially hazardous. Named after (1862) Apollo discovered in 1932, they have semi-major axes of >1 AU and perihelion distances of <1.017 AU. There are over 10,500 known Apollos. See also *Amor asteroids, Aten asteroids* and *Atira asteroids*.
Argument of the Perihelion (ω)	For the *orbit* of a *comet* or *meteoroid stream*, for example, the angle from the *ecliptic* to the line connecting the Sun to the *perihelion*.
Ascending node (Ω)	For the *orbit* of a *comet* or *meteoroid stream*, for example, the point at which the comet or stream intersects the Earth's orbital plane travelling from south to north. See also *descending node*.
Asteroid	See *minor planet*.
Asteroid belt	A band of rocky debris between the orbits of Mars and Jupiter, believed to be the remnants of the planet building era approximately 4.56 Gyr ago. Most of the *meteorites* found on Earth originated in the asteroid belt.
Astronomical Unit (AU)	The average distance from the Sun to the Earth, approximately 150 million kilometres.

Aten asteroids	Aten-type asteroids have orbits that cross that of the Earth and are therefore potentially hazardous. Named after (2062) Aten discovered in 1976, they have semi-major axes of <1 AU and aphelion distances of >0.983 AU. There are over 1800 known Atens. See also *Apollo asteroids, Amor asteroids* and *Atira asteroids*.
Atira asteroids	Atira-type asteroids have orbits that lie entirely within the Earth's orbit. Named after (163693) Atira discovered in 2003, there are more than 110 known members. See also *Apollo asteroids, Amor asteroids* and *Aten asteroids*.
Atlantic Gap	Name given to the paucity of observations as a *meteor shower radiant* passes over the Atlantic.
Bolide	An exceptional bright *fireball* that usually explodes.
Brownlee particles	See *dust (interplanetary)*. Named after Don Brownlee, who pioneered research into this material.
Circumpolar	A radiant, or star, is said to be circumpolar when it is always above the observer's horizon. To determine if a radiant is circumpolar, subtract the observer's latitude from 90° and anything with a declination greater than the result will be circumpolar. For example, for someone observing from 40°N: 90° − 40° = 50°. So any radiant with a declination of more than 50° will never set on the observer's horizon.
Comet	A comet is essentially a large, dirty snowball composed of water ice and solid grains. Many spend much of their lives in the deep freeze of the outer reaches of the Solar System. Comets can be separated into several types including: • Short period or *Jupiter-family comets* with orbital periods of less than 20 years • *Halley-type comets* with orbital periods of between 20 and 200 years • Long period comets with orbital periods in excess of 200 years • Interstellar comets which are interlopers from beyond the Solar System and, unless their trajectory is radically altered by a close approach to one of the planets, or the Sun, simply pass through the system
Comet designations	There are six key designations for *comets* that identify what type of object they are:

Prefix	Explanation
P/	Periodic comet i.e. a comet with an *orbital period* of 200 years or less.
C/	A non-periodic comet. Most non-periodic comets actually have orbital periods of more than 200 years.
X/	A comet for which no reliable orbit could be calculated.
D/	A comet that has disappeared. This could be due to a close encounter with a planet that has radically changed the comet's orbit, or a comet that has fragmented, or one that has impacted with a planet or the Sun.
A/	An object that was previously thought to be a comet but has subsequently been shown to be a *minor planet*.
I/	An interstellar comet or object.

Contamination	Where an observational record of a shower's activity erroneously includes meteors from another shower. It is sometimes incorrectly referred to as pollution.
Cosmic dust	See *dust (interplanetary)*.
Counterglow	English term for the *gegenschein*.

Appendix D: Glossary, Symbols and Abbreviations

Culmination	The highest point of a *radiant* above the horizon. In the Northern Hemisphere, radiants culminate due South. In the Southern Hemisphere, radiants culminate due North. For a *shower* displaying meteors over several hours, the higher the radiant, the greater the detectable activity.
Descending node (℧)	For the *orbit* of a *comet* or *meteoroid stream*, the point at which the comet or stream intersects the Earth's orbital plane travelling from north to south. See also *ascending node*.
Dust (interplanetary)[a]	Finely divided solid matter with particle sizes in general smaller than *meteoroids*, moving in or coming from interplanetary space.
Dust trail	The trail of debris left behind as a *comet sublimates* and disintegrates. Dust trails are components of *meteoroid streams*.
Earthgrazers	*Meteors* that skim through the upper reaches of the atmosphere, almost parallel with it. Earthgrazers will often appear near the horizon and pass overhead producing long, bright meteors. Some have been known to skip back into space.
Ecliptic (Plane of)	The plane of the Earth's orbit extended into space. The path of the ecliptic is often marked on star atlases and runs through each of the zodiacal constellations. The Sun appears to travel along the ecliptic.
Ejecta	Solid material ejected from the surface of an object either during an impact, or due to thermal destruction and outgassing.
Exoplanet	A planet that lies outside of the Solar System. Most exoplanets are in orbit around stars but there are also exoplanets wandering through interstellar space, having been gravitationally ejected from their own planetary system.
Falling star	Popular name for a *meteor*.
Filament	A rope-like concentration of *meteoroids* in a *meteoroid stream* caused by planetary perturbations.
Fireball	A *meteor* with a visual magnitude brighter than m_v −4.0.
Fireball type	A classification system for *fireballs* based on their origins:

Type	Origin	Percentage
I	Ordinary chondrite meteorites	29%
II	Carbonaceous chondrite meteorites	33%
IIIa	High-density comets	29%
IIIb	Low-density comets	9%

Fusion crust	A thin, glassy crust that forms on the surface of a meteorite in the final stages of atmospheric flight. It is caused by the surface minerals melting.
Gaseous giant	One of the large, outer planets that are made mostly of gas: Jupiter, Saturn, Uranus and Neptune.
Gegenschein	A roughly oval, dim glow directly opposite the Sun. The gegenschein is part of the *Zodiacal Dust Cloud*. It is sometimes known by its English name of counterglow.
Halley-type comet (HTC)	Comets that have orbital periods of 20–200 years.
Hourly rate (HR)	The actual number of *meteors* an observer sees during a 1-h period. This has to be scaled up to calculate the *zenithal hourly rate*.
Inclination (i)	The angle between the Earth's orbital plane and that of a *comet* or *meteoroid stream*.
Interplanetary dust particle (IDP)	See *dust (interplanetary)*.
Jupiter-family comet (JFC)	Comets whose orbits have been strongly influenced by Jupiter. They have *orbital periods* of 20 years or less and *inclinations* of up to 30°. See also *Halley-type comet*.

Long period comet	A *comet* with an *orbital period* of more than 200 years. See also *Short period comet*.
Longitude of the ascending node (Ω)	For the orbit of a *comet* or *meteoroid stream*, the angle from the *First Point of Aries* to the *ascending node*.
Magnitude (m_v, m_p)	The brightness of an astronomical object. The faintest stars visible to the naked eye are typically m_v +6.5, the brightest is the Sun at m_v −26.7. Camera and video equipment can extend the photographic magnitude down to about m_p +8. Average magnitude is indicated by a bar over the m: $\bar{m}_v$, $\bar{m}_p$.
Major shower	A *meteor shower* where the *zenithal hourly rate* usually exceeds 9 meteors an hour at maximum.
Major stream	A *meteoroid stream* that results in a *major shower*.
Mesosphere	That part of the atmosphere between about 50 and 100 km above the Earth's surface. Most *meteors* occur in the mesosphere, while meteoric *dust* may contribute to the formation of *noctilucent clouds*.
Meteor[a]	The light and associated physical phenomena (heat, shock, ionization), which result from the high speed entry of a solid object from space into a gaseous atmosphere.
Meteor shower[a]	A group of *meteors* produced by *meteoroids* of the same *meteoroid stream*.
Meteor storm	A *meteor storm* occurs when there are too many meteors to be counted manually.
Meteoric dust	The dust left behind after a *meteor* event.
Meteoric region	The part of the atmosphere in which *meteors* occur. This is, for the most part, between 80 and 120 km altitude but can vary.
Meteoric smoke[a]	Solid matter that has condensed in a gaseous atmosphere from material vaporized during the *meteor* phase. Meteoric smoke particles (MSP) are typically less than 100 nanometres in size.
Meteorite[a]	Any natural solid object that survived the *meteor* phase in a gaseous atmosphere without being completely vaporized. The object becomes a meteorite once the meteor phenomena cease and it falls to Earth as a dark and cooling body. On a body without an atmosphere, 'meteorites' are called 'impact debris' because there cannot be a meteorite without there first being a meteor—although there can be a micrometeorite without a meteor!
Meteoroid[a]	A solid natural object of a size roughly between 30 micrometres and 1 metre, moving in or coming from interplanetary space.
Meteoroid stream[a]	A group of *meteoroids* that have similar *orbits* and a common origin.
Micrometeorite	A *meteorite* smaller than 1 millimetre.
Micrometeoroid	See *dust*. The IAU discourages the use of this term.
Minor planet	Debris dating back to the formation of the Solar System 4,650 Ma ago. The debris is largely composed of silicates and metals such as nickel-iron, and range in size from metres to hundreds of kilometres. Some objects that appear to be minor planets may in fact be defunct *comets* that have lost their volatile material and so no longer produce the tails for which comets are famous. Minor planets are also commonly called asteroids.
Minor shower	A *meteor shower* where the *zenithal hourly rate* is usually less than 10 meteors an hour at maximum.
Minor stream	A *meteoroid stream* that results in a *minor shower*.
Noctilucent cloud	Very high-altitude cloud, 75–85 km above the Earth, whose formation may be partly due to meteoric *dust*. They can only be seen about 4 weeks either side of 21 June in parts of the Northern Hemisphere, or 21 December in parts of the Southern Hemisphere.

Non-gravitational perturbations	Processes that alter the orbital characteristics of *comets* and *meteoroids* that are not associated with gravity. The *outgassing* of a comet produces jets that can either increase or slow its orbital speed. Meteoroids can spiral into the Sun because of the *Poynting-Roberston Effect* but can also be blown out of the Solar System by the *solar wind*.
Oort cloud	A theoretical reservoir of *comets* some 2000 to at least 50,000 AU from the Sun. Also known as the Öpik–Oort cloud.
Orbit	The path around the Sun of an object such as a *meteoroid* or *comet*.
Orbital period (P)	The time it takes for an astronomical object to make a complete journey around the Sun. See also *long period comet* and *short period comet*.
Outburst	A sudden increase in the number of *meteors* from a *meteor shower*, often caused by *filaments* or clumps of material.
Outgassing	The release of gases from a comet's surface due to sublimation of its ices.
Pacific Gap	Name given to the paucity of observations as a *meteor shower radiant* passes over the Pacific.
Perihelion (q)	The closest point in an orbit to the Sun.
Period (P)	See *orbital period*.
Perseids (PER)	A *major shower* which reaches maximum in the second week of August each year.
Persistent train	A glowing train left behind by a bright *meteor*. Persistent trains often last several seconds.
Plane of the ecliptic	See *ecliptic*.
Population index (r)	An estimate of the ratio of the number of *meteors* in subsequent magnitude classes. The value of r usually ranges from 2.0 to 3.5.
Retardation point	During the fall of a meteorite, the altitude at which the cosmic velocity is exhausted and the meteorite falls to Earth only under the influence of the planet's gravity.
Poynting-Robertson effect	A mechanism that causes *cosmic dust* particles to spiral in towards the Sun. The particle absorbs radiation from the Sun and then re-radiates at a slightly different wavelength, causing a tangential drag on the dust particle resulting in an ever-decreasing orbit.
Radiant	The point on the sky from which shower *meteors* appear to radiate. It is an optical illusion.
Rock comet	This term is being superseded by *active asteroid*.
Semi-major axis	A common but incorrect term for the semi-axis major. Half the longest diameter of an elliptical orbit.
Semi-minor axis	A common but incorrect term for the semi-axis minor. Half the shortest diameter of an elliptical orbit.
Shooting star	Popular name for a *meteor*.
Short period comet	A *comet* with an *orbital period* of 200 years or less. See also *long period comet*.
Shower	See *meteor shower*, *major shower* and *minor shower*.
Solar longitude $\lambda_\odot$	A measure of the Sun's position in the sky calculated in degrees from the *Vernal Equinox*. As the solar longitude does not suffer the irregularities of a civil calendar (e.g. leap years), it can be used as a more stable time system. This is particularly useful when trying to determine the time of maximum activity of a *meteor shower*.
Solar wind	A stream of charged particles, mostly electrons, protons and alpha particles, released from the Sun's corona (its upper atmosphere). Solar wind exerts a force on Solar System bodies, but it is only significant for small particles such as *meteoroids*.
Sporadic meteor	A *meteor* that does not appear to be associated with any of the known *meteor showers*.
Storm	See *meteor storm*.

Sublimation	The process by which ice turns directly to gas without going through a liquid phase. Sublimation is a prominent feature of *comets*.
Superbolide[a]	A *meteor* brighter than absolute visual magnitude −17 (at 100 km altitude).
Terrestrial planet	One of the inner planets: Mercury, Venus, Earth and Mars.
Trail	See *dust trail*.
Train[a]	The light or ionization left along the trajectory of the *meteor* after it has passed.
Wake	The short-lived trail of a *meteor*. Wakes range from 1 metre to 1 km at a height of 100 km.
Vernal Equinox	A point on the celestial sphere at which the ecliptic intersects the Celestial Equator. It is at this point that the Sun appears to cross from south to north of the Equator. Currently, the Vernal Equinox occurs around March 21. It is often referred to as the First Point of Aries, although the Equinox is now actually in Pisces, having drifted there as a result of the Earth gyrating on its axis.
Zenith attraction	Because the Earth's gravity changes the direction and velocity of incoming meteoroids, the radiant appears to be displaced towards the zenith.
Zenithal hourly rate (ZHR)	The number of *meteors* an observer is likely to see if the *radiant* is at the zenith, the limiting magnitude is m_v +6.5 and there is no cloud cover. It is a theoretical value that is rarely achieved in practice. See *hourly rate*.
Zodiacal band (ZB)	A faint extension to the *Zodiacal Light* on both sides of the Sun. The two bands merge to form the *gegenschein*.
Zodiacal dust cloud (ZDC)	A cloud of cosmic *dust* that spreads out from the Sun angled at about 3° to the plane of the Solar System. The dust particles are typically in the order of 30–100 micrometres with an average mass of just 150 micrograms.
Zodiacal light (ZL)	A cone-shaped glow stretching from the Sun along the ecliptic. The zodiacal light is due to sunlight reflected from dust particles in the *zodiacal dust cloud*.
a	Semi-major axis (or, more correctly, the semi-axis major)
A/	Asteroid previously believed to be a comet
AHY	α-Hydrids
AMO	α-Monocerotids
AMS	American Meteor Society
AND	Andromedids
ARC	April ρ-Cygnids
AU	Astronomical Unit (150,000,000 km)
AUD	August Draconids
AVB	α-Virginids
b	Semi-minor axis (or, more correctly, the semi-axis minor)
BAA	British Astronomical Association
BHY	β-Hydrusids
BTA	Daytime β-Taurids
C/	Non-periodic comet
CAN	c-Andromedids
CAP	α-Capricornids
COM	Comae Berenicids
CTA	χ-Taurids
D/	Disappeared comet
DAD	December α-Draconids
DAQ	δ-Aquarids
DKD	December κ-Draconids
DLM	December Leonis Minorids

Appendix D: Glossary, Symbols and Abbreviations 343

DPC	December φ-Cassiopeiids
DRA	October Draconids
DSV	December σ-Virginids
EAU	ε–Aquilids
EGE	ε-Geminids
EPG	ε-Pegasids
ERI	η-Eridanids
ETA	η-Aquarids
FAN	49 Andromedids
FIDAC	Fireball Data Centre
GDR	July γ-Draconids
GEM	Geminids
GUM	γ-Ursae Minorids
HR	Hourly rate
HTC	Halley-type comet
HVI	h Virginids
HYD	σ-Hydrids
i	Inclination of an *orbit*
I/	Interstellar comet or object
IAU	International Astronomical Union
IDP	Interplanetary dust particle
IMO	International Meteor Organisation
JFC	Jupiter-family comet
JLE	January Leonids
JPE	July Pegasids
JXA	July ξ-Arietids
KCG	κ-Cygnids
KUM	κ-Ursae Majorids
LBO	λ-Boötids
LEO	Leonids
LMI	Leonis Minorids
LUM	λ-Ursae Majorids
LYR	April Lyrids
MON	December Monocerotids
MP	Minor planet
m_p	Photographic magnitude
$\overline{m}_p$	Average photographic magnitude
m_v	Visual magnitude
$\overline{m}_v$	Average visual magnitude
NDA	Northern δ-Aquariids
NEO	Near-Earth Object
NIA	Northern ι-Aquariids
NOA	Northern October δ-Arietids
NPI	Northern δ-Piscids
MSP	Meteoric smoke particles
NTA	Northern Taurids
NUE	ν-Eridanids
OCC	October Capricornids
OCT	October Camelopardalids
OCU	October Ursae Majorids

OER	o−Eridanids
ORI	Orionids
ORN	Northern χ-Orionids
ORS	Southern χ-Orionids
OSE	ω-Serpentids
P	Orbital period
P/	Periodic comet
PAU	Piscis Austrinids
PCA	ψ-Cassiopeiids
PER	Perseids
PHO	December Phoenicids
PPS	φ−Piscids
PPU	π Puppids
PSU	ψ Ursae Majorids
PUP	Puppid-Velids
Q	Aphelion
q	Perihelion
r	Population index
QUA	Quadrantids
RPU	ρ-Puppids
SLD	Southern λ-Draconids
SLE	σ-Leonids
SDA	Southern δ-Aquariids
SDO	Scattered Disc Objects
SOA	Southern October δ-Arietids
SPI	Southern δ-Piscids
SPO	Sporadic meteor
SSE	σ-Serpentids
STA	Southern Taurids
THA	November θ−Aurigids
URS	Ursids
UT	Universal Time
X/	Comet for which an orbit has not been calculated
XDR	ξ-Draconids
ZHR	Zenithal hourly rate
ZB	Zodiacal band
ZCS	ζ-Cassiopeiids
ZDC	Zodiacal dust cloud
ZL	Zodiacal light
ZPE	Daytime ζ-Perseids
$\lambda_\odot$	Solar longitude
α	Right ascension
δ	Declination
Δα	Drift in the right ascension of a radiant
Δδ	Drift in the declination of a radiant
ω	Argument of the Perihelion
Ω	Longitude of the ascending node

[a] Definition by IAU Commission F1 on Meteors, Meteorites and Interplanetary Dust. These replaced earlier definitions by IAU Commission 22 on Meteors and Meteorites

Appendix E: Wind Chill Chart

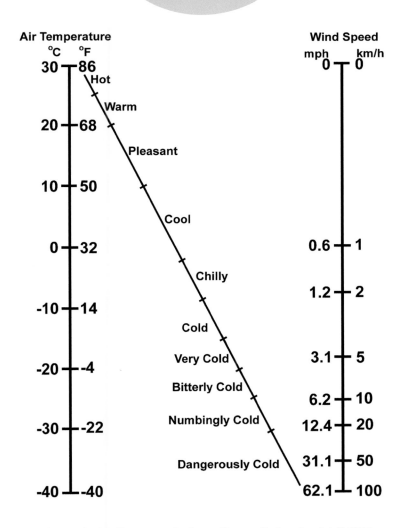

Appendix F: Visual Meteor Observation Form

Appendix F: Visual Meteor Observation Form

Visual Meteor Observation Form
Please use Coordinated Universal Time (UTC)

Page ___ of ___

Date			Start	End	Duration	- Breaks	= Obs. Time
yyyy	mm	dd-dd	hh:mm	hh:mm	hh:mm	hh:mm	hh:mm

Observer: _____ Location: _____ Lat: _____ Long: _____

Limit. Mag.: _____ Moon: ___ d Sky Conditions: _____

Time	Mag	Shower	Notes

Appendix G: Magnitude Comparison Charts

Andromeda

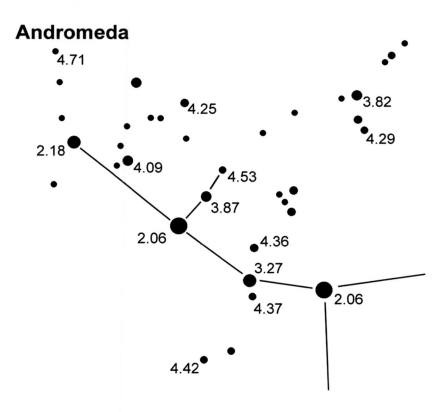

350 Appendix G: Magnitude Comparison Charts

Apus
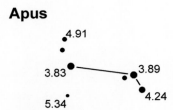

Aquila
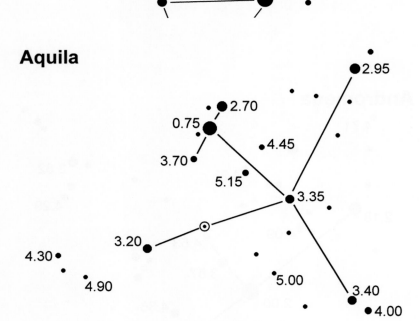

Ara
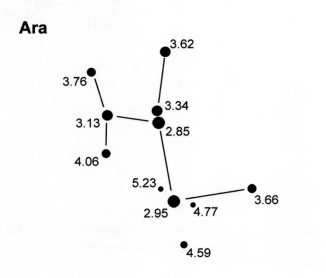

Appendix G: Magnitude Comparison Charts 351

Aries

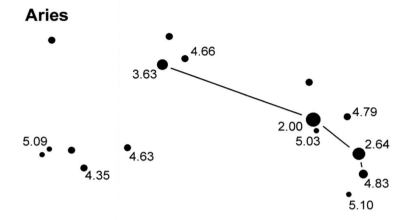

Auriga

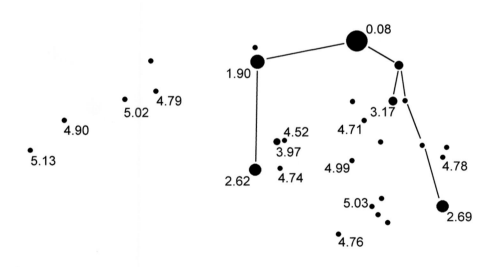

Appendix G: Magnitude Comparison Charts

Boötes

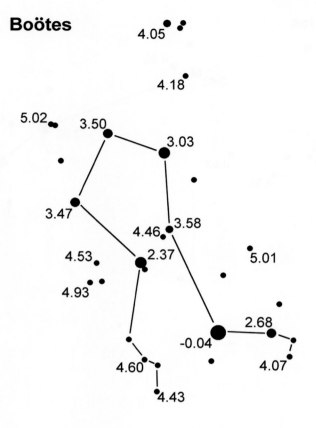

Cancer

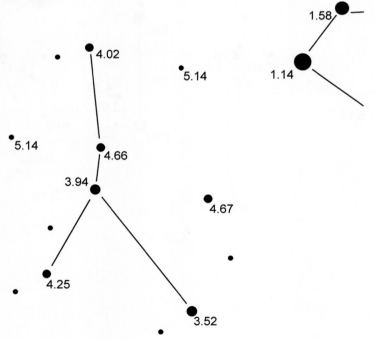

Appendix G: Magnitude Comparison Charts 353

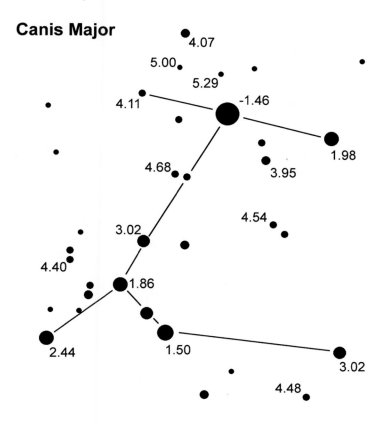

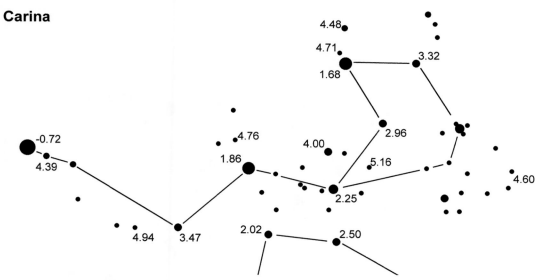

Cassiopeia

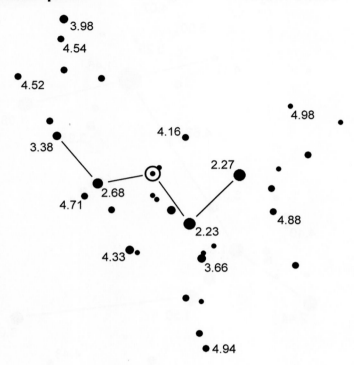

Centaurus and Crux

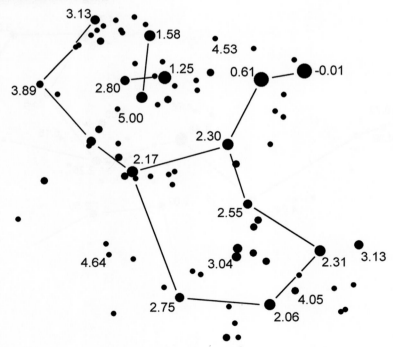

Appendix G: Magnitude Comparison Charts

Cepheus

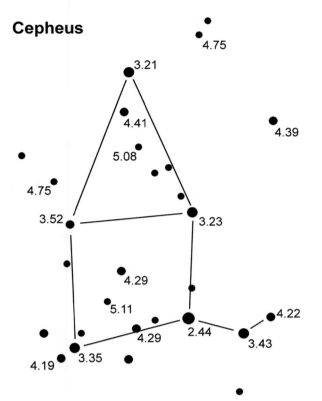

Cetus

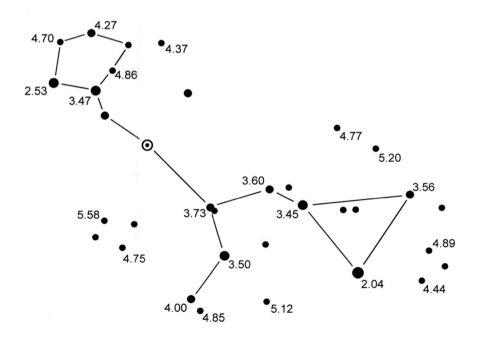

Appendix G: Magnitude Comparison Charts

Corona Borealis

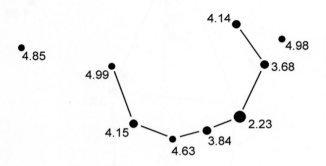

Cygnus

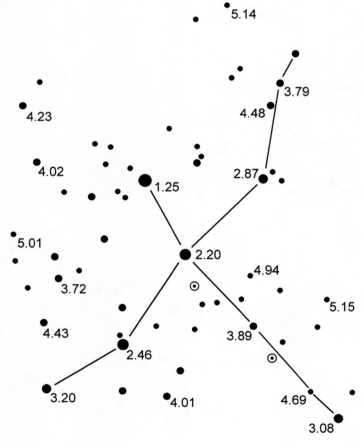

Appendix G: Magnitude Comparison Charts 357

Gemini

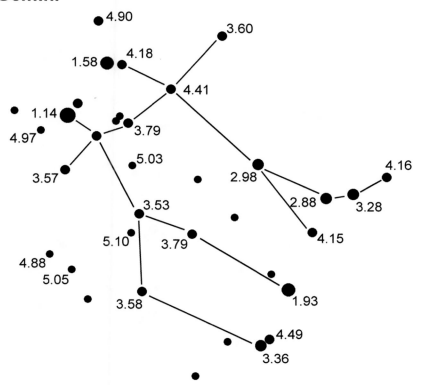

Grus

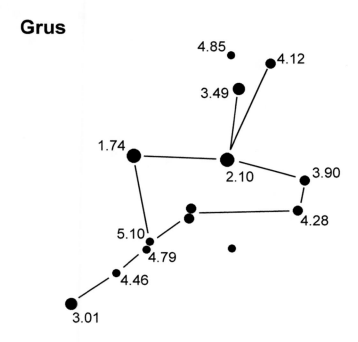

Hercules and Lyra

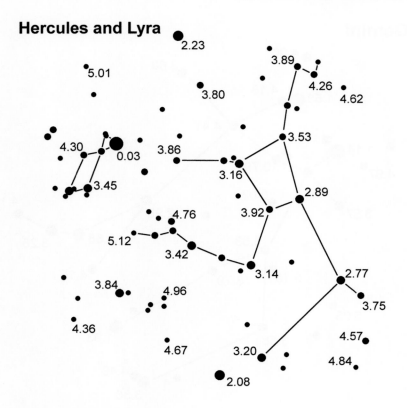

Leo

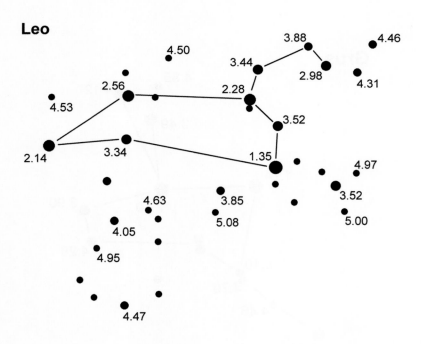

Appendix G: Magnitude Comparison Charts

Ophiuchus

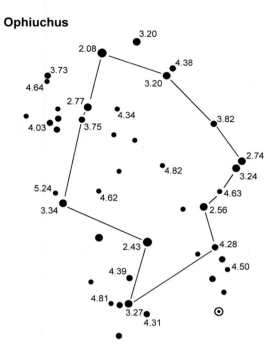

Orion

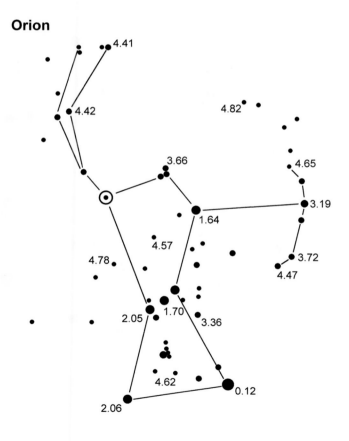

Appendix G: Magnitude Comparison Charts

Pavo

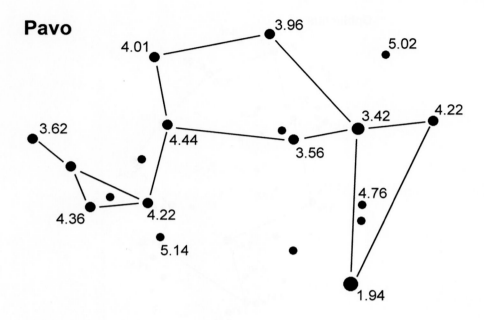

Pegasus

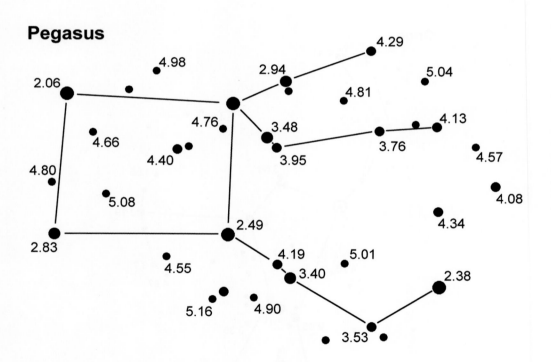

Appendix G: Magnitude Comparison Charts 361

Perseus

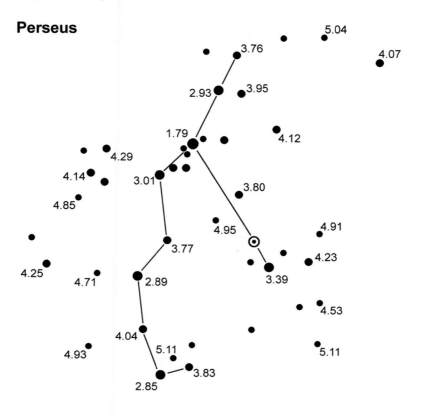

Phoenix

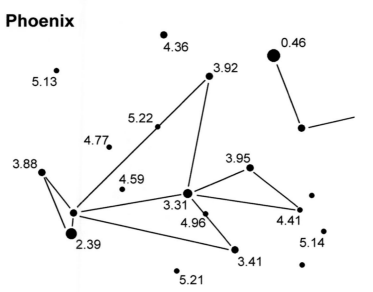

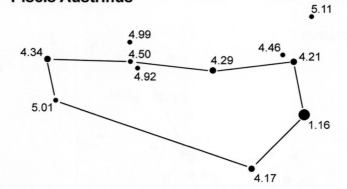
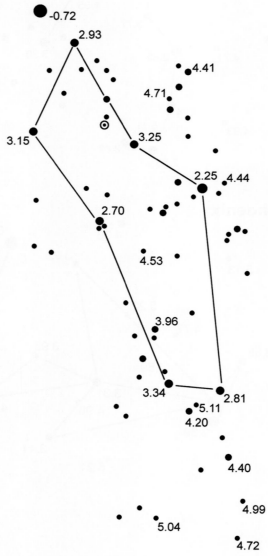

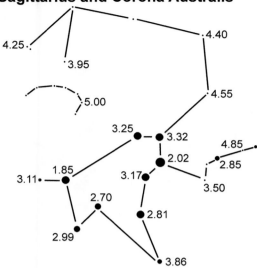

Sagittarius and Corona Australis

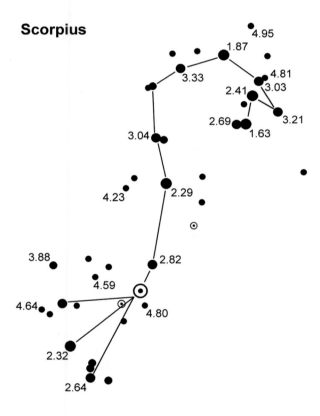

Scorpius

364 Appendix G: Magnitude Comparison Charts

Taurus

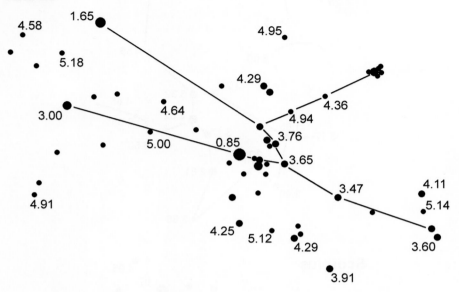

Triangulum Australe

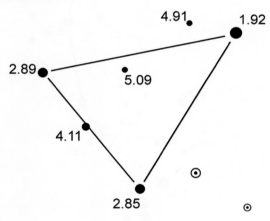

Appendix G: Magnitude Comparison Charts 365

Ursa Major

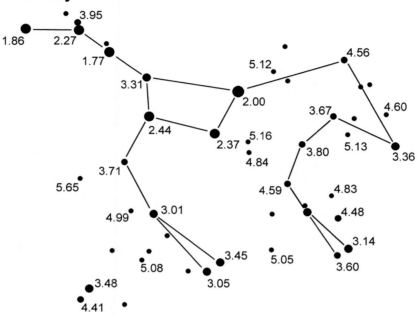

Ursa Minor

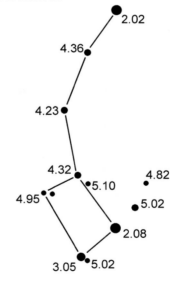

Appendix G: Magnitude Comparison Charts

Vela

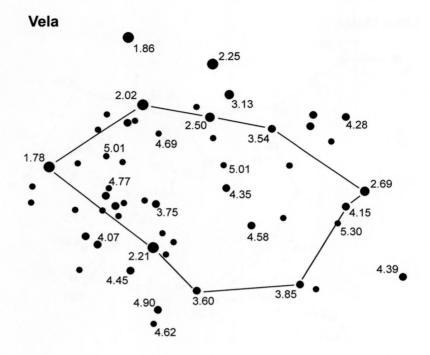

Virgo

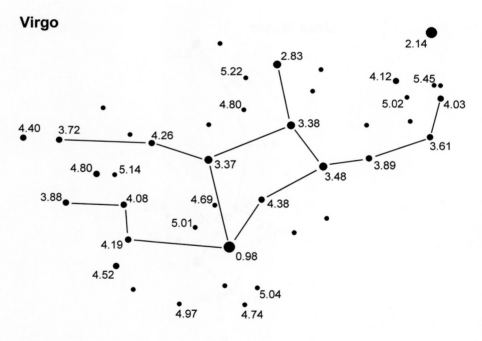

Appendix H: Gnomonic Plotting Charts

These plotting charts are available as downloads for printing in both US and Metric A4 formats from: https://doi.org/10.1007/978-3-030-76643-6_4.

Charts A8a to A8r are for each of the 10 major meteor showers. Each shower has two charts at approximately right angles to one another to provide maximum sky coverage.

The remaining charts, H.8s to H.8zc, can be used for minor showers. Each is centred on an easily recognizable constellation.

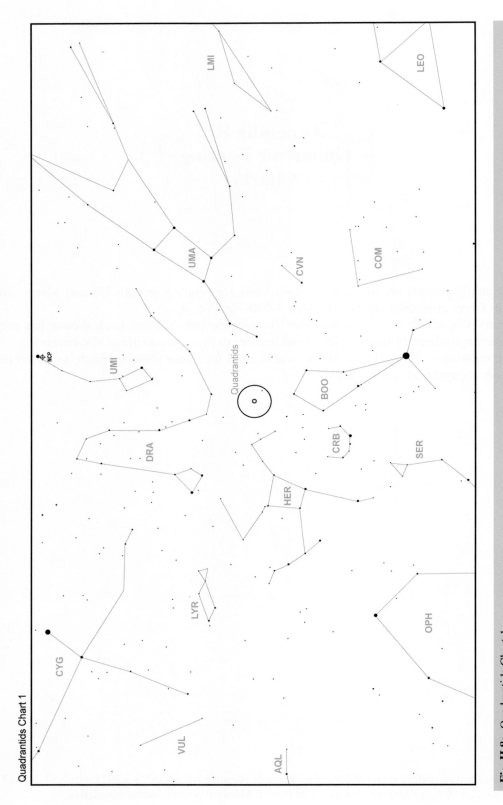

Fig. H.8a Quadrantids Chart 1

Appendix H: Gnomonic Plotting Charts 369

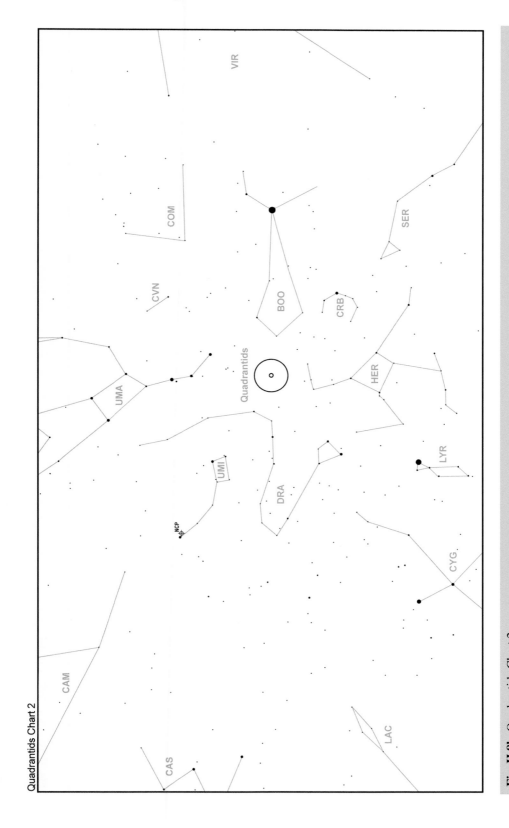

Fig. H.8b Quadrantids Chart 2

370 Appendix H: Gnomonic Plotting Charts

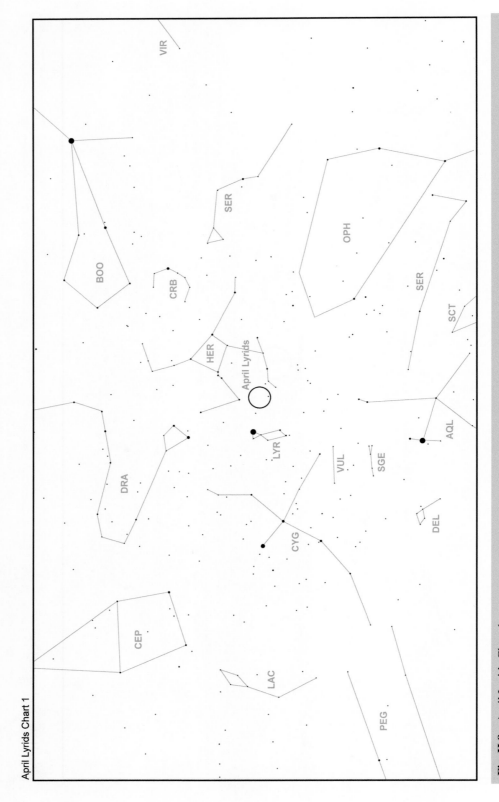

Fig. H.8c April Lyrids Chart 1

Appendix H: Gnomonic Plotting Charts

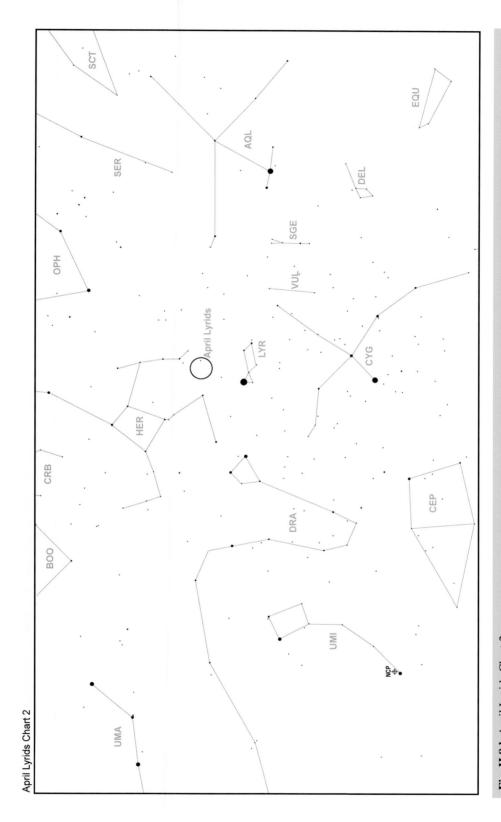

Fig. H.8d April Lyrids Chart 2

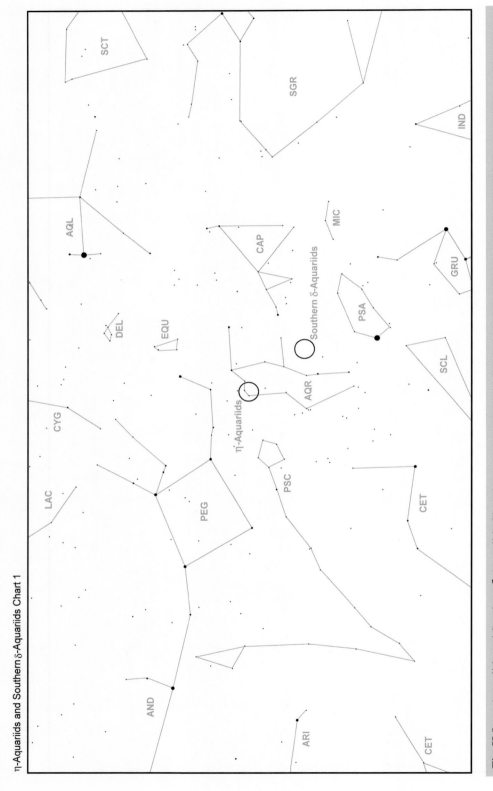

Fig. H.8e η-Aquariids and Southern δ-Aquariids Chart 1

Appendix H: Gnomonic Plotting Charts 373

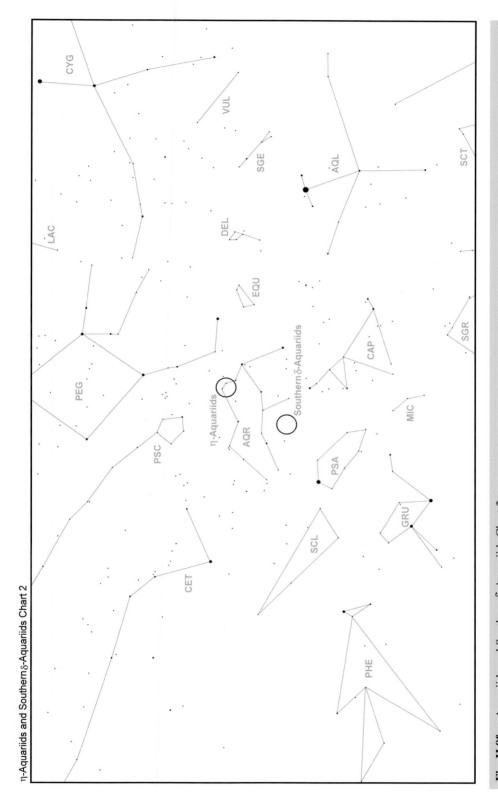

Fig. H.8f η-Aquariids and Southern δ-Aquariids Chart 2

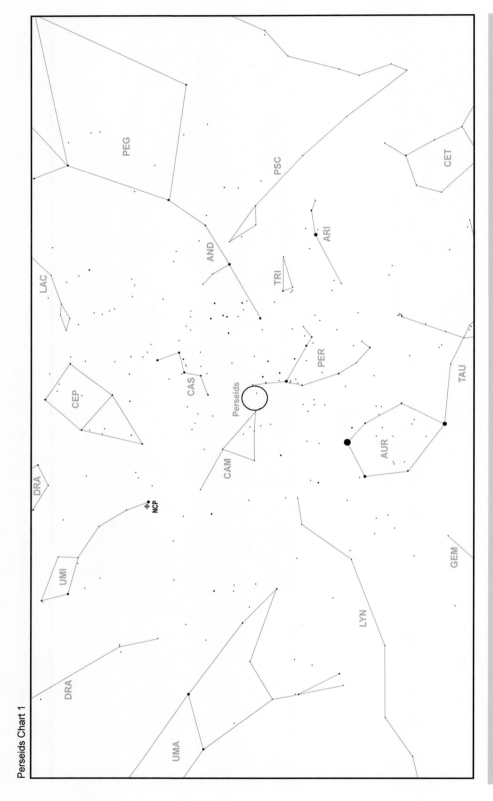

Fig. H.8g Perseids Chart 1

Appendix H: Gnomonic Plotting Charts

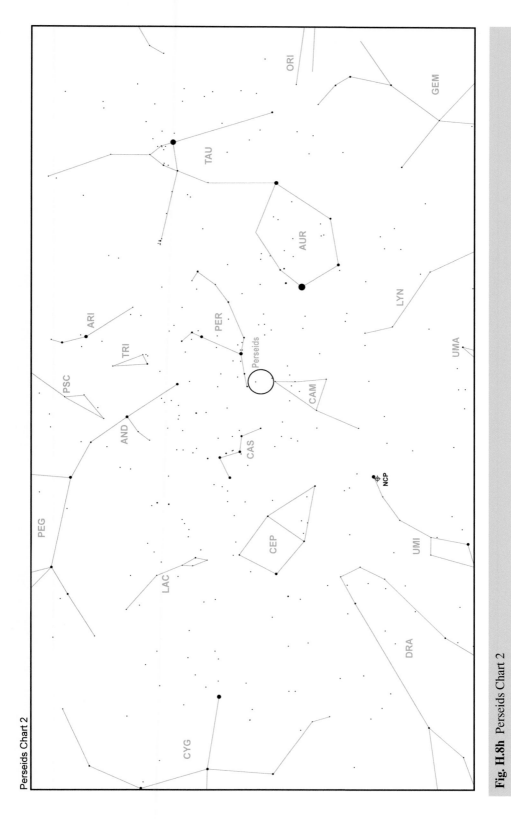

Fig. H.8h Perseids Chart 2

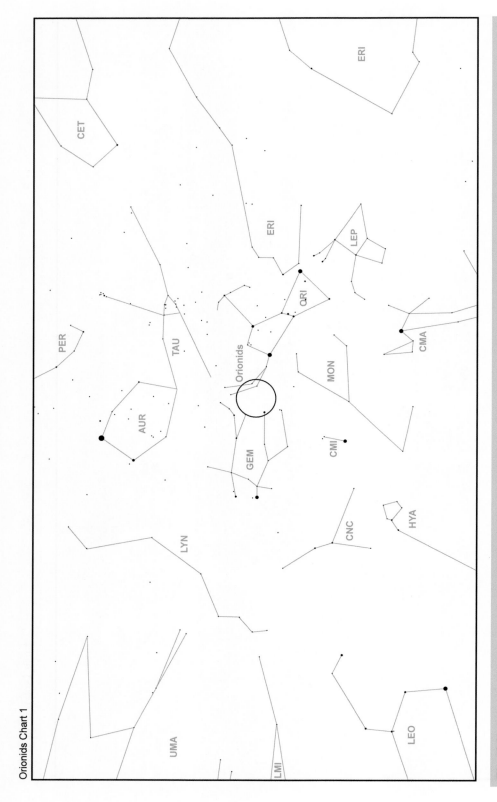

Fig. H.8i Orionids Chart 1

Appendix H: Gnomonic Plotting Charts

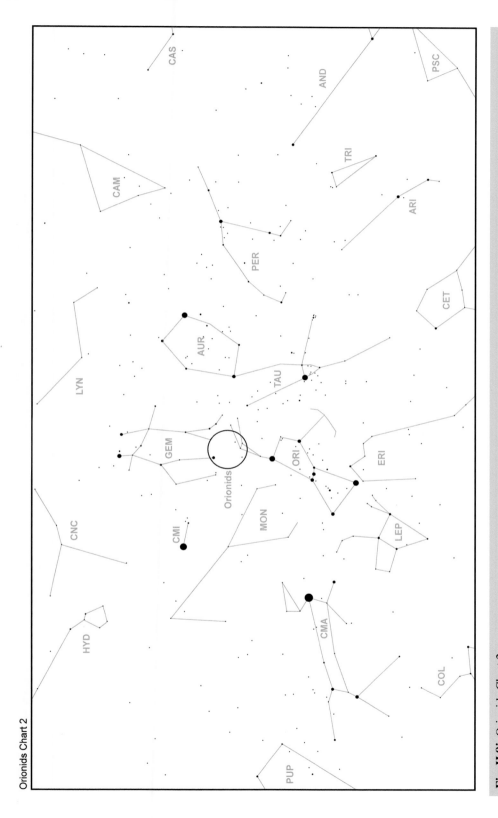

Fig. H.8j Orionids Chart 2

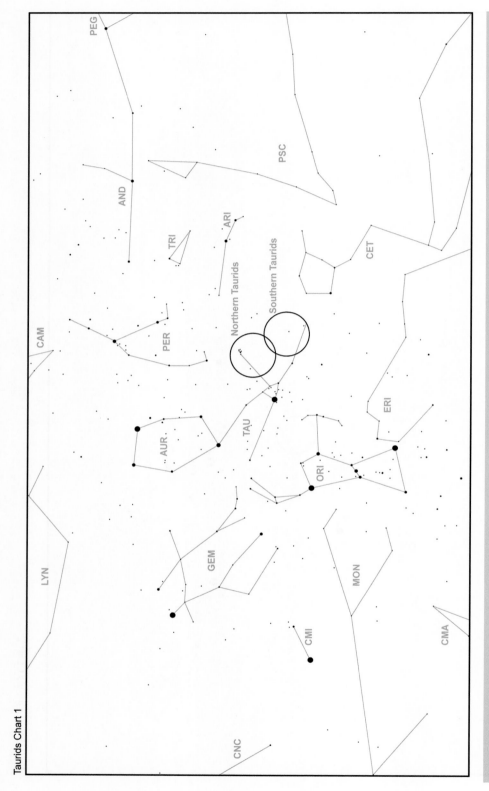

Fig. H.8k Taurids Chart 1

Appendix H: Gnomonic Plotting Charts 379

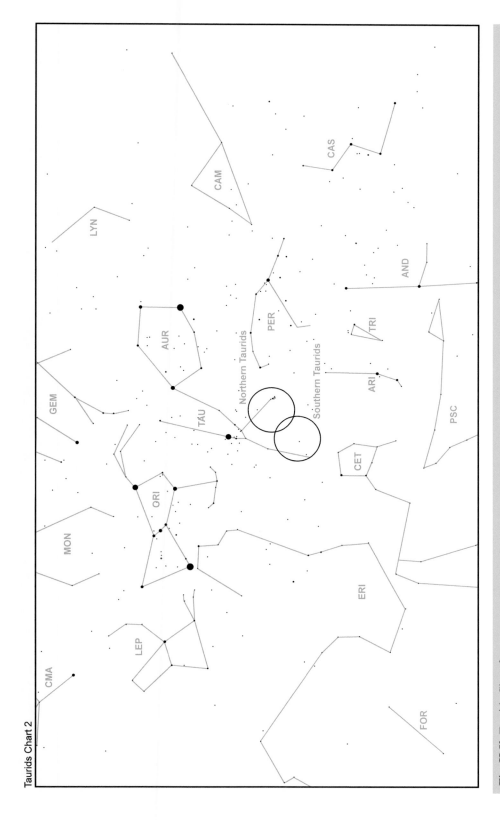

Fig. H.81 Taurids Chart 2

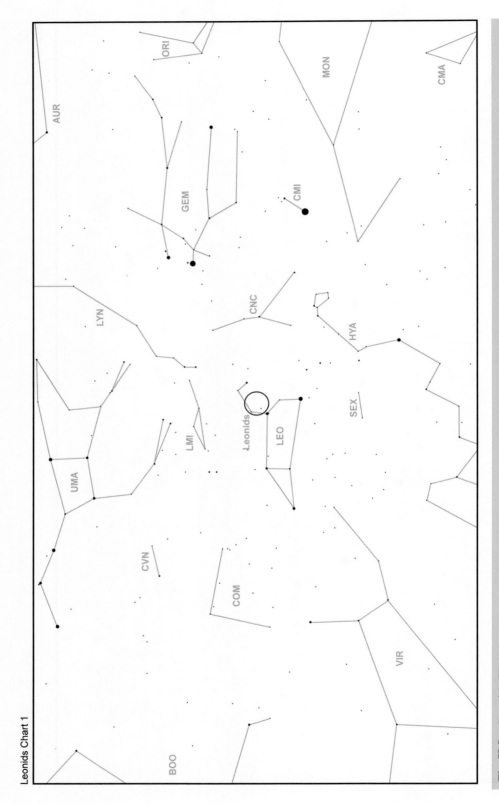

Fig. H.8m Leonids Chart 1

Appendix H: Gnomonic Plotting Charts 381

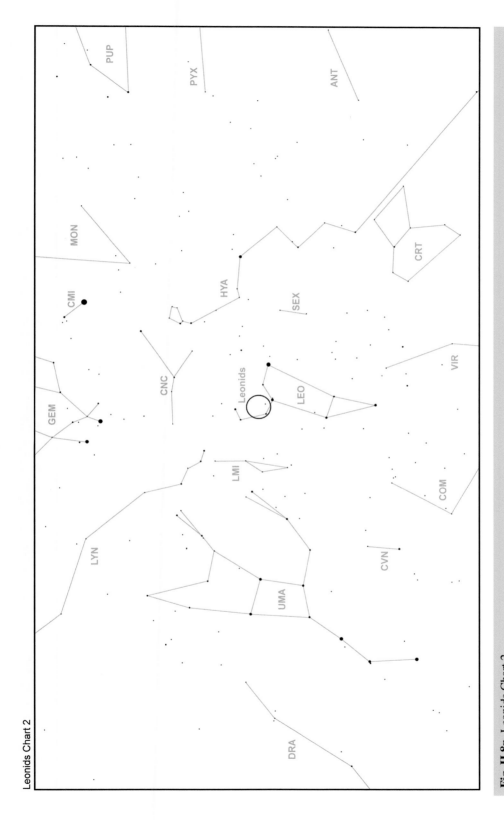

Fig. H.8n Leonids Chart 2

382 Appendix H: Gnomonic Plotting Charts

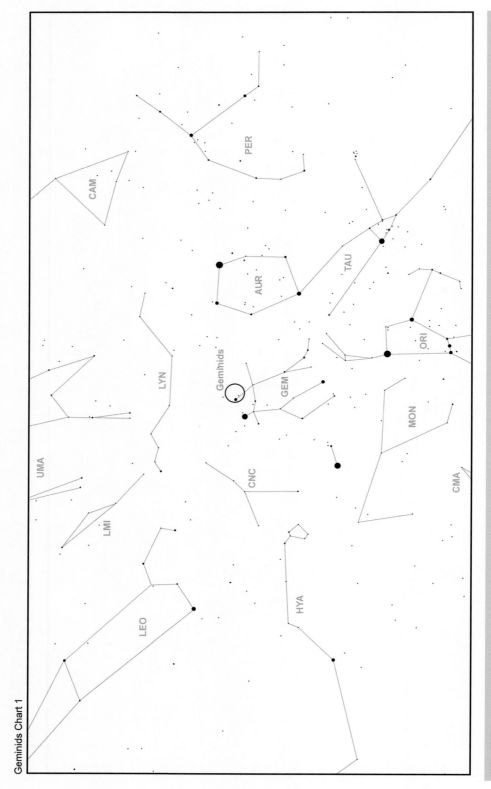

Fig. H.80 Geminids Chart 1

Appendix H: Gnomonic Plotting Charts 383

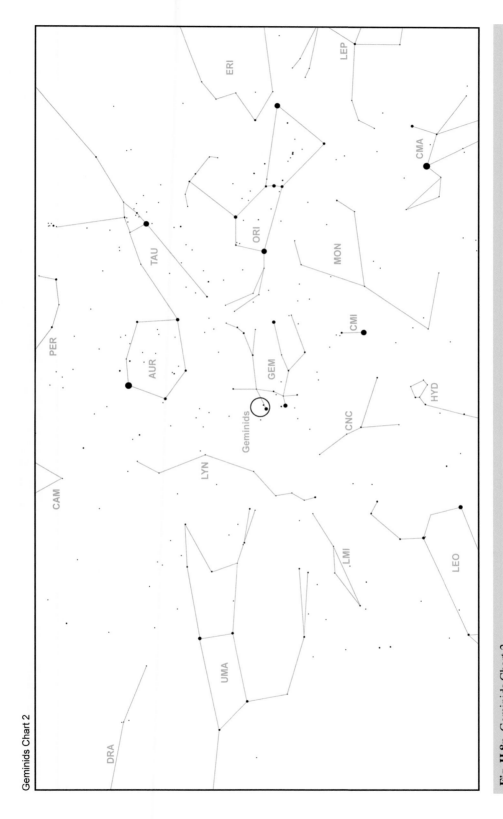

Fig. H.8p Geminids Chart 2

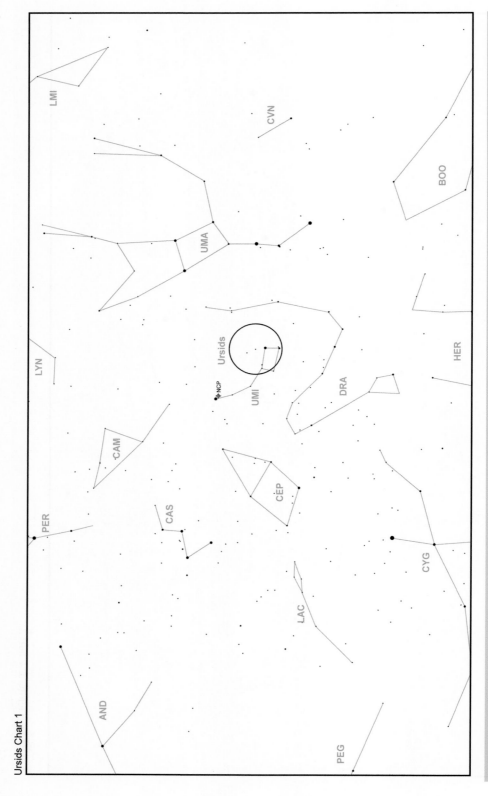

Fig. H.8q Ursids Chart 1

Appendix H: Gnomonic Plotting Charts

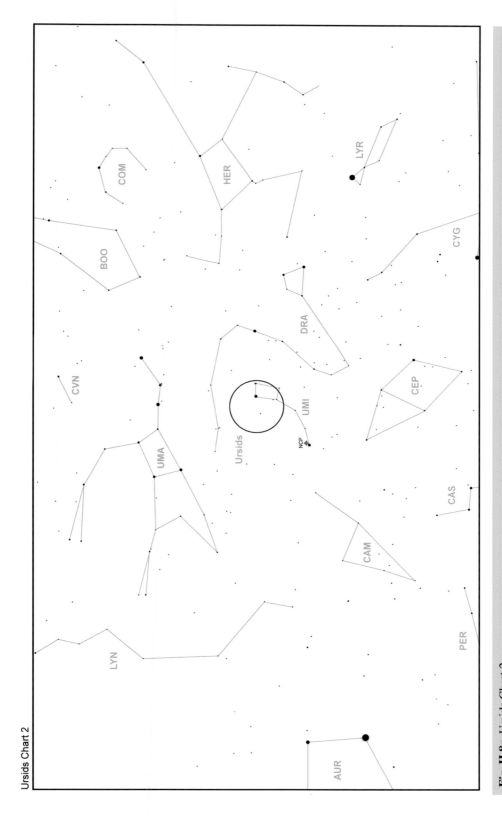

Fig. H.8r Ursids Chart 2

386　　　　　　　　　　　　　　　　　　　　Appendix H: Gnomonic Plotting Charts

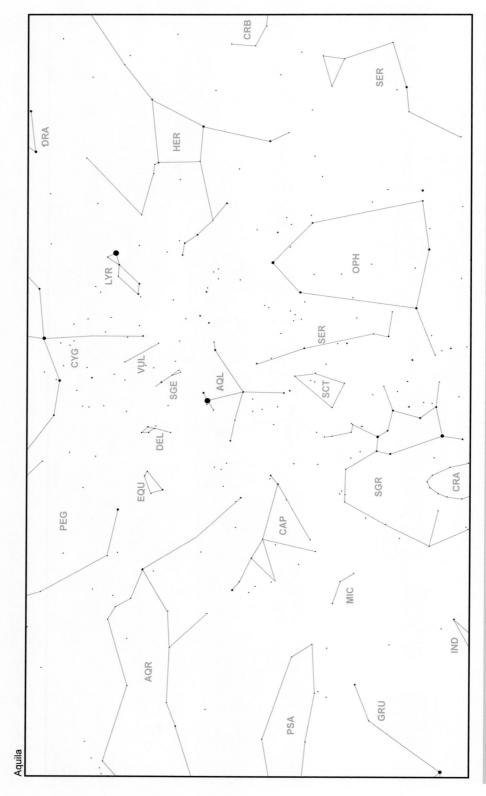

Fig. H.8s Aquila

Appendix H: Gnomonic Plotting Charts 387

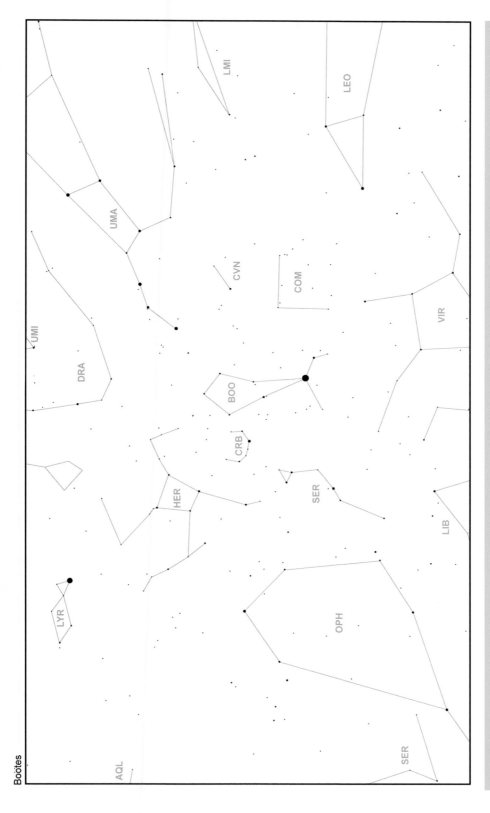

Fig. H.8t Boötes

388 Appendix H: Gnomonic Plotting Charts

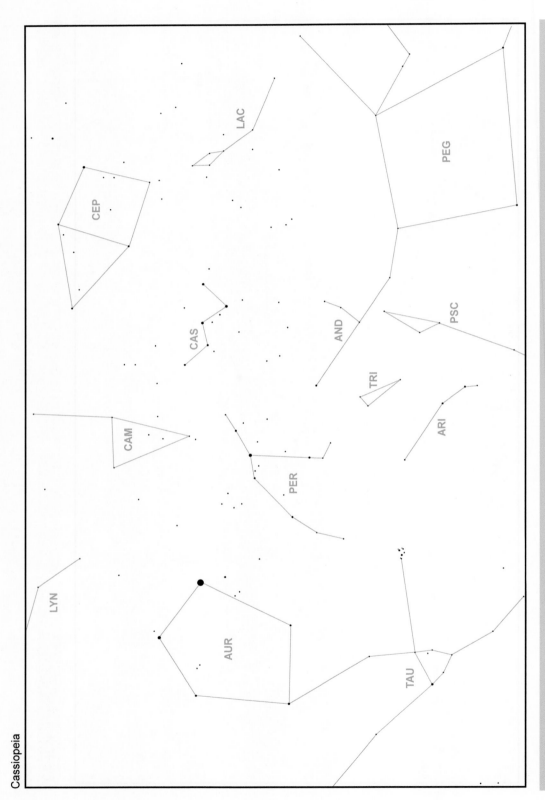

Fig. H.8u Cassiopeia

Appendix H: Gnomonic Plotting Charts 389

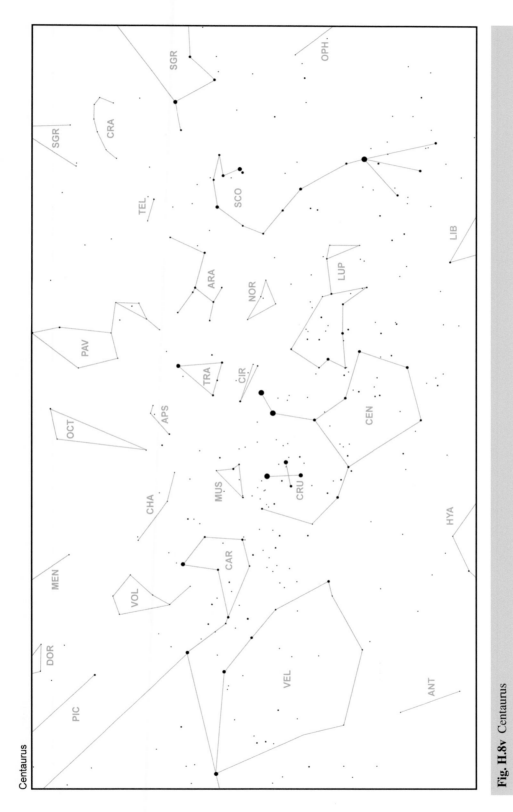

Fig. H.8v Centaurus

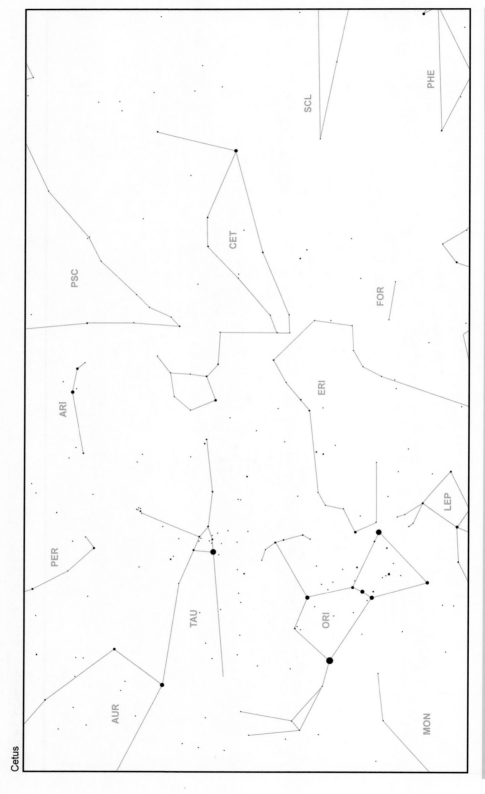

Fig. H.8w Cetus

Appendix H: Gnomonic Plotting Charts

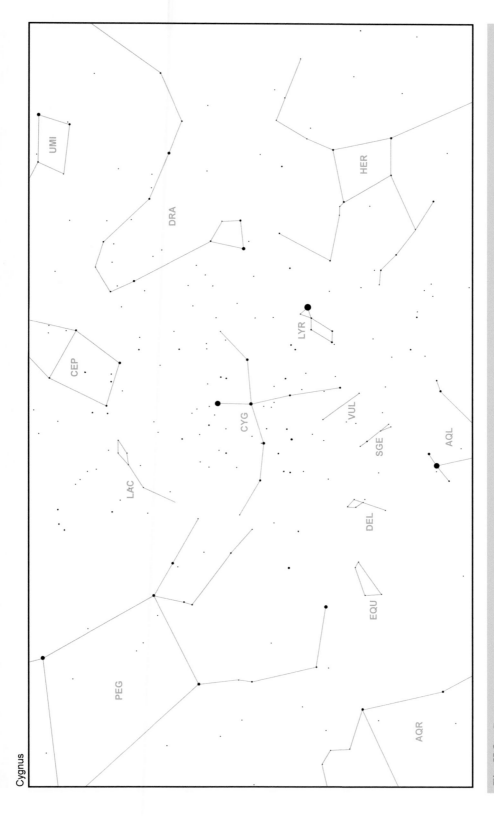

Fig. H.8x Cygnus

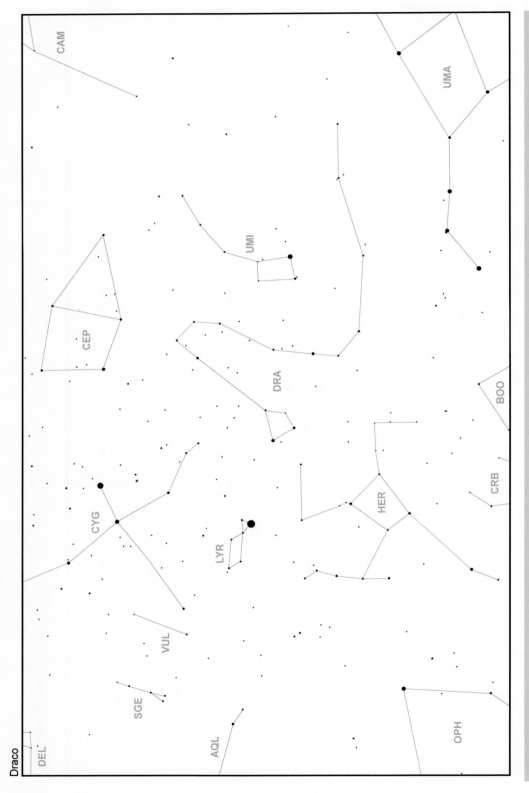

Fig. H.8y Draco

Appendix H: Gnomonic Plotting Charts

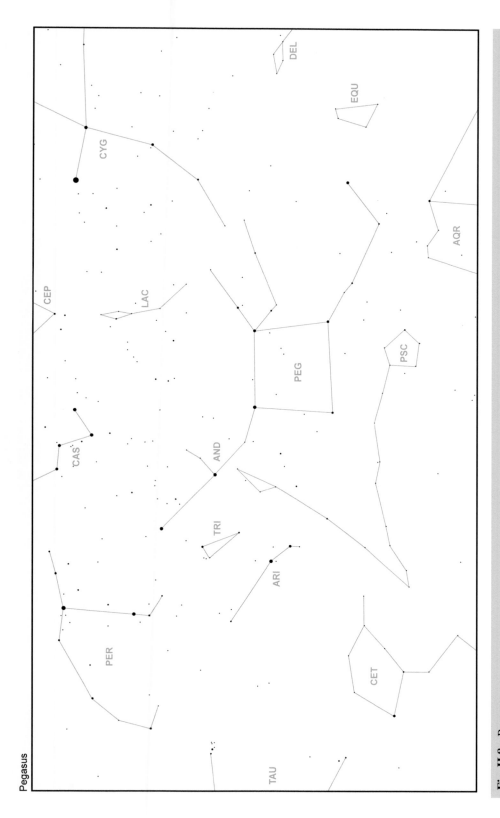

Fig. H.8z Pegasus

394 Appendix H: Gnomonic Plotting Charts

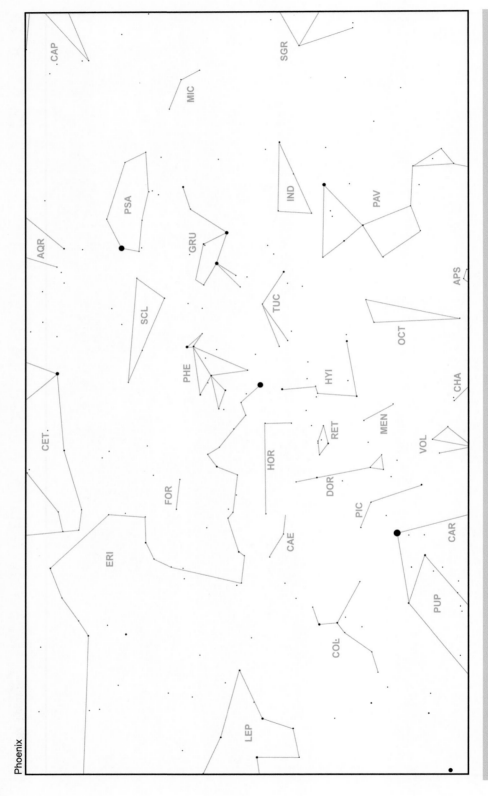

Fig. H.8za Phoenix

Appendix H: Gnomonic Plotting Charts

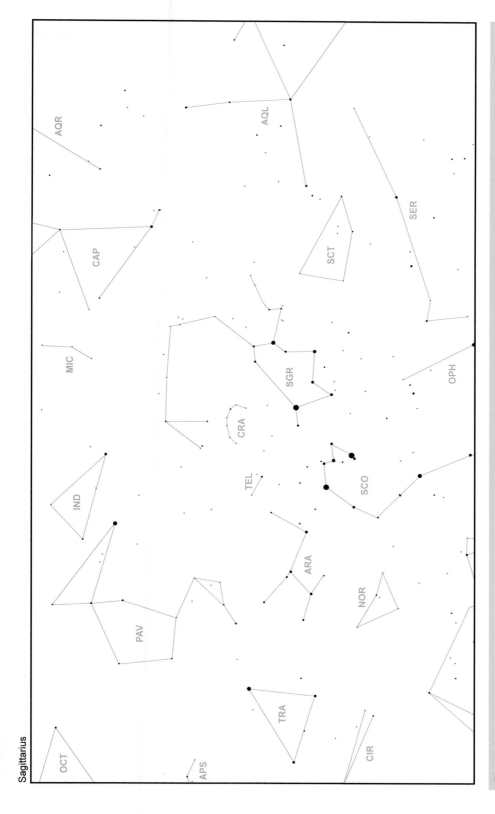

Fig. H.8zb Sagittarius

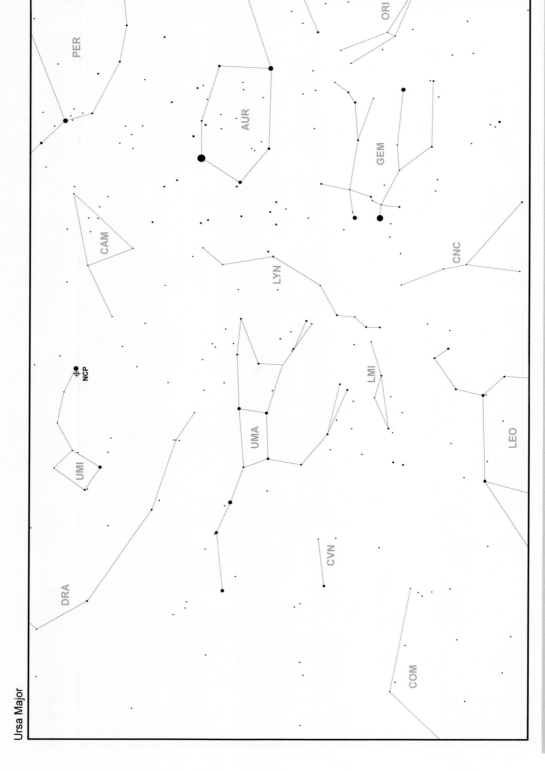

Fig. H.8zc Ursa major

Appendix I: Minor Meteor Showers

Table I.1: Minor Meteor Showers

Many parameters uncertain. Calendar dates are approximate. Use solar longitude $\lambda_\odot$
*These showers are discussed in more detail in the text

Shower	IAU Code	Start $\lambda_\odot$	Max $\lambda_\odot$	End $\lambda_\odot$	At Max. α	At Max. δ	V_g (km/s)	ZHR
o-Leonids	OLE	Jan 1 281°	Jan 8 288°	Jan 28 308°	$09^h 36^m$ 144°	+7°	41.5	<1
κ-Cancrids	KCA	Jan 8 287°	Jan 10 289°	Jan 15 294°	$09^h 12^m$ 138°	+9°	47.3	<1
γ-Ursae Minorids	GUM	Jan 9 288°	Jan 19 298°	Jan 22 301°	$15^h 12^m$ 228°	+67°	30	3
α-Centaurids*	ACE	Jan 31 311°	Feb 8 319.2°	Feb 20 331°	$14^h 00^m$ 210°	-59°	58	6
γ-Normids	GNO	Feb 25 336°	Mar 15 354°	Mar 28 007°	$15^h 56^m$ 239°	-50°	56	5
α-Virginids	AVB	Mar 10 350°	Apr 7-18 17°- 28°	May 6 46°	$13^h 36^m$ 204°	-11°	17.7	5-10
σ-Leonids	SLE	Apr 8 18°	Apr 15 25°	April 28 38°	$13^h 28^m$ 202°	+3°	19	1
April ρ-Cygnids	ARC	Apr 11 21°	Apr 22? 32°?	May 8 48°	$21^h 24^m$ 321°	+45.5°	41.4	<1
π-Puppids	PPU	Apr 15 25°	Apr 23 33.5°	Apr 28 38°	$7^h 20^m$ 110°	-45°	15	Var.
h-Virginids	HVI	Apr 20 30°	Apr 28-30 38°- 40°	May 4 44°	$13^h 40^m$ 205°	-11°	18.4	<1
η-Lyrids*	ELY	May 3 43°	May 8 48°	May 14 54°	$19^h 08^m$ 287°	+44°	43	3
ε-Aquilids	EAU	May 4 44°	May 21-22 61°	May 27 67°	$19^h 18^m$ 289.5°	+18°	31.2	<1
φ-Piscids	PPS	Jun 8 77°	Jul 4 102°	Aug 2 130°	$02^h 06^m$ 32°	+31°	67	<1
June Boötids*	JBO	Jun 22 91°	Jun 27 95.7°	Jul 2 100°	$14^h 56^m$ 224°	+48°	18	Var.
c-Andromedids	CAN	Jun 24 93°	Jul 9 106°	Jul 21 119°	$02^h 10^m$ 32.4°	+48.4°	57.8	<1
ε Pegasids	EPG	Jun 28 97°	Jul 11 109°	Jul 13 111°	$22^h 28^m$ 337°	+16°	28.4	<1
ζ-Cassiopeiids	ZCS	Jul 2 100°	Jul 15 113°	Jul 27 124°	$00^h 56^m$ 14°	+53°	57	1
July ξ-Arietids	JXA	Jul 2 100°	Jul 13 111°	Aug 1 129°	$02^h 44^m$ 41°	+10.2°	69	1
α-Capricornids*	CAP	Jul 3 101°	Jul 25-30 122°- 127°	Aug 15 142°	$20^h 28^m$ 307°	-10°	22.7	5
July Pegasids	JPE	Jul 3 101°	Jul 17 115°	Aug 5 133°	$23^h 40^m$ 355°	+13°	63.3	1
ψ-Cassiopeiids	PCA	Jul 4 102°	Jul 21 119°	Aug 7 135°	$02^h 08^m$ 32°	+73°	42.1	1
49-Andromedids	FAN	Jul 5 103°	Jul 20 118°	Aug 13 140°	$01^h 36^m$ 24°	+48°	60.2	1
Piscis Austrinids*	PAU	Jul 15 113°	Aug 1 129°	Aug 10 138°	$22^h 44^m$ 341°	-22.8°	43.1	5
Northern δ-Aquariids	NDA	July 16 114°	Aug 11 139°	Sep 9 167°	$23^h 06^m$ 346.4°	+1.4°	39.1	3

Appendix I: Minor Meteor Showers

Shower	Code	Start	Max	End	RA/Dec	Dec	V∞	ZHR
July γ-Draconids	GDR	Jul 24 / 121°	Jul 27 / 125°	Aug 1 / 129°	18ʰ 41ᵐ / 280.2°	+50.9°	27.4	Var.
κ-Cygnids*	KCG	Aug 3 / 131°	Aug 17 / 145°	Aug 24 / 152°	18ʰ 59ᵐ / 284.7°	+59°	23	3
η-Eridanids	ERI	Aug 2 / 130°	Aug 9 / 137°	Sep 16 / 176°	02ʰ 58ᵐ / 44.4°	-12.2°	64	<1
Northern ι-Aquariids	NIA	Aug 11 / 139°	Aug 18 / 146°	Sep 10 / 168°	22ʰ 17ᵐ / 334.3°	-5°	29.9	<5
August Draconids	AUD	Aug 13 / 141°	Aug 15 / 143°	Aug 19 / 147°	18ʰ 08ᵐ / 272.1°	+59°	19.2	1
β-Hydrusids	BHY	?	Aug 16 / 144°	?	02ʰ 25ᵐ / 36.3°	-74.5°	22.8	<1
ν-Eridanids	NUE	Aug 24 / 151°	Sep 24 / 181°	Nov 16 / 234°	08ʰ 28ᵐ / 127°	+16	67	<1
Aurigids*	AUR	Aug 28 / 155°	Sep 1 / 158.6°	Sep 5 / 162°	06ʰ 04ᵐ / 91°	+39°	66	5
Northern δ-Piscids	NPI	Sep 3 / 161°	Sep 17 / 175°	Sep 21 / 178°	00ʰ 29ᵐ / 7.3°	+6.9°	65.7	<1
September ε-Perseids	SPE	Sep 5 / 162°	Sep 9 / 166.7°	Sep 21 / 178°	03ʰ 12ᵐ / 48°	+40°	64	5
χ-Cygnids	CCY	Sep 5 / 162°	Sep 15 / 172°	Sep 25 / 182°	20ʰ 00ᵐ / 300°	+31°	19	3
Southern δ-Piscids	SPI	Sep 8 / 166°	Sep 24 / 182°	Sep 21 / 178°	01ʰ 17ᵐ / 19.2°	+1°	46.6	<1
October Capricornids	OCC	Sep 21 / 178°	Oct 2 + 9 / 189° + 196°	Oct 12 / 199°	20ʰ 12ᵐ / 303°	-09°	12	Var.
Southern October δ-Arietids	SOA	Sep 22 / 179°	Oct 10 / 197°	Oct 27 / 214°	02ʰ 10ᵐ / 32.6°	+9.6°	21	<1
Northern October δ-Arietids	NOA	Sep 22 / 179°	Oct 16 / 203°	Oct 22 / 209°	02ʰ 27ᵐ / 36.7°	+19°	62	<1
October Camelopardalids	OCT	Oct 4 / 191°	Oct 5 / 192.6°	Oct 9 / 196°	10ʰ 56ᵐ / 164°	+79	46.6	5
October Draconids	DRA	Oct 6 / 193°	Oct 8 / 195.4°	Oct 10 / 197°	17ʰ 28ᵐ / 262°	+54	21	Var.
δ-Aurigids	DAU	Oct 10 / 196°	Oct 12 / 198°	Oct 18 / 204°	05ʰ 36ᵐ / 84°	+44	64	2
Leonis Minorids	LMI	Oct 12 / 199°	Oct 24 / 211°	Nov 5 / 223°	10ʰ 48ᵐ / 170°	+37°	62	2
October Ursae Majorids	OCU	Oct 14 / 201°	Oct 15 / 202°	Oct 19 / 206°	09ʰ 43ᵐ / 145.8°	+63.8°	38.8	<1
o-Eridanids	OER	Oct 16 / 203°	Nov 5 / 223°	Nov 24 / 242°	03ʰ 00ᵐ / 45°	-04°	29	1
χ-Taurids	CTA	Oct 20 / 207°	Nov 4 / 222°	Nov 17 / 235°	03ʰ 32ᵐ / 53°	+25°	41	1
ξ-Draconids	XDR	Oct 21 / 208°	Oct 23 / 210°	Oct 24 / 211°	11ʰ 21ᵐ / 170.3°	+73°	35.8	<1
Andromedids	AND	Oct 26 / 213°	Nov 6 / 224°	Nov 17 / 235°	00ʰ 38ᵐ / 10°	+24°	19	<1
λ-Ursae Majorids	LUM	Oct 27 / 214°	Oct 28 / 215°	Oct 29 / 216°	10ʰ 24ᵐ / 156°	+49°	61	<1
Southern λ-Draconids	SLD	Nov 1 / 219°	Nov 3 / 221°	Nov 4 / 222°	10ʰ 48ᵐ / 162°	+68°	49	<1
κ-Ursae Majorids	KUM	Nov 3 / 221°	Nov 7 / 225°	Nov 10 / 228°	09ʰ 49ᵐ / 147°	+45°	66	<1
ρ-Puppids	RPU	Nov 10 / 228°	Nov 14 / 232°	Nov 20 / 238°	08ʰ 23ᵐ / 126°	-25°	58	<1

Shower	Code	Begin	Max	End	RA/Long	Dec	V	ZHR
November Orionids*	NOO	Nov 13 231°	Nov 28 246°	Dec 6 254°	06ʰ 04ᵐ 91°	+16	43	3
Southern χ-Orionids	ORS	Nov 13 231°	Dec 2 250°	Dec 21 269°	05ʰ 29ᵐ 82°	+18°	26	1
α-Monocerotids	AMO	Nov 15 233°	Nov 21 239.32°	Nov 25 243°	07ʰ 48ᵐ 117°	+0.9°	63	Var.
November θ-Aurigids	THA	Nov 17 235°	Nov 26 244°	Dec 1 249°	06ʰ 13ᵐ 93°	+35	33	<1
December σ-Virginids	DSV	Nov 22 239°	Dec 21 269°	Jan 25 269°	12ʰ 58ᵐ 194°	+8°	66	1
Northern χ-Orionids	ORN	Nov 23 241°	Dec 12 260°	Dec 18 266°	05ʰ 52ᵐ 88.1°	+25.7°	29	<1
December Monocerotids	MON	Nov 27 245°	Dec 9 257°	Dec 20 268°	06ʰ 40ᵐ 100°	+8°	41	3
Phoenicids*	PHO	Nov 28 246°	Dec 2 250°	Dec 8 256°	01ʰ 12ᵐ 18°	-53°	18	Var.
December α-Draconids	DAD	Nov 30 248°	Dec 8 256°	Dec 15 263°	13ʰ 34ᵐ 204°	+58	44	1
Puppids-Velids*	PUP	Dec 1 249°	Dec 7? 255°	Dec 26 274°	08ʰ 12ᵐ 123°	-45	40	9
December φ-Cassiopeiids	DPC	Dec 1 249°	Dec 6 254°	Dec 8 256°	01ʰ 36ᵐ 24°	+50	16	Var.
December κ-Draconids	DKD	Dec 2 250°	Dec 3 251°	Dec 7 255°	12ʰ 29ᵐ 187°	+70	41	2
ψ-Ursae Majorids	PSU	Dec 2 250°	Dec 5 253°	Dec 10 258°	11ʰ 16ᵐ 169°	+42	62	1
σ-Hydrids*	HYD	Dec 3 251°	Dec 9 257°	Dec 21 269°	08ʰ 20ᵐ 125°	+2°	62	7
December Leonis Minorids	DLM	Dec 6 254°	Dec 20 268°	Feb 4 315°	10ʰ 40ᵐ 160°	+30°	63	5
σ-Serpentids	SSE	Dec 7 255°	Dec 27-28 275°-276°	Jan 12 292°	16ʰ 16ᵐ 244°	-1.7°	45.5	<1
Comae Berenicids	COM	Dec 12 260°	Dec 16 264°	Dec 23 271°	11ʰ 40ᵐ 175°	+18°	63.5	3
α-Hydrids	AHY	Dec 17 266°	Jan 1 280°	Jan 17 296°	08ʰ 19ᵐ 124.9°	-7.0°	43.2	1
c-Velids	CVE	Dec 19 268°	Dec 28 276°	Jan 6 285°	09ʰ 14ᵐ 138°	-54°	39	1
Volantids	VOL	Dec 31 279°	Jan 2 281°	Jan 1 280°	07ʰ 56ᵐ 119°	-75°	28.1	9
January Leonids	JLE	Dec 31 279°	Jan 4 283°	Jan 8 287°	09ʰ 51ᵐ 147.7°	+24.1°	55.2	<1
λ-Boötids	LBO	Dec 31 279°	Jan 16 295°	Jan 17 296°	14ʰ 40ᵐ 220°	+43°	41	<1

Appendix J: Major Meteor Showers

Table J.1: Major Meteor Showers							
IAU No.	IAU Code	Shower Name	Duration (date)[1] (Solar longitude λ☉)	Maximum (date)[2] (Solar longitude λ☉)	ZHR[3]	R.A.[4] α	Decl[5] δ
0010	QUA	Quadrantids	Dec 28 274° / Jan 12 274°	Jan 4/5 291°	91	$15^h\,18^m$ 229°	+49.5°
0006	LYR	April Lyrids[6]	Apr 14 24° / Apr 30 40°	22 Apr 32.32°	18	$18^h\,07^m$ 272°	+33.1°
0031	ETA	η-Aquarids	Apr 19 29° / May 28 68°	May 6/7 46.2°	83	$22^h\,32^m$ 338°	−0.8°
0005	SDA	Southern δ-Aquarids	Jul 12 109° / Aug 19 147°	Jul 29/30 126.5°	28	$22^h\,44^m$ 341°	−16°
0007	PER	Perseids[7]	Jul 18 115° / Aug 25 153°	Aug 12/13 140°	130	$03^h\,11^m$ 48°	+58°
0002	STA	Southern Taurids	Sep 10 168° / Nov 20 238°	Nov 5/6 223°	5	$03^h\,33^m$ 53°	+12.9°
0008	ORI	Orionids	Oct 3 190° / Nov 7 220°	Oct 22/23 209°	26	$06^h\,24^m$ 96°	+15.7°
0017	NTA	Northern Taurids	Oct 20 207° / Dec 10 258°	Nov 12/13 230°	5	$03^h\,57^m$ 59°	+22.3°
0013	LEO	Leonids	Nov 6 224° / Nov 30 248°	Nov 18 236°	17	$10^h\,17^m$ 154°	+21.4°
0004	GEM	Geminids	Dec 4 252° / Dec 17 265°	Dec 14 262.2°	100	$07^h\,27^m$ 112°	+32.3°
0015	URS	Ursids	Dec 17 265° / Dec 26 274°	Dec 23 271°	15	$14^h\,28^m$ 217°	+75.4°

[1], [2] Approximate dates. Use the solar longitude value for greater accuracy.
[3] Zenithal hourly rate (2010-2020 average):
[4] Right ascension in hours and minutes with degrees underneath.
[5] Declination.
[6] Often just called the Lyrids.
[7] Sometimes called the August Perseids.

Author Index

A
Alcock, George Eric Deacon (1912–2000), 58, 292
Almond, Mary, 122, 199
Arago, François (1786–1853), 80
Araki, Genichi, 292
Aristotle (384–322 BCE), vii
Arlt, Rainer, 120
Arter, T.R., 83
Asher, David J., ix
Astapovič, Igor' Stanislavovich (1908–1976), 203

B
Backhouse, Thomas William (1842–1920), 198
Baeker, Carl Wilhelm (1819–1882), 83
Bečvář, Antonín (1901–1965), 272, 273
Benzenberg, Johann (1777–1846), vii, 80, 175
Biela, Wilhelm von (1782–1856), 6, 288
Biot, Jean-Baptiste (1774–1862), 80
Blaauw, R., 251
Blanpain, Jean–Jacques (1777–1843), 322, 323
Bonpland, Aimé Jacques Alexandre (1773–1858), 226
Booth, David, viii
Brandes, Heinrich Wilhelm (1777–1834), vii, 175
Briggs, Robert E., 34
Brown, Peter G., 199
Brucalassi, Antonio (1717–1866), 60

C
Campbell-Brown, Margaret, 104
Cassini, Giovanni Domenico (1625–1712), 19
Ceplecha, Zdeněk (1929–2009), ix, 272
Chasles, Michel (1793–1880), 80
Childrey, Joshua (1623–1670), 19

Chladni, Ernst Florens Friedrich (1756–1827), vii
Clap, Thomas (1703–1767), vii
Clube, S.V.M., 199
Cooke, William, 199
Cooper, Tim, 33, 292
Corder, Henry (1855–1944), 80

D
Davidson, Martin (1880–1968), ix, 299
Davies, John K., 33, 250
Denning, William Frederick (1848–1931), viii, 60, 80, 102, 122, 145, 173, 176, 198, 272, 273, 299, 306, 308, 312
Denza, Francesco (1834–1894), 299
Drummond, Jack D., 292
Dubietis, Audrius, 120
Duillier, Nicolas Fatio de (1664–1753), 19

E
Edgeworth, Kenneth Essex (1880–1972), 26
Egal, Auriane, 104
Elkin, William Lewis (1855–1933), viii

F
Falb, Rudolph (1838–1903), viii, 103, 176

G
Galle, Johann (1812–1910), 81
Goldschmidt, Hermann Mayer Salomon (1802–1866), 27

Goodall, William, ix
Gravnik, Mikael, 251
Green, Simon F., 33, 250
Greg, Robert Philips (1826–1906), 250
Guillemin, Amédée (1826–1893), 289
Guth, Vladimir (1905–1980), 310

H

Hajduk, Anton (1933–2005), 102
Hamid, Salah E., 60, 123
Hasegawa, Ichiro (1928–2016), 61
Hawkins, Gerald Stanley (1928–2003), ix
Heis, Eduard (1806–1877), viii, 122, 198
Herrick, Edward Claudius (1811–1862), 80
Herschel, Alexander Stewart (1836–1907), viii, 80, 102, 103, 176, 250
Hey, James, ix
Hindley, Keith, 81, 198
Hoffmeister, Cuno (1892–1968), 199, 304, 310, 315, 324
Hughes, David W., 61, 103, 146
Humboldt, Friedrich Wilhelm Heinrich Alexander von (1769–1859), vii, 226

J

Jacchia, Luigi Giuseppe (1910–1996), 34, 124
Jenniskens, Peter, 61, 78, 272, 296, 299
Jewitt, David, 323

K

Kämtz, Ludwig Friedrich (1801–1867), 249
Kegler, Ignatius (1680–1746), 145
King, Alphonse, 122
Kirkwood, Daniel (1814–1895), viii, 28, 145
Knopf, O.H.J., 199
Konkoly-Thege, Miklós de (1842–1916), 296
Kornoš, Leonard, 81
Krafft, Georg Wolfgang (1701–1754), vii
Kresák, Ľubor, 199
Kuiper, Gerard Peter (1905–1973), 26–28, 36

L

Lindblad, Bertil A., 81, 83
Le Verrier, Urbain Jean Joseph (1811–1877), 227, 228
Lovell, Sir Bernard (1913–2012), 102, 104
Lowe, Edward Joseph (1825–1900), 306
Lunsford, Robert, 315, 317
Lyttleton, Raymond Arthur (1911–1995), 30
Lyytinen, Esko, 272

M

Machholz, Donald, 61, 123
MacLennan, Eric, 251
Marchenko, Valerie, 199

Marsden, Brian G. (1937–2010), 145, 146
Marsh, B.V., 250
Martynenko, Vasily Vasilevich (1930–2000), 144
Maskelyne, Nevil (1732–1811), vii
Masterman, Stillman, 60, 63
McBride, Neil, 61, 103, 146
McCrosky, Richard Eugene (1924–2012), 319
McIntosh, Bruce Andrew (1929–2015), ix, 61, 102, 123
McIntosh, Ronald Alexander (1904–1977), ix, 102, 122
McKinley, D.W.R., 124
McNaught, Robert H., ix
Méchain, Pierre François André (1744–1804), 199, 273
Mellish, John Edward (1886–1970), 323
Messier, Charles (1730–1817), 199, 288
Miller, F., 250
Montaigne, Jacques Leibax (1716–1785?), 288
Montanari, Geminiano (1633–1687), vii, 176
Moser, Danielle E., 199
Muller, Richard, 36
Musschenbroeck, Petrus van, vii, 144

N

Nagaoka, Hantaro (1865–1950), ix
Nakamura, K., 308
Neumayer, Georg Balthazar von (1826–1909), viii, 122, 124
Newton, Hubert Anson (1830–1896), viii, 102, 103, 227

O

Olbers, Heinrich Wilhelm Matthias (1758–1840), 227
Olivier, Charles Pollard (1884–1975), viii, 80, 102, 103, 145, 176, 306
Olmsted, Denison (1791–1859), viii, 226
Oort, Jan Hendrik (1900–1992), 28
Öpik, Ernst Julius (1893–1985), 28
Oppolzer, Theodor von (1841–1881), 81, 227

P

Pape, C.F., 81
Parsons, John, ix
Piazzi, Giuseppe (1746–1826), 26
Porubčan, Vladimír, 81
Posen, A., 318
Prentice, John Philip Manning (1903–1981), 58, 173
Puzio, Michael, 290

Q

Quetelet, Adolphe (1796–1874), viii, 144, 176

R

Ridley, Harold Bytham (1919–1995), 322
Roggemans, Paul, 145
Russell, Henry Norris (1877–1957), 83

Author Index

S
Sato, M., 104
Schaeffer, G.C., 144
Schafer, John, ix
Schiaparelli, Giovanni (1835–1910),
　viii, 81, 145
Schmidt, Johann Friedrich Julius (1825–1884), 102
Secchi, Angelo (1818–1878), 288
Shain, Charles Alexander (1922–1960), 322
Skellett, Albert M., ix
Southworth, Richard, ix
Stepanek, J., 310
Stewart, Gordon, ix
Štohl, J., 199
Swift, Lewis A. (1820–1913), viii, 145

T
Teichgraeber, Arthur, 310
Tempel, Ernst Wilhelm Liebrecht
　(1821–1899), 227
Terentjeva, Alexandra Konstantinovna,
　122, 315
Thatcher, Albert E. (1829–?), 81, 310
Tolian, Athanasia, 251
Trouvelot, Étienne Léopold (1827–1895), 229
Turco, E.F., 312
Tuttle, Horace Parnell (1837–1923), 81, 145, 146, 227,
　272, 273
Tupman, George Lyon (1838–1922), viii, 102, 104, 199,
　203, 299

V
Vaubaillon, Jérémie J., 296

Venter, S.C., 322
Verniani, Franco, 34
Vida, Denis, 104
Vlcek, J., 310
Vratnik, A. Wantanabe, Junichi, 310

W
Wiegert, Paul, 104
Weinek, Ladislaus (1848–1913), viii
Weiss, A.A., 323
Weiss, Edmond (1827–1917), 81
Weryk, Robert, 199
Whipple, Fred Lawrence (1906–2004), 30, 33, 60, 123,
　199, 250, 299, 323
Williams, V., 304, 322
Wilson, Fiametta, 60
Wood, Jeff, 103
Wright, Frances Woodworth (1897–1989),
　124, 203

X
Xuanye, Aisin Gioro (1654–1722), 145

Y
Ye, Quanzhi, ix, 33, 290, 292
Yongzheng (Yinzhen), Aisin Gioro (1722–1735), 145
Youssef, Mary N., 60

Z
Zezioli, G., 198
Znojil, V., 174

Subject Index

A
Accretion, 16
Active asteroids, ix, 18, 19, 25, 33, 35, 251, 290, 337, 341
American Meteor Society (AMS), ix, 41, 47, 80, 312, 342
Angular distance, 43, 44
Antihelion/anthelion, 18, 20, 198, 199, 296
Ascending node, 57, 103, 146, 173, 337, 339, 340, 344
Asteroid belt
   masses, 25, 26
Asteroids, 337
   (1) Ceres, 25, 251
   (2) Pallas, 227
   (4) Vesta, 25, 227
   (10) Hygiea, 25
   (21) Lutitia, 27
   (1221) Amor, 339
   (1862) Apollo, 339
   (2062) Aten, 340
   (3200) Phaethon, 33, 61, 250, 251
   (36108) Haumea, 26
   (101955) Bennu, ix, 251, 290
   (136199) Eris, 26
   (136472) Makemake, 26
   (163693) Atira, 340
   1973 NA, 61
   2002 XM_{35}, 319
   2003 EH_1, 33, 61
   2003 WY_{25}, 324
   2010 LU_{108}, 319
Astronomical Society of South Africa, 33, 292, 322
Athens Observatory, 102
Atlantic Gap, 198, 338

B
Bilk Observatory, 80
Bolides, 12, 48, 308, 338
British Astronomical Association (BAA), viii, 47, 80, 81, 198, 299, 322, 342
Brussels Observatory, viii, 60, 144, 176

C
Carbon dioxide, 15
Centaurs, 28, 35
Chinese annals, vii, 323
Chinese royal astrologers, 80
Comet
   coma, 31, 323
   dirty snowball model (*see* Whipple model)
   Halley-type, 28, 323, 339
   icy conglomerate model (*see* Whipple model)
   Jupiter family, 61, 339
   long period, 28, 81, 310, 341
   named, 61, 145, 199
      1737 II, 146
      1783 W1, 62
      1811 I, 31
      1819 IV, 322
      1861 I, 81
      1862 III, 81, 145
      1866 I, 227
      1892 T1, 62
      1917 I, 323
      Barnard-Boattini, 62
      Biela, 6, 288
      Blanpain, 322, 323
      Brorsen, 62
      Brorsen-Metcalf, 62

Comet (*cont.*)
    C/1490 Y$_1$, 61, 62
    C/1743 X$_1$, 31
    C/1811 F$_1$, 31
    C/1860 D$_1$, 62
    C/1917 F$_1$, 323
    C/1939 B$_1$, 62
    D/1819 W$_1$, 323
    D/1978 R$_1$, 202
    De Cheseaux, 31
    Encke, 30, 199, 288
    Flaugergues, 31
    Giacobini-Zinner, ix, 6, 299, 303
    Hale-Bopp, 30
    Halley, vii, 6, 30, 100, 103, 146, 171, 199, 288
    Haneda-Campos, 202
    Kiess, 81
    Kozik-Peltier, 62
    Liais, 62
    Machholz 1, 62
    Mellish, 323
    Pigott-LINEAR-Kowalski, 62
    Pons-Brooks, 62
    Pons-Winnecke, 6, 306
    Sarabat
    Stephan-Oterma, 62
    Swift-Tuttle, viii, 81, 145, 146
    Tempel-Tuttle, ix, 227
    Thatcher, 81, 310
    Tuttle, viii, 145, 227, 273
  nucleus, 30–32, 61, 323
  outgassing, 30, 288, 341
  rocket effect, 30
  rubble pile, 30
  sandbank model, 30
  short period, 28, 31, 288, 340, 341
  sublimate/sublimation, 30, 31, 33, 251
  tail, 31
  Whipple model, 30
Coordinated Universal Time (UTC), 2, 43, 45
Counterglow, 18, 338, 339
Czech Astronomical Society, 310

D
Dark adaption, 43
Daylight Saving Time (D.S.T.), 2
Descending node, 81, 103, 303, 337, 339
Dirty snowball model, 30
Dust particles
  mass infall to Earth, 1, 15
Dwarf planet, 26

E
Earthgrazers, 102, 339
Ecliptic, 12, 18, 20, 28, 81, 146, 173, 337, 339, 341, 342

Edgeworth–Kuiper belt, 26
Epoch, 2
Exoplanets, 27, 339

F
Filaments, ix, 32, 81, 144, 145, 176, 339, 341
Fireball
  smoke train, 12
  sounds, 16, 17, 48

G
Gegenschein, 18, 338, 339, 342
Great Leonid Storm, viii, 80, 226
Great Meteor Procession, 17
Greenwich Mean Time (GMT), 2, 43, 299

H
Haleakala Observatory, 323
Harvard Meteor Survey, 80, 318
Harvard Observatory, ix, 323
Helion radiant
HMS Prince Consort, 102
Hungarian Meteor and Fireball Network, 310

I
Icy conglomerate model, 30
Infra-Red Astronomy Satellite (IRAS), 33, 250, 292
International Astronomical Union (IAU), xv, 6, 12, 32, 33, 54, 55, 287, 290, 292, 315, 327, 331, 340, 343, 344
International Meteor Organization (IMO), ix, xv, 2, 47, 48, 55, 103, 120, 122, 175, 199, 270, 304, 315, 343
Italian Meteoric Society, 102

J
Japanese annals, 272
Jodrell Bank, 102, 122, 142, 199, 250
Jupiter-family comets (JFCs), 28, 338

K
Kirkwood gaps, 28, 145
Korean annals, 226, 249
Kuiper belt
  Kuiper belt objects (KBO), 26

L
Lagrangian Points
l'École Militaire Observatory
Limiting magnitude (lm), 6, 40, 47, 50, 54, 342

Subject Index

Local time (LT), 2, 20, 41, 43, 54, 57, 58, 78, 100, 120, 142, 174, 224, 246, 312, 318
Lowell Observatory Near-Earth-Object Search (LONEOS), 63

M

Machholz Complex, 61
Magnitudes, xi, 1, 8–9, 12, 15, 17, 34, 35, 43, 45, 48, 50, 52, 54, 55, 60, 80, 100, 120, 144, 171, 173, 176, 198, 224, 227, 246, 270, 304, 308, 310, 318, 322, 339–343, 349–366
Major showers, xi, 5, 33–35, 50, 53–55, 57, 58, 198, 250, 326, 340, 341
Mesosphere, 15, 340
Meteor
　ablation, 13
　annual variation
　antihelion sources, 198
　definition, 12
　diurnal variation
　energy, 15
　heights, vii
　Helion source, 20, 199, 203
　metal atoms, 15
　North Apex source, 20
　North Toroidal source, 20
　persistent train, 60, 100, 120, 299, 310, 318
　plasma, 13, 15
　South Apex source, 20
　South Toroidal source, 20
　sporadics, 25, 35, 80
　trains, 304, 322
　velocities, 15, 102, 292, 304
　wake, 12, 30, 122, 246
Meteor showers
　calendar, 38, 53, 171
　class, 324
　first recorded, 80
　named
　　α (alpha) Aurigids, 310
　　α (alpha) Capricornids, 6, 34, 122, 296, 299
　　α (alpha) Centaurids, 304
　　α (alpha) Monocerotids, 224
　　α (alpha) Virginids, 82, 103, 342, 398
　　Andromedids, 6, 32, 288
　　April Lyrids, 38, 78–81, 331
　　April ρ (rho) Cygnids, 82, 103, 343, 398
　　August Draconids, 123, 147, 342, 399
　　Aurigids, 81, 310
　　β (beta) Hydrusids, 123, 147, 342, 399
　　β (beta) Piscids, 122
　　β (beta) Taurids, 198, 199
　　Bielids, 6, 32, 202, 288
　　c Andromedids, 123, 147, 342, 398
　　χ (chi) Taurids, 174, 201, 226, 342, 399
　　Comae Berenicids, 248, 272, 332, 333, 342, 400
　　Daytime Arietids, 61, 122
　　Daytime β (beta) Taurids, 200
　　Daytime ζ (zeta) Perseids, 200
　　December α (alpha) Draconids, 201, 248, 342, 400
　　December κ (kappa) Draconids, 202, 248, 342, 400
　　December Leonis Minorids, 62, 248, 272, 400
　　December Monocerotids, 249, 318
　　December φ (phi) Cassiopeiids, 202, 248, 400
　　December Phoenicids, 202, 248, 320
　　December σ (sigma) Virginids, 248, 272, 400
　　δ (delta) Aquariids, 55, 60, 61, 100–103, 120–123, 312
　　Draconids, 6
　　ε (epsilon) Aquilids, 103, 398
　　ε (epsilon) Geminids, 246–251
　　ε (epsilon) Pegasids, 123, 147, 398
　　η (eta) Aquariids, 6, 38, 100–103, 173, 176, 331
　　η (eta) Eridanids, 123, 147, 399
　　η (eta) Lyrids, 292
　　γ (gamma) Ursae Minorids, 62, 398
　　Geminids, xiii, 17, 33, 38, 61, 173, 175, 246–251, 287, 288, 296, 318, 323, 324
　　Giacobinids, 6, 299
　　Halleyids, 6, 103
　　h-Virginids, 82, 103, 398
　　July γ (gamma) Draconids, 123, 147, 399
　　July Pegasids, 123, 147, 398
　　July ξ (xi) Arietids, 123, 147
　　κ (kappa) Cygnids, 32, 296, 299
　　κ (kappa) Ursae Majorids, 174, 201, 226, 399
　　λ (lambda) Taurids, 198
　　λ (lambda) Ursae Majorids, 174, 201, 399
　　Leonids, 15, 34, 80, 100, 224–227
　　Leonis Minorids, 174, 201, 399
　　Northern χ (chi) Orionids, 200, 248, 318, 400
　　Northern δ (delta) Aquariids, 61
　　Northern δ (delta) Piscids, 200, 399
　　Northern ι (iota) Aquariids, 123, 147, 343, 399
　　Northern October δ (delta) Arietids, 200, 203
　　Northern Taurids, 198, 199
　　November Orionids, 249, 318, 324
　　November θ (theta) Aurigids, 201, 226, 400
　　ν (nu) Eridanids, 201, 226
　　October Camelopardalids, 174, 201, 399
　　October Capricornids, 174, 201, 399
　　October Draconids, 299, 304
　　October Ursae Majorids, 174, 201, 399
　　o (omicron) Eridanids, 174, 201, 226, 399
　　Orionids, 6, 102, 103, 171–176, 249, 318, 324
　　Perseids, 6, 32, 34, 142–146, 171, 198, 251, 287, 288, 299, 315
　　φ (phi) Piscids, 123, 147, 398
　　Phoenicids, 34, 320, 323
　　π (pi) Puppids, 82, 103, 398
　　Piscis Austrinids, 122, 312
　　Pons-Winneckids, 6, 306
　　ψ (psi) Cassiopeiids, 123, 147, 398
　　ψ (psi) Ursae Majorids, 202, 248, 400
　　Puppids-Velids (PUP), 324

Meteor showers (*cont.*)
  Quadrantids, 6, 32–34, 38, 54, 55, 58–61, 100, 122, 287, 288
  ρ (rho) Puppids, 174, 201, 226, 399
  September ε (epsilon) Perseids, 315, 318
  September Perseids, 315
  σ (sigma) Hydrids (HYD), 318, 324
  Southern χ (chi) Orionids (ORS), 249, 318, 324
  Southern δ (delta) Aquariids, 4, 61, 120–123
  Southern δ (delta) Piscids, 200, 399
  Southern λ (lambda) Draconids, 174, 201, 399
  Southern October δ (delta) Arietids, 203, 399
  Southern Taurids, 33, 198, 224
  Taurids, 33, 53, 175, 196–199, 224
  Ursids, 34, 270–273
  ξ (xi) Draconids, 174, 201, 399
  ζ (xi) Perseids, 198, 315
  ζ (zeta) Cassiopeiids, 123, 147, 398
  ζ (zeta) Perseids, 196, 198, 317
  49 Andromedids, 123, 147, 398
Meteoric dust, 12, 30, 340
Meteoric smoke particles (MSP), 12, 340, 343
Meteorite
  Chelyabinsk, 14
  cosmic velocity, 341
  crushing strength, 13
  energy, 13, 15, 246
  impacts, 340
  injury, 14
  middlesbrough, 14
  point of retardation, 13
  terminal impact velocity
Meteoroids
  masses, 18, 246
Meteoroid stream
  age, 176
  dissipation, 19
  masses, 103
Minor planet, *see* Asteroids
Minor showers, xi, xiii, 5, 6, 33, 34, 38, 50, 53, 60, 122, 251, 287–327, 340, 341, 367
Moonlight, 41, 54, 58, 78

N
NASA Ames Research Center, 78
National Association of Planetary Observers, 318
Near-Earth Objects (NEO)
Nemesis, 36
New Zealand Astronomical Society, 122
Noctilucent clouds, 15, 340
Non-gravitational forces, 31, 287

O
Olbers Paradox, 227
Ondřejov Observatory, 308
Oort cloud, 28, 29, 341
Öpik–Oort Cloud, *see* Oort Cloud

Orbits
  eccentricity, 57
  ellipse/elliptical, 29
  energy, 30
  inclinations, 28, 81, 323, 339
  periods, 27–29, 36, 81, 272, 323, 339
  resonance, 28, 272
Outbursts, ix, 32, 34, 35, 47, 53, 78, 103, 270, 290, 299, 304, 306, 308, 310, 315, 322, 323, 341

P
Pacific Gap, 198, 341
Panoramic Survey Telescope and Rapid Response System (Pan-STARRS), 323
Perihelia/perihelion, 28, 32, 57, 60, 122, 145, 176, 250, 251, 290, 309, 323
Persistent trains, 45, 60, 100, 120, 299, 341
Planets
  Earth, 12, 16
  Jupiter, 12, 27, 339
  Mars, 13, 25, 36, 199, 250
  Mercury, 25, 36, 199, 250
  Neptune, 339
  Saturn, 339
  Uranus, 339
  Venus, 12, 25, 48, 102, 199
Poynting-Robertson effect, 31, 246, 287, 341

Q
Quadrans Muralis, 6, 32, 58, 60
Quadrantid Complex, 122, 123

R
Radiant
  antihelion, 296
  average radiant, 4, 54, 304, 306
  helion, 20, 199
  radiant diameter, 44, 54, 326
  radiant drifts, viii, 55, 102, 199, 312
  radiant points, 4, 54, 315
  radiant sizes, 44, 304
  stationary radiants, 176
RAMBo, 308
Red light, 43
Retardation point, 199, 341
Rock comet, 251, 341
Rosetta, 27, 32

S
Sandbank, 30
Scattered disc
  Scattered disc objects (SDO), 28
SETI Institute, 78
Smithsonian, 145
Smoke train, 12

Subject Index

Solar longitude, 2, 53, 55, 58, 196, 341, 344
Solar system, 1, 2, 12, 16, 18, 19, 25–31, 35, 36, 56, 287, 288, 290, 338–342
Solar wind, 19, 31, 341
Sonneberg Observatory, 310
Spanish Meteor Society, 317
Sporadics, 3, 20, 32, 45, 50, 52, 324, 326, 341, 344
Stefanik Observatory, 310
Steward Observatory, 292
Storms, vii, ix, 1, 47, 80, 224, 226, 227, 288, 303, 340, 341
Swarm Years, 198, 199

T
Train, 12, 43, 48, 144, 246, 308, 341, 342
Trans-Neptunian Objects (TNOs), 28
Tunguska, 199

V
Vienna Observatory, 81
Visual magnitude (m_v), 8, 12, 34, 55, 173, 198

W
Wakes, 12, 30, 122, 246, 342
Western Australia Meteor Society (WAMS), 103, 304, 322
Whipple model, 30

Y
Yale University, 60, 80

Z
Zenithal Hourly Rate (ZHR), ix, xi, xiii, 5–8, 33, 40, 44, 50, 52–55, 58, 60, 78, 100, 120, 142, 144–146, 171, 173, 176, 196, 198, 224, 227, 246, 250, 270, 272, 287, 288, 292, 296, 299, 303, 304, 306, 308, 310, 312, 315, 318, 320, 322, 324, 339, 340, 342, 344
Zenith attraction, 4, 120, 342
Zodiacal band, 18, 342, 344
Zodiacal dust cloud (ZDC)
    mass, 18, 19
Zodiacal light, 12, 18, 19, 342, 344

Printed by Printforce, the Netherlands